Geoarchaeology

Geoarchaeology
Earth Science and the Past

edited by

D.A. Davidson
and M.L. Shackley

Westview

First published in 1976 in the United Kingdom by
Gerald Duckworth and Co. Ltd., London, England

Published 1976 in the United States of America by
Westview Press, Inc.
1898 Flatiron Court
Boulder, Colorado 80301
Frederick A. Praeger, Publisher and Editorial Director

Library of Congress Cataloging in Publication Data
Main entry under title:

Geoarchaeology: earth science and the past.

Presented at a symposium on the theme
'Sediments in archaeology' held at the
University of Southampton Dec. 15-16, 1973.
    Includes indexes.
    1. Sediments (Geology) – Congresses.
    2. Geology, Stratigraphic – Quaternary – Congresses.
    3. Archaeology – Congresses.
    I. Davidson, Donald A.
    II. Shackley, Myra L.
QE471.2.G46   930.1'01'55   76-25224
ISBN 0-89158-635-0

Filmset by Specialised Offset Services Ltd., Liverpool
and printed and bound in Great Britain by
Redwood Burn Limited, Trowbridge & Esher

CONTENTS

SECTION 3 TERRESTRIAL ENVIRONMENTS

SECTION 4 BIOLOGICAL SEDIMENTS

PREFACE

The papers in this volume were presented at a symposium on the theme 'Sediments in archaeology'. An important current trend in archaeology is the application of techniques culled from the physical and biological sciences to archaeological problems. The organisers held the view that the full potential contribution of the earth sciences to archaeology has never been realised, even though much pioneering work has been done in recent years. The initial intention was to hold an informal meeting for discussion of this theme, but the response to a circular indicated that there was a large number of people engaged in this type of research. It was therefore decided to invite contributions for a weekend symposium. The response to and interest in archaeological sediments gave strong evidence that the production of a volume of the symposium papers would be much appreciated. Types of sediment were interpreted in a broad sense to include not only geological sediments, but also soils and organic remains. Participants were drawn from a variety of academic disciplines including geography, geology, geomorphology, pedology, physics, archaeology and other environmental sciences. The common concern was the reconstruction of former environments using specialist techniques. As a result the symposium had a distinct unity of feeling, and this is reflected in the title of this volume. 'Geoarchaeology' is a term which Professor A.C. Renfrew uses in his introduction, meaning the study of archaeology using earth science techniques.

The meeting was divided into four sections. In Section 1, which was on techniques, attention was first focussed on field recording and later concentrated on suitable laboratory methods and their applications. This was followed by two sections concerned with applications to particular sedimentary environments: Section 2 on coastal and lacustrine environments and Section 3 on terrestrial environments. The final Section dealt specifically with biological sediments and their environmental interpretation. It is hoped that this collection of the papers is a representative selection of research themes in geoarchaeology and that the overall effect is an illustration of their relevance.

Grants were made towards the cost of the symposium by the Royal Society and the British Academy and grateful acknowledgment is made of this financial assistance. Support was also obtained from St. David's University College, Wales and from the University of Southampton. The organisers wish to express their appreciation to the following people for their help in preparing and organising the symposium: Mr. and Mrs. A. Connor, Mrs. C.E. Davidson, Mrs. N. Fergusson, Mrs. R. Mercer, Professor A.C. Renfrew, and Mrs. A. Tabesfar. Thanks are extended to the chairmen of symposium sessions: Professor V.B. Proudfoot, Professor A.C. Renfrew, Professor A. Straw, Mr. J.M. Wagstaff, and Mr. J.J. Wymer. Contributors are warmly thanked for participating in the meeting and for submitting their papers within a tight schedule. Finally the editors are very pleased to acknowledge the enthusiasm and constructive help received from the publishers.

Donald A. Davidson   Myra L. Shackley

Lampeter   Southampton

LIST OF PARTICIPANTS†

J. Bintliff (Leckhampton, Corpus Christi College, University of Cambridge)

P.C. Buckland (Doncaster Museum, Doncaster, Yorks.)

J.A. Catt (Pedology Department, Rothamsted Experimental Station, Harpenden, Herts.)

G.H. Cheetham (Department of Geography, University of Reading)

D.A. Davidson (Department of Geography, St. David's University College, Lampeter, Wales)

C. Delano Smith (Department of Geography, University of Nottingham)

G.W. Dimbleby (Institute of Archaeology, 31-4 Gordon Square, London WC1)

J.G. Evans (Department of Archaeology, University College, Cardiff)

F.G. Fedele (Istituto e Museo di Antropologia, Università di Torino, Italy)

I. Graham (Rheinisches Landesmuseum, 53 Bonn. Colmantstr. 14-16, West Germany)

W.G. Jardine* (Department of Geology, University of Glasgow)

D.A. Jenkins (Departments of Biochemistry and Soil Science, University College of North Wales, Bangor)

R.L. Jones (Department of Geography, Lanchester Polytechnic, Coventry)

A. Kirby* (Department of Psychology, University of Leeds)

M.J. Kirkby (Department of Geography, University of Leeds)

H. Laville* (Laboratoire de Géologie du Quaternaire et Préhistoire, Université de Bordeaux, 33405 Talence, France)

S. Limbrey (Department of Ancient History and Archaeology, University of Birmingham)

A. Morrison (Department of Archaeology, University of Glasgow)

I.A. Morrison* (Department of Geography, High School Yards, University of Edinburgh)

B. Proudfoot (Department of Geography, University of St. Andrews, Fife, Scotland)

A.C. Renfrew (Department of Archaeology, University of Southampton)

F.R. Schweitzer (South African Museum, P.O. Box 61, Cape Town, South Africa)

M.L. Shackley (Institute of Archaeology, University of Oxford)

A. Straw (Department of Geography, University of Exeter)

*Authors who were unable to attend
†The meeting was held in the University of Southampton on 15-16 December 1973.

A.J. Tankard* (South African Museum, P.O. Box 61, Cape Town, South Africa)

A.H. Weir (Pedology Department, Rothamsted Experimental Station, Harpenden, Herts.)

T.J. Wilkinson (24 Swallowcroft, Eastington, Stonehouse, Glos.)

J.Ll. Williams (Department of History, University College of North Wales, Bangor)

J.J. Wymer (Department of Anatomy, University of Chicago, U.S.A.)

INTRODUCTION

Colin Renfrew
Archaeology and the earth sciences

Archaeologists have always been ready to exploit the natural sciences in order to make the material remains of the past yield more information about human activities and human history. And workers in the sciences, for at least two centuries, have been willing to use their special skills in this way. Sir Isaac Newton, already in 1690, had employed his mathematical and astronomical expertise to calculate the precise date of astronomical events reported by the Greeks, and used these calculations to amend the chronology of ancient Egypt. Halley in 1720 utilised what was then known of petrology to comment on the composition and age of the stones of Stonehenge. Yet it is only in the past two or three decades that the archaeologist and the scientist have gone beyond the confines of their own speciality, working together to bridge their mutual ignorance of each others' disciplines and to broaden the limited perspectives which so often keep the archaeological text and the specialist report rigorously segregated at different ends of any publication.

The past twenty years have, as a result, seen the emergence of at least two new disciplines. Neither pure science nor pure archaeology, they sometimes arouse suspicions on both sides, but at their best they are able to draw upon the accumulated skills of two different traditions of scholarship and research. In the first place, the application of the techniques of the physical sciences to the problems of the remote sensing of archaeological features, to the dating of artefacts and to their chemical analysis, led to the emergence of groups of specialists engaged in what is today called *archaeometry*.

That discipline is now practised by a number of workers, all trained in the physical sciences but devoting their skills specifically to archaeological problems, to the extent that they are no longer simply scientific advisers or outside specialists aiding archaeological research, but archaeological researchers themselves.

A second focus of research has been the application of the biological sciences to the documentation of the flora and fauna exploited by early man, especially as food resources, and of the plant and animal environment in general. Again groups of specialists have worked towards the application of the appropriate scientific techniques to archaeology to the extent that they are no longer simply zoologists or botanists as such. Their skills are directed towards the solution of archaeological problems, problems which they themselves help to define, rather than acting merely as advisers called in by the archaeologist. Conscious of the new unity of their discipline, they now often refer to this endeavour as *bio-archaeology*.

These two disciplines can only very loosely be subsumed under the general heading 'archaeological science', for in fact they hold little in common. The field of research activity of each and the methods employed in them are remote. Discussions of plant remains, for instance, would look out of place if published in *Archaeometry* or the *MASCA Newsletter*, and the principal bond which archaeometry and bio-archaeology have in common is archaeology itself.

Traditionally the earth sciences enjoyed a much closer relationship with archaeology than these relative newcomers, indeed they assisted at the very birth of the subject in the early days when the antiquity of Man was first established. In recent decades, however, their application has been primarily in the field of Pleistocene studies and students of post-Palaeolithic man have made less frequent use of geomorphology or of soil science than of physics or metallurgy. Only in the past ten years or so has interest revived, and the relevance of geological processes working in geologically recent times been more fully appreciated. The conference which this volume records amply indicated both the wide field of problems and the broad range of methods for approaching them within the scope of what perhaps may now be called 'geoarchaeology'. For it seems that we have now reached a threshold of awareness comparable to that achieved some time ago among colleagues practising in the physical and biological sciences. We have been developing, or assisting in the development of, what amounts to the emergence of a new discipline. This discipline employs the skills of the geological scientist, using his concern for soils, sediments and landforms to focus these upon the archaeological 'site', and to investigate the circumstances which governed its location, its formation as a deposit and its subsequent preservation and life history. This new discipline of geoarchaeology is primarily concerned with the context in which archaeological remains are found. And since archaeology, or at least prehistoric archaeology, recovers almost all its basic data by excavation, every archaeological problem starts as a problem in geoarchaeology.

The nature of geoarchaeology

Geology has always played a major role in archaeological theory and practice. The principle of stratigraphic succession, upon which all archaeological excavation rests, is itself a construct of the geologist. Sir Charles Lyell concerned himself in his *Antiquity of Man*, published in 1863, both with the geological context of archaeological remains, and with the nature, and above all the rate of the processes of sedimentary deposition of the deposits in which those remains were found. A direct line of work links Lyell with F.E. Zeuner, whose *Dating the Past* in 1946 summarised a century of research. But geology or geomorphology at that time, when brought to bear upon archaeology, was primarily concerned with questions of chronology – indeed Zeuner used the term 'geochronology' to describe his work. It is only in the past two decades that archaeology has been rid of this overriding concern by the development of isotopic dating methods, so that deposition rates or the world-wide correlation of supposed 'pluvials' are much less fascinating than they used to be. There are of course still major questions relating to world-wide climatic changes which remain to be solved, and this volume contains an important contribution to one of them, the question of the chronology of eustatic sea-level changes in Holocene times, but today most geoarchaeo-logical research is focussed upon the site.

An archaeological 'site' is a preserved locus of human activity. The term implies that human activity took place at the location in question, that it left some material trace (possibly in the form of artefacts), and usually that this trace became buried so that some indications of it have been preserved. The most important archaeological sites are settlements – places of residence – but any traces of activity, if found in their original position, are sufficient to constitute a site: the notion of a site without artefacts is not a new one. In special circumstances even the stipulation that the remains be preserved in their original position is not essential.

1. *The site and its position in time*

The determination of the stratigraphic and hence chronological position of the site is the traditional concern of geoarchaeology. And indeed it remains a primary concern today. Occasionally sites of relatively recent date are amenable to study of this kind, such as the settlement destroyed in the Late Bronze Age eruption of Santorin, whose stratigraphic context was first systematically discussed in 1879 by F. Fouqué in *Santorin et ses éruptions*. In general, however, remains of earlier formation are more amenable to geological study in this kind of way. The special feature of this approach is that the deposit which actually contains the archaeological remains need not necessarily be any more interesting than the strata underlying or overlying it, for the deposit is studied for its context within a sequence of geological processes. It remains true, however, that broad geographical comparisons fall within the scope of geoarchaeology and the important work of correlating, for instance, glacial and interglacial periods in different areas, is still far from complete. Such problems are no longer new but they remain central to Palaeolithic studies.

2.   The formation of the site

It is only very recently that archaeologists, more self-critical now about problems of sampling and the interpretation of evidence, have become specifically concerned with the processes by which artefacts and other materials are lost, broken or discarded in such a way as to find a place in the archaeological record. Matching a new concern for detailed distributional studies is an awareness of the need to understand more clearly the essentially geological processes (albeit modified by human action) by which these materials come to be buried at all.

Archaeological excavation has always involved stratigraphic interpretation, but in the classical sense this means little more than establishing relative sequence. Soil horizons are today increasingly studied, but there has still been exceedingly little research into the processes of formation of midden deposits, from which the majority of artefacts on settlement sites in fact come. Dr. D.A. Davidson's paper in this volume is indeed one of the first studies to focus on processes of formation and erosion on tell mounds. Many of the questions raised are applicable also to the deep deposits found on urban sites.

The archaeologist spends much of his time digging, and most of what he removes from the ground may be described as 'sediments'. Yet very rarely is the nature and origin of this 'dirt' itself studied by the scientific means available. Increasingly it is subjected to flotation and fine sieving by the bio-archaeologists, to obtain seeds, terrestrial molluscs, small bones and now insect remains. But the potential offered by particle-sized analysis, chemical analyses and other methods of the geomorphologist has yet to be fully exploited.

3.   Dispersal processes

It is an archaeological truism: 1) that few artefacts used on a site are generally lost or discarded, 2) that few of these are buried, 3) that few of those buried are preserved, 4) that only a proportion of those preserved lie in the areas selected for excavation, and 5) that not all of those preserved in those areas are in fact recovered. At each stage there are problems of differential preservation. And the difficulties of sampling even that which is now present are substantial, as Professor B. Proudfoot's interesting discussion of phosphate analysis incidentially makes clear. Numbers 2, 3 and 4 are the product of geomorphological and sedimentological processes as much as of human agency. Indeed the elegant paper by Dr. A. Kirkby and Professor M.J. Kirkby is one of the first to investigate this question, and illustrates the extent to which purely geomorphological processes will change distributions which might erroneously have been supposed to be the result primarily of human patterning.

4.   The site and its environment

The ecological approach widely practised in archaeology today implies that the reconstruction of the early environment is a major concern of the excavator. In the past archaeologists have often assumed that gross changes in the environment, like those illuminatingly discussed in Professor A. Straw's paper, are in general to be expected only for Palaeolithic sites. As a result the environmental approach has focussed primarily on changes of flora and fauna. Increasingly, however, the archaeologist is being taught by the geomorphologist that the changes in landform occurring in geologically recent times can be of substantial importance. In this volume, for instance, Miss C. Delano Smith working in Italy and Mr. J. Bintliff in Greece discuss major changes in landscape and site environment which have occurred within the past few thousand years. Such changes must be' of real concern to anyone working within an environmental or ecological framework.

The reconstruction of the original environment of the site offers in addition hope of interpretation of a different kind. For the environment enjoyed by the occupants of a site is to some extent a matter of choice, change of environment being brought about by change in site location. Archaeologists are now increasingly realising how a study of the location of sites can be used to reveal a broader strategy of exploitation by their original inhabitants. Studies of site location thus bring into closer relationship the geomorphologist and physical geographer on the one hand, and the human geographer and anthropologist on the other.

For too long archaeologists have regarded the sites they dig as simply stratigraphic successions, with destruction phases and construction phases, without reaching any real understanding of just how these successive episodes actually produced the solid evidence unearthed. The digging archaeologist has always been proud to call himself a 'dirt archaeologist', to distinguish himself from the mere armchair variety. But until recently the 'dirt' has been mere waste, tiresomely intercalated between the artefacts sought.

Today we are coming to realise that this itself may be the real 'pay dirt', promising detailed insights into the formation of the 'cultural' strata themselves, and containing microfaunal and microfloral remains which can give a surprisingly full and detailed picture of the local environment.

This potential can only be realised, however, if fieldworkers perceive simultaneously both the archaeological aims and the technical methods appropriate for their realisation. And this perception can be made effective only by close collaboration actually in the field, by specialists who understand each other's perspective and language, by a real meeting of disciplines. The questions to be answered are not of archaeology alone, nor of geomorphology, but of geoarchaeology.

Section 1
Techniques in Geoarchaeology

M.L. Shackley

The Danebury project: an experiment in site sediment recording

1.   The hillfort at Danebury

The Iron Age hillfort of Danebury,(Hampshire, England) is of medium size and is exceptionally well preserved. It is situated some 5 km north-west of Stockbridge, at 160 m O.D. on the Upper Chalk. The massive univallate defensive earthworks enclose an area of some 13 ha and are roughly circular. There is evidence suggesting initial Beaker and Bronze Age occupation of the site, but the first defence of the hilltop is assigned to the fifth century B.C. The fort was remodelled in the fourth century B.C. in a design that stayed virtually constant until a hasty refitting on the eve of the Roman invasion.

1.1 The excavations

Danebury belongs to Hampshire County Council, who in 1969 sponsored a trial excavation directed by Professor B.W. Cunliffe, to test the archæological potential of the fort in view of a proposal to plant it with trees. The results showed that the fort had been intensively occupied. With this in mind a broadly based research programme was instigated, within the 'rescue' framework, the object of which was totally to excavate the defended area. Such a proposal will take at least five years of continuous excavation which began in the spring of 1974, sponsored by the newly formed Danebury Trust. The Danebury project will also feature within a wider scheme aimed at studying the evolution of the Wessex landscape (Bowen and Cunliffe, 1973).

With the exception of some early structures, all phases of the occupation of the fort were associated with numerous pits dug into the chalk, and these pits varied in style and shape. By the end of the 1971 field season, 117 pits and nearly 1000 post-holes had been examined, yielding a settlement pattern of five rows of rectangular timber buildings on lines parallel to the rampart. An earlier phase of occupation associated with smaller four-post structures could be recognised from discrete examples in the 1972 excavations. A later phase, mainly recognised from pits, was also evident. During the 1972 season, 1600 m² were examined in the centre of the fort to reveal a total of 266 pits dating from the fourth to the first centuries B.C. Finds from the site, coming chiefly from the pits, included pottery groups and many small objects, as well as large quantities of animal and human bone, charred grain, seeds and charcoal.

1.2 The pits

The function of the large numbers of pits occurring at Iron Age sites has been a matter of speculation for some years. Bersu (1940) was the first to postulate the theory that most were intended for grain storage, although he suggested that others were designed for the storage of other foods such as nuts or berries. Some certainly have specialised functions, including clay puddling or burial, but the majority were clearly used for storing grain. Bersu (1940), Bowen and Wood (1968) and Reynolds (1967, 1969) have shown by experiment that grain may be stored perfectly well in chalk cut pits, either lined or unlined, for several years. The grain will germinate satisfactorily as long as the pits are efficiently sealed and all rodents excluded. There is, however, evidence to suggest that the pit turns 'sour' after an unspecified number of years.

At Danebury the pits were abandoned after a time, and then either filled with rubbish and occupation material, or left open and allowed to silt naturally. It is clear, therefore, that since the bulk of the archaeological material from the site is recovered from the pits an interpretation of the pit sedimentation patterns is necessary if the development and chronology of the site is ever to be fully understood.

Over 500 pits had been excavated by the end of the 1973 season, but it is estimated that the total excavation of the site may yield in the order of 5000 more. It was therefore decided that a computer based system for recording the pit sediments and their inclusions was necessary, a pilot scheme being initiated during the 1973 season. Since the natural siltings are composed principally of chalk weathering products, random sampling for laboratory analysis would not have been profitable, the information obtainable from routine testing of such samples being minimal. Few questions were asked that could not be answered by adequate field description supplemented by more selective laboratory analyses.

2.   *The recording system*

Due to the rapid excavation of the pits, the fillings of which are half-sectioned and drawn before being removed, it was necessary to record the sedimentation pattern from sections similar to that shown in Fig. 1 although further observations could, of course, be made as the second segment was being removed. Since it is neither possible nor desirable to record all the characteristics of a sediment, those chosen were selected as being likely to give the maximum relevant information for the questions that required answering. Most data were recorded in the field directly on to computer coding sheets, together with information supplied by the bone specialist, Mrs. A.

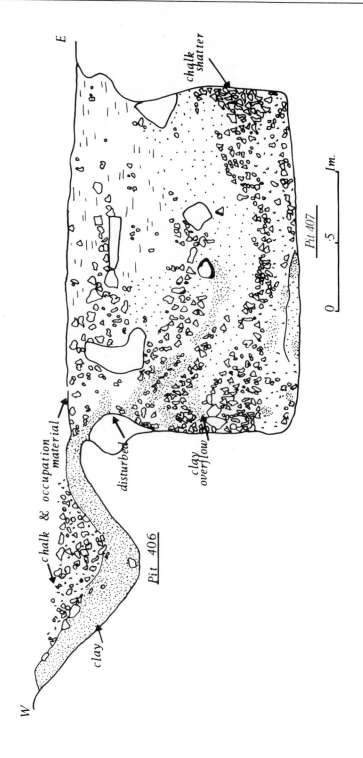

Figure 1. Sedimentation pattern of pits 406-407.

Grant, who was also present on site. This minimised data loss in transference and optimised use of laboratory time. In the early stages of the project it was found that simple coding systems used, for example, in describing the shape of a pit or layer, had to be devised off the site and added later. This procedure is not to be recommended since it is difficult to make such observations after the excavation. Now that the coding systems are established the recording should progress much more smoothly.

2.1 Pit coding sheets

Dr. J.D. Wilcock very kindly collaborated with the writer on the computing side of the project, since the handling program forms part of the PLUTARCH system (Wilcock and Shackley, 1974). Computer coding sheets were used for the recording of data, each row on the sheet representing a computer card, and each column on the sheet a single character on the card. Fig. 2 illustrates the standard layout of a pit coding sheet, complete with hypothetical data. The first three cards of each record give standard information that may be printed on to the sheet, since it remains constant for each pit record. Numerical data are recorded in structured data fields, with adequate space left for general comments. A request to the computer concerning pit shape, for example, would result in a print-out of the dimensions (maximum depth, base diameter and pit volume) as well as the general description and the shape code. Note was taken of the basal angle of the pit walls and the general character of the sedimentation pattern of each pit, which was exactly located on the fort by means of a six figure site grid reference, tied in to the larger National Grid system. These data may later be recovered as histograms, for dimensions, or as computer drawn maps of the distributions of particular types or sizes of pits. The date of the excavation was recorded as a decimal number for later retrieval together with other parameters, a computer routine being used that recognises all valid forms of date. Mrs. Grant provided information on the quantity of bone found in each pit, as well as a general coding for special features such as the presence of human bone or articulated animal skeletons.

The system design is very flexible and any amount of information may be changed or added at a later date. Particulars yet to be recorded include the detailed pottery dating sequence, and the presence or absence of other features such as wattle, daub, cobb, charcoal or toolmarks. Work on the coding of these data is in progress.

2.2 Layer coding sheets

In addition to the standard 20 cards for each pit a further 16 cards were used to record more detailed information about each layers. A hypothetical example is shown in Fig. 3. These recorded the characteristics of the layer, including codes for the sediment type, and

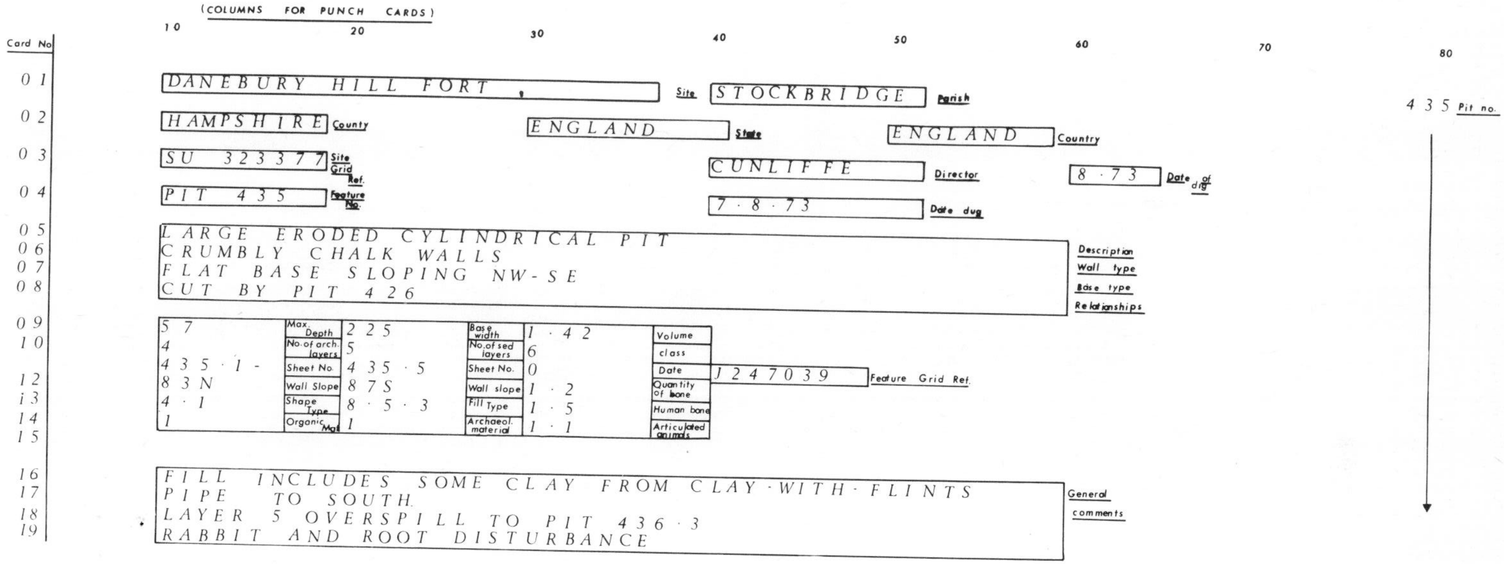

(COLUMNS FOR PUNCH CARDS)

10   20   30   40   50   60   70   80

DANEBURY HILL FORT   Site STOCKBRIDGE   Parish   ENGLAND Country   4 3 5 Pit no.

HAMPSHIRE County   ENGLAND State

SU 3 2 3 3 7 7 Site Grid Ref.   CUNLIFFE Director   8 . 7 3 Date dug

PIT 4 3 5 Feature No.   7 · 8 · 73 Date dug

LARGE ERODED CYLINDRICAL PIT   Description
CRUMBLY CHALK WALLS   Wall type
FLAT BASE SLOPING NW-SE   Base type
CUT BY PIT 426   Relationships

5 7   Max. Depth 2 2 5   Basal width 1 · 4 2   Volume
4   No. of arch. layers 5   No. of used layers 6   class
4 3 5 · 1 -   Sheet No. 4 3 5 · 5   Sheet No. 0   Date
8 3 N   Wall Slope 8 7 S   Wall slope 1 · 2   Quantity of bone   1 2 4 7 0 3 9 Feature Grid Ref.
4 · 1   Shape Type 8 · 5 · 3   Fill Type 1 · 5   Human bone
1   Organic Mat. 1   Archaeol. material 1 · 1   Articulated animal

FILL INCLUDES SOME CLAY FROM CLAY-WITH-FLINTS   General comments
PIPE TO SOUTH TO PIT 436 · 3
LAYER 5 OVERSPILL TO PIT 436 · 3
RABBIT AND ROOT DISTURBANCE

Card No
0 1
0 2
0 3
0 4
0 5
0 6
0 7
0 8
0 9
1 0
1 2
1 3
1 4
1 5
1 6
1 7
1 8
1 9

Figure 2. Pit coding sheet.

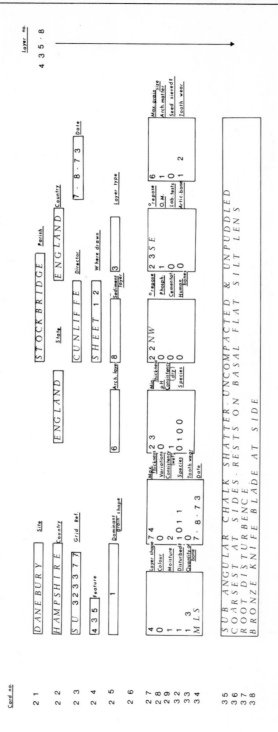

Figure 3. Layer coding sheet.

for the layer shape. Thickness, angle of repose, grain size, colour, moisture content, consistency and cementation were all noted, and pH, phosphate and humus contents were determined when required. Bone records for each layer included information about the species and numbers present, together with notes on the human bones and an animal age factor. Space was left for general comments and the layer number was printed in the last eight columns of each card. The number of cards used for recording the pits excavated in the 1973 season exceeded 8000, and will rapidly become unmanageable as the excavation progresses. It is therefore planned to store all the data permanently on magnetic tape.

3. Preliminary results

3.1 Shape classification

Several classifications of the shape of Iron Age pits already exist, the first being that of Bersu (1940) for the site of Little Woodbury. Within his excavated area 190 pits were examined, most having been used for grain storage; on the basis of aerial photographs a total of 360 was predicted for the whole settlement. Bersu divided his pits into two classes by the shape of their cross-sections in the horizontal plane, and further subdivided them into six shape types. He identified the presence of two major forms, the 'cylindrical' and the 'beehive' types, which have since been found at many other sites. His classifications were made simply on morphological grounds without reference to the sedimentation pattern of the pits, and several of his categories, notably his type F ('barrel' shaped pits) are difficult to identify elsewhere.

The Danebury classification (Fig. 4) is generally similar to that of Bersu, but differs in detail. Its validity will later be tested using the PLUTARCH system. The 'beehive' and 'cylindrical' forms are represented both by fresh and eroded examples, and other minor varieties such as the 'boat-shaped' pit can be clearly identified. There is a special category for pits of distinctive form but not belonging to either of the two main types, a prime example of which is the inverted conical clay puddling pit (Fig. 1), examples of which have also been recovered from the site of Hod Hill (Richmond, 1968). At Danebury an attempt is being made to see whether the pit size, shape or sedimentation pattern tends to vary with time, and if specific types of pit had been dug at different periods (Ellison and Drewett, 1971), although the pottery dating sequence which forms the basis for this sort of chronological work is not yet completed. There is evidence to suggest that the ratio of 'beehive' to 'cylindrical' pits (approximately 5:1) remains constant. The distinction between 'beehive' and 'cylindrical' pits is occasionally difficult to see, although they vary both in basal angle and in sedimentation pattern.

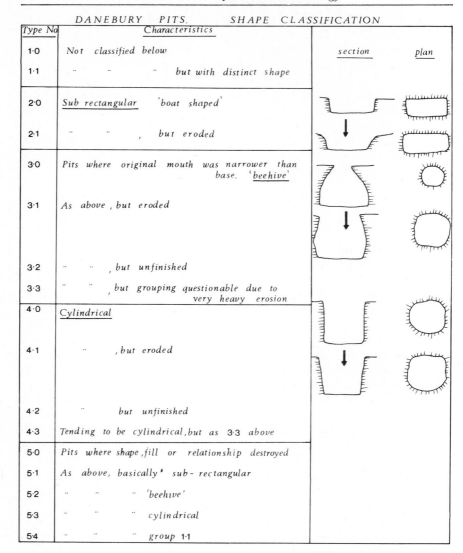

Figure 4. Classification of Danebury pits according to their shape.

3.2 Pit size

Bersu (1940) noted that the 'cylindrical' pits (his types A and B), tended to be shallow, less than 0.5 m deep. His eroded 'beehives', on the other hand, were generally between 1 and 2 m deep. At the site of Gussage All Saints (Wainwright, 1973) 'beehive' pits up to 2 m deep were found, and 'cylindrical' pits 2 to 3 m deep. The Iron Age settlement of Tollard Royal (Wainwright, 1968) had 50 per cent of its pits less than 1 m deep, a figure compatible with the Little Woodbury calculation of 49 per cent less than 1 m deep (Bersu, 1940). A primary question, therefore, is to obtain a picture of the variations in pit size with shape at Danebury, and to examine the chronological progress of the correlations produced.

It seems probable that a close relationship exists between size, shape and function, and this again is a hypothesis that is being tested. Tentative evidence from Danebury suggests an increase of pit capacity with time, although this may be due to the pressure put upon the space available for pit digging.

3.3 Pit sedimentation patterns

It was found possible to make models of the natural filling pattern of the pits by observation. In the 'beehive' examples, which have a basal angle of between 70 to 80°, the primary deposit is generally a domed chalk wash, followed by a dejection cone of chalk shatter. Toolmarks are often visible up to the point where the angle changes, that is in the unweathered zone (Fig. 5). The broadening of the lip by weathering is characteristic, but in uneroded specimens (a fine example of which occurred in 1973, buried by the rampart siltings soon after its abandonment), the lip may be very narrow. In the 'cylindrical' pits (Fig. 5) with basal angles of between 85 to 90°, the first deposit is generally a flat layer of chalky wash, the rest of the fill typically consisting of a little weathered chalk and a great deal of washed and tumbled-in material. The quantity and variety of the archaeological and faunal material is much greater in the 'cylindrical' pits than in the 'beehive' type, due to the greater importance of washed-in material as opposed to sediments produced by erosion of the sides. Fig. 1 shows an anomalous sedimentation pattern where the fill of a large 'cylindrical' pit has been affected by the overspill from the nearby clay-puddling pit (Richmond, 1968) which cuts it. The clay in pit 406 had an acid pH reading, a positive field test for phosphates and a rich microfauna and flora.

A primary question is the rate of fill of the pits. Wainwright (1973) noted that many of the pits at Gussage had been filled in one operation with occupation material and rubbish, and that most had then been sealed and levelled. Bersu (1940) noted that the filling of the Little Woodbury pits was rapid, since no palaeosols or other features had accumulated to indicate a period of stillstand. At Barley in Hertfordshire all the pit fills were deliberate (Cra'ster, 1961), but the present season at Danebury shows that a pit filled entirely deliberately was the exception rather than the rule. This may, however, be due to area specialisation. In naturally silted pits observation has shown that chalk weathering products accumulate very rapidly. Even during the very dry summer of 1973 chalk cut pits left open would silt with fine-grained basal wash deposits at a rate of several centimetres per week. Chalk shatter deposits also form fast, especially in 'beehive' pits where the shape and lip configuration are particularly susceptible to erosion. Pitt-Rivers (1898) sectioned the ditch of Wor Barrow, which he had left open after an excavation, and

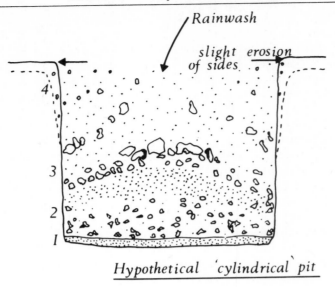

Rainwash

slight erosion
of sides

4

3

2

I

Hypothetical 'cylindrical' pit

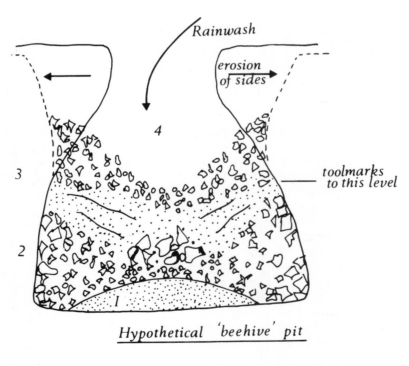

Rainwash

erosion
of sides

4

3   toolmarks
to this level

2

I

Hypothetical 'beehive' pit

Figure 5. Models of two different types of pit. Numbers refer to infill phases.

estimated that it had filled to a depth of 70 cm with chalk erosion products in 4 years. The Overton Down experiment (Crabtree, 1971; Coles, 1973; Jewell, 1963; Jewell and Dimbleby, 1968; Proudfoot, 1967) stressed the speed at which chalk shatter deposits were formed, and provided useful statistics on their accumulation rate.

In some of the larger pits there are several major dejection cones of shatter, separated by wash laminae, which indicate that the deposits probably formed over several seasons. However, in the smaller pits there seems little reason to suppose that there needed to be much more than a year between the abandonment of the pit and its complete silting, and if the pottery dating sequence is sufficiently tight it may be possible to offer some suggestions on the number and density of pits that could theoretically have been open at any one time. A common feature of the naturally silted pits is a central hollow in the concavity of the dejection cone, which is often rich in molluscan remains.

Bearing in mind potential complications such as the influence of topography, it should be possible to make inferences on the rate of fill of the pits, which would affect the interpretation of included archaeological material. From a sedimentological point of view the variations in angle of rest of the chalk shatter deposits is an interesting feature of the site, and it may be possible to include this in theoretical models of chalk weathering as an extension of the work of Scheideggar (1961).

Conclusions

Apart from the research lines indicated above it is hoped that the project will yield information concerning the distribution of faunal and archaeological material on the site, in relation to position within pits and in sediments of different types at different periods. Further work is being done on the development of silting models and the differential preservation of organic materials in different sediments. Changes in the faunal and artifact assemblages in pits of different shapes, types and functions will also be noted, as will changes in the amount of human bone recovered. At present it seems likely that it will be possible to comment on the seasonality of pit digging and pit abandonment.

The system used (PLUTARCH) is sufficiently flexible to cope with the field recording methods outlined above, and with the results of selective laboratory analyses, and with any future amendments. If data are to be accurately recorded from such a large and valuable site, a computerised system is obviously necessary, and there seems little reason why a version of this system should not become standard practice for recording excavations of this type. The project will yield information of value fundamental to the study of the sediments and to an understanding of the complex relationships between the sediments and other factors affecting the interpretation of the site as a whole.

Acknowledgments

The writer would like to thank Professor B.W. Cunliffe for permission to work on his site, and Dr. J.D. Wilcock, Mr. D.W. Startin and Mrs. A. Grant for their help and co-operation, fundamental to the design and execution of this pilot project.

REFERENCES

Bersu, G. (1940) 'Excavations at Little Woodbury, Wiltshire. Part 1: The settlement as revealed by excavation', *Proc. prehist. Soc.* 6, 30-112.

Bowen, H.C. (1967) 'Corn storage in antiquity', *Antiquity* 41, 214-15.

Bowen, H.C. and B.W. Cunliffe (1973) 'The evolution of the landscape', *Ant. J.* 53, 9-13.

Bowen, H.C. and P.J. Fowler (1966) 'Romano-British rural settlement in Dorset and Wiltshire', in C. Thomas, ed. *Rural Settlement in Roman Britain*.

Bowen, H.C. and P. Wood (1968) 'Experimental storage of corn underground and its implications for Iron Age settlements', *Bulletin of the University of London Institute of Archaeology* 7, 1-14.

Coles, J. (1973) *Archaeology by Experiment*. London.

Crabtree, K. (1971) 'The Overton Down experimental earthwork, Wiltshire 1968: geomorphology of the area'. *Proceedings of the University of Bristol Speleological Soc.* 12, 237-44.

Cra'ster, M.D. (1961) 'The Aldwich Iron Age settlement at Barley, Herts.', *Proc. Cambridge Antiq. Soc.* 54, 22-46.

Cunliffe, B.W. (1971) 'Danebury, Hampshire: first interim report on the excavations', *Ant. J.* 51, 240-53.

Ellison, A. and P. Drewett (1971) 'Pits and postholes in the British Iron Age: some alternative explanations', *Proc. prehist. Soc.* 37, 183-195.

Jewell, P.A. (1963) *The Experimental Earthwork on Overton Down, Wiltshire, 1960*, publication of the British Association for the Advancement of Science.

Jewell, P.A. and G.W. Dimbleby (1968) 'The experimental earthwork on Overton Down, Wiltshire: the first four years', *Proc. prehist. Soc.* 32, 313-42.

Pitt-Rivers, A.H. (1898) *Excavations in Cranbourne Chase*, 4 (private publication).

Proudfoot, V.B. (1967) 'Experiments in archaeology', *Sci. J.* 3, 59-64.

Reynolds, P.J. (1967) 'Experiments in Iron Age agriculture', *Trans. Bristol and Gloucester archaeol. Soc.* 86, 60-73.

Reynolds, P.J. (1969) 'Experiments in Iron Age agriculture, part 2', *Trans. Bristol and Gloucester archaeol. Soc.* 88, 29-32.

Richmond, I.A. (1968) *Hod Hill 2: Excavations carried out between 1951 and 1958*. London.

Scheideggar, A.E. (1961) *Theoretical Geomorphology*. Berlin.

Wainwright, G.J. (1968) 'The excavation of a Durotrigian farmhouse near Tollard Royal in Cranbourne Chase, south England', *Proc. prehist. Soc.* 34, 102-48.

Wainwright, G.J. and M. Spratling (1973) 'The Iron Age settlement at Gussage All Saints', *Antiquity* 42, 109-30.

Wilcock, J.D. and M.L. Shackley (1974) 'The recovery of information from Iron Age pits: The Danebury project and PLUTARCH,' *Computer Applications in Archaeology* 2, 82-90.

DISCUSSION

Miss C. Burek asked two questions about the method of recording information. In answer to the first part the speaker stated that the time taken to record information was variable since at the beginning of the experiment the need for developing coding systems for particular features tended to slow down progress. However the recording system became faster to use as the operator became more familiar with it, and once the majority of information was recorded on the site this also increased recording speed by cutting out a data transference stage. Miss C. Burek's second question concerned the adopted procedure if there was any doubt about details, and the speaker replied that in this case a blank was recorded on the appropriate card and a note made on the comment card.

Dr. M.S. Tite asked whether the study had provided evidence for methods by which rubbish pits were filled. In reply the speaker said that there seemed to be no overall patterns, but that the pits appeared mainly to have been filled by scraping off the surrounding area and dumping it into the pit. There was evidence for included burnt material, but little for the actual burning of rubbish in the pits. Mr. J.J. Wymer enquired whether there was any evidence for cess pits on the site, and Dr. Shackley answered that none had been recognised yet, although there were examples known from similar hillforts. Dr. J.G. Evans commented that molluscan evidence might be a useful guide to the rate of pit infill and to the identification of stillstand phases. Mr. P. Buckland suggested a similar use for insect remains although these are not generally well preserved in calcareous material.

F.G. Fedele

Sediments as palaeo-land segments: the excavation side of study[1]

> We can say with confidence that systems theory *does* provide a new way of looking at old problems, and therefore it will almost certainly lead to important new insights. (Doran, 1970, p. 294)

Sediments on archaeological sites have usually been considered embarrassing material concealing the objects of prime importance. The recognition that sediments have to be considered a subsidiary and optional source of interesting information to parallel other lines of evidence about the past came much later. After some twenty years of gradual refinement, it is now possible to take an integrated view of sedimentological studies in archaeology. Today a formalisation of these studies is badly needed. The goal is still far distant and the task apparently unattempted, although the increasing wealth of available technical aids makes the need more acute.

The case for conceptualisation is proposed in this paper which presents a tentative suggestion of ways to approach this subject. It is easy to maintain that sediment studies should be carried out as explicitly and objectively as possible. However, formalisation goes a step further, as a means of emerging from amorphism towards consciousness and rational design. It casts case studies and individual contributions in the context of epistemological perspective and provides frameworks to integrate them into a viable whole.

The basis of 'archaeological' sediments may be viewed as merely one facet of a complex multidimensional phenomenon. There is essential unitary congruence between the 'sediments' or 'deposits' and the other dimensions of cultural-environmental information systems. In the first section of this paper attention is given to the conceptual background to these types of studies, and this is followed by a discussion of the problems. Only an explicit appraisal of the theory underlying a discipline or a subdiscipline allows a comprehensive realisation of research needs and strategies. The latter section of this paper is devoted to aspects of the interactions between sediment studies and excavation techniques. An attempt is made to describe some relatively simple field procedures that can profitably be applied to a wide range of situations, in keeping with the previous theoretical assumptions. Examples are simply outlined and kept to a minimum.

1.   *Towards a general theoretical model*

1.1 A human ecosystem model

Sediments of potential archaeological significance should be considered within the comprehensive conceptual framework of a 'human ecosystem' (*HE*). By human ecosystem is meant a dynamic system (Ashby, 1965; Klir, 1972) comprising the total environmental context of given cultural entities, the cultural (or socio-cultural) entities themselves, and the interrelated networks of their mutual

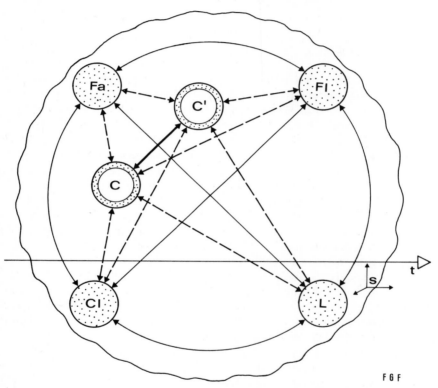

Figure 1. A static and simplified model of a Human Ecosystem (HE), presenting:
(i)   a three-dimensional, highly interconnected territory or 'space' (s);
(ii)   the dynamic equilibrium between the 'human' component in the HE (given Culture System, C, and the summation effect of alien culture systems, C', connected to C by 'cultural coactions') and the non-human 'environment' component;
(iii)   the organisation of the (non-human) Environment System under four essential subsystems ('biotic' – Fa, fauna; Fl, flora; and 'abiotic' – Cl, climate; L, land), connected to each other by 'environmental coactions' and to C and C' by 'interactions';
(iv)   the time dimension (arrow t), implying a 'diachronic' state trajectory of the ecosystem.
(Adapted from Fedele, in press, a)

behaviours (Fedele, in press, a). Subordinate entities are systems in their turn with internal systemic organisation. In this sense the environment embodies all the attributes external to a culture system and their varying and successive states/values in time and space.

The validity of this as well as of other system constructions might be subject to dispute on general grounds (Clarke, 1971, Appendix A), but at present it seems a reasonable and heuristic hypothesis in the light of current data and problems. A schematic model of a human ecosystem is shown in Fig. 1; it is basically inspired by D.L. Clarke's model and nomenclature, in view of their appealing clarity (Clarke, 1968, ch. 3; Stjernquist, 1971; Fedele, 1973a, and in press, a).[2]

Such a system is actually a total ecosystem and might be named accordingly. Naming it a 'human' ecosystem stresses anthropological perspective. This emphasises the peculiar position that man has been occupying in the ecosystem ever since his specialised hunter-gatherer stage of cultural evolution (Braidwood, 1960). Man as a cultural species deliberately and effectively manipulates, disturbs, or pollutes ecosystems, creating irregularities which can be observed in the archaeological record (Stjernquist, 1971). A very special attitude is also possessed by the human observer with respect to cultural accomplishments in his total 'World'. In particular, many material effects of man-terrain interactions ($C \leftrightarrow L$) are among the best perceivable in the archaeological record and are of primary concern in archaeological fieldwork (Binford, 1964).

According to this conceptual framework, the study of archaeological sediments should be co-ordinated in the perspective of the human palaeoecological evidence and its overall research strategy. It ceases, therefore, being subsidiary to some other objective; instead archaeological sediments become a central study in their own right. The term 'sediments' is used in this paper to include everything referring to the ground under archaeological or palaeoecological investigation. In more particular contexts other terms such as 'deposit', 'terrain' (cf. Davidson, 1972) and 'soil' may be used.

1.2 Land and palaeo-land

A basic entity in a HE is 'land' (D.L. Clarke's 'geology'; cf. G. Clark's 'habitat', 1952). Preference is given to land (L) as a technical term for the earth's crust surface supporting biotic components and behaviours, with both its physical and spatial attributes (L_s, where s indicates the space properties of the subsystem). Biotic components include human components. Soil, for example, has been defined from the cultural viewpoint as the physical base on which man lives and works and with which he interacts in energy flow (Hyams, 1952; Hole *et al.*, 1973, p. 230).

Human palaeoecology refers by definition to 'past' systems, or rather to systems which move through successive series of states in archaeological time-trajectories. When the significance of former ecosystems is assessed with reference to man, the term 'palaeo-land' (PL) can be suggested for a specific land subsystem at one point in time in a specific human ecosystem of the past ($PL_{s,t}$). Subscript t indicates time. 'Past' is not a key word for chronological reasons alone, but also for the fundamental side-effects of the elapsing of time ('ageing') and of system transformations. At the upper limit of the time arrow, palaeo-land and land obviously merge into one.

The following variables, interrelationships, and research topics may be considered in this context.

(a) Time implications of the PL, either relative, absolute, or chronometric (Oakley, 1969), and its trends and trajectory through time. Chronological operations are not discussed in this paper.

(b) The mutual oscillatory interrelationships between the PL and the other entities in the ecosystem, especially man. Interactions between the PL and the culture systems, and environmental coactions within the environmental system (Fig. 1), might be distinguished. Let

$HE_{s,t}$ be a specific human ecosystem in time and space, or the summation effect of all cultural plus environmental networks,

$PL_{s,t}$ be the palaeo-land participating in the same ecosystem,

(f)   be a 'buffering' factor accounting for the regulatory capacity of the palaeo-land subsystem,

then a two-way or equilibrium relationship such as is provided by this coupling can be assumed:

$$PL_{s,t} \leftrightharpoons (f).HE_{s,t} . \tag{1}$$

The detailed plot of coactions and interactions of the PLs represents the bulk of investigation and interpretive reasoning in environmental archaeology, or human palaeoecology. The corresponding trend seems to be true in modern geography, mainly as a result of locational analysis (Haggett, 1965; Chorley *et al.*, 1967).

(c) The relationships between an *original* PL and its *extant* relics, in other words, the proportions, degree, and modality of preservation of the PL. PL relics are a function of time, or of events in time:

$$PL_{s,t_p} = f_s \left(\Sigma \varepsilon_s, \Delta t_p \right) \tag{2}$$

where

t_p refers to the observer's present,

PL_{s,t_p} then represents the extant relics of the original PL,

$$PL_{s,t_o} \ (PL_{s,t_p} \subset PL_{s,t_o}),$$

$\Sigma \varepsilon$ is the summation of the relevant events in the time interval,

$$\Delta t_p = t_p - t_o$$

t_o refers to the original point in time or 'zero time'.

In this model, f_s marks a most complex 'segmenting function'; this topic will be expanded below.

1.3 Sediments as palaeo-land segments

It may be anticipated that PLs are in general preserved in different and incomplete ways. For instance, it is widely agreed that palaeosols can occur in relict, buried, and exhumed status according to the modality of occurrence (Ruhe, 1965, 1967). The lack of integrity of the palaeo-environmental record has three inter-related facets: (a) 'fragmentation', or breakage and disjunction into parts; (b) 'dispersion', or separation and dispersal of fragments; and (c) 'incompleteness', or lack of integrity of the fragments themselves. Let $\Delta PL_{s,\Delta t_p}$ represent the total 'loss' of the PL subsystem, measuring its 'non-integrity' after the relevant events in the interval Δt_p:

$$\Delta PL_{s,\Delta t_p} = PL_{s,t_o} - PL_{s,t_p} ; \tag{3}$$

then

$$PL_{s,t_p} = PL_{s,t_o} - \Delta PL_{s,\Delta t_p} = f_s(\Sigma \epsilon_{s,\Delta t_p}). \tag{4}$$

From a general standpoint, sediments at archaeological sites may be viewed as a class of palaeo-land remnants, or 'segments'. The term 'segment' is a technical substitute for fragment — anything that is generated when a continuum collapses. Segment is interpreted as a relatively incomplete yet 'oriented' entity conceivable as a sub-unit of a former whole, with both spatial and temporal delimitations. A segment is oriented in that it keeps some stored 'memory' of the structured continuum to which it belonged. It appears to resemble a sample. Nevertheless the difference is substantial, because samples are culturally controlled fractions, purposefully selected and representative (*Am. Geol. Inst.*, 1960, 1962; see section 2.1), while for that matter segments are more or less culturally uncontrolled. Finally, segments must be often viewed as extrinsic sub-units, as they did not possess any individual identity before the disintegration of the whole.

These qualities also apply to PL segments. Sedimentary bodies down to 'sediment units', regardless of their definition and scale,[3] represent actual parts of past landscapes or at least reflect them indirectly. They are unique information reservoirs on the PLs, and according to relationships (4) and (1) they potentially relate to the broader ecosystems (see relationship 6). A central task for sedimentologists is therefore to detect and investigate in the recognisable PL segments the HE interrelationships as were reflected in the PL whole.

To develop this argument, let 'deposit' designate any local time-oriented set comprising a finite number of seriated sediment units as built up during a given interval Δt. Such a complex sedimentary body represents a higher-power reservoir of ordered evidence and potential information about past ecosystems. At archaeological sites, say at 'palaeo-environmental' sites, *stratified palaeo-land segments* are encountered; there are superposed segments of successive palaeo-land states connected with transforming ecological factors and circuits.

If

S_{t_n} is a sediment unit $\equiv$ a PL_{s,t_n} segment, $\in PL_{s,t_P}$

$D_{\Delta t}$ is a 'deposit', corresponding to the time interval $\Delta t = t_n - t_o$; n being a positive integer if time-units are stratigraphically expressed,

subscript s and t are pointers of distribution in a space-time framework, as above,

then by virtue of (1), (2) and (4) the system

$$S_{t_n} \in PL_{s,t_P} = f_s \left(\Sigma \epsilon_{s,\Delta t_P} \right) \qquad (5)$$

$$S_{t_n} \Rightarrow PL_{s,t_n} \Leftarrow (f) \,.HE_{s,t_n} \qquad (6)$$

$$D_{\Delta t} = (S_{t_o}, \ S_{t_1}, \ S_{t_2}, \ldots S_{t_n}) \qquad (7)$$

may be considered a symbolic depiction of the inventory of aims and problems to be pursued in the study of archaeological sediments.

Some of the problems in the operation of the above model have already been discussed; several others must also be examined and these are as follows.

(a) Two main implications of the lack of integrity of the palaeo-environmental record deserve mention (see relationship 5); they apply to all disparate populations of palaeo-environmental data. The first one results from dealing with fragments or segments, not with the whole units, or with residual parts, not with 'organisms'. The raw material for the archaeologist and the sedimentologist occurs as scattered and incompletely preserved parts of former units (Childe, 1956; Piggott, 1965, 1973; Clarke, 1968; Fedele, in press, a). Even where there is almost 'entire' preservation, inherent limits of practical research allow only segments or samples of the unit to be studied (Binford, 1968; Fedele, 1973b). The relationships of a segment to its possible whole (here a sediment unit and the 'implied' PL) resemble those of a sample to its parent population, and have to be thoroughly assessed. Causal agencies must be looked for. Why did segmentation occur? Why and how did a part survive when another was destroyed? The kinds and activity patterns of 'segmenting' agencies constitute a primary field of inquiry; it seems possible to think in terms of a 'segmenting function' f_s as symbolised in (2) and (5).

(b) The second consequence is procedural; the whole must be viewed through its segments. Problems and attitudes in research must be transposed from the theoretical unit to its surviving segments. The previous discussion on the implications of the PLs may be similarly replicated in the context or at the level of discrete PL segments. Some difficulties may arise as to the discrepancy in properties between wholes and segments. Some properties are attributable to a whole but are not evenly inscribed in all or most of the individual parts, especially in a systemic and polythetic distribution model of compound entities (Sokal *et al.*, 1963; Clarke, 1968). Implications and converse implications are not always recognisable or clear-cut. (On theoretical problems of properties and definitions, the work of Achinstein (1971) is relevant.) Moreover, the precise structure and function of the 'living' unit has often been masked or erased in the disjunct fragments of the available record and appears difficult to detect.

The bearing of the segment on the interpretation of the whole is a general and crucial point. Its relevance in inference making and inference reliability is still underestimated, and it heavily affects any interpolation or extrapolation which is drawn from the factual evidence. The danger of extrapolating from the data secured at a single point to inferences of a much wider scope (local, regional, or even super-regional scales) is well known, yet may be difficult to avoid due to its subtlety. The limits of generalisation and inference should be evaluated and explicitly stated in each case.

(c) The local and/or superlocal mechanisms responsible for the accumulation of sediment units in a given terrain (see relationship 7) provide the basis for the interpretation of deposits. The following elements must be analysed to describe and understand a deposit: (i) sediment units, as to sedimentation cycle and their intrinsic attributes; (ii) vertical discontinuities, or stratigraphic boundaries; (iii) horizontal discontinuities and juxtapositions; (iv) stratification (the origins and parameters of superposition); and (v) secondary transformations, both physical and chemical, including biotic-cultural products (the post-depositional changes of the 'fossil' sediments becoming gradually included within successive PLs).

Sedimentation is a complex cycle usually starting from the separation of the particles from the parent rock and ending in the last deposition and immobilisation. It includes the sources from which sediments are derived, transportation and deposition, the changes which take place, the depositional environments, the multiple environmental conditions, and the textures and structures developed in the final deposition. The exhaustive study of sediment units requires all these steps and mechanisms to be taken into account (Sparks, 1971). The origins of the bedded structure of a deposit by discontinuities and superposition may be examined in the light of the

diachronic variation and partition of sedimentary events. Intimate
attention should be regularly given to the specific agencies and
processes within the specific palaeo-morphologic conditions of each
site (or *locus* within a site) and site period. This is most valid for caves
because cave sedimentation (Schmid, 1969; Butzer, 1972) is very
sensitive to the space relationships between sedimentation points and
the geomorphological environment of the cave; the latter is subject to
significant alteration by vault collapses, entrance obstructions, and
similar effects (*cf.* Fedele *et al.*, 1974).

2. · *The excavation-phase study of sediments*

2.1 A model

While several different techniques at different points of investigation
are used today, mainly in the laboratory, the case for a fuller
treatment of archaeological sediments at the very moment of
excavation is stressed. This need may be expressed in two ways: (i)
operational, in other words to carry out sediment analyses, or
'geological' studies in general, in an integrated field strategy; (ii)

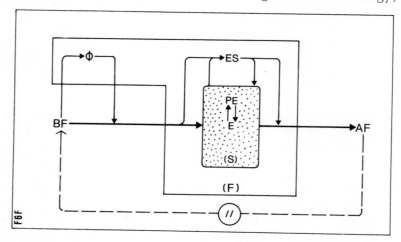

Figure 2. A general operational model for archaeological research activities.

BF	Before-field operations – survey; planning of research; organisation of project; ϕ, checking of plan in the field
F	Fieldwork
S	Site operations (stippled)
E	(Formal) excavation operations
PE	'Para-excavation' operations – integrative operations within the site, such as site survey and mapping of site features
ES	Extra-site operations – integrative investigations outside the site
AF	After-field operations – outside field studies; laboratory work; report preparation, etc.
(II)	a new research-cycle – the behavioural feed-back stimulated by the outcome of the original cycle

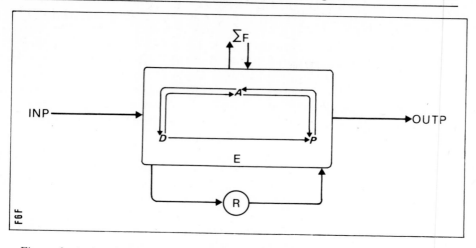

Figure 3. Archaeological fieldwork: a model for excavation operations, within the domain of site and field operations.

INP	input of system E: 'plan', or research objectives and constraints
OUTP	output of system E: 'results', or materials and observations
E	Excavation operations (see Fig. 4)
D	Digging department
P	Processing department
A	Analysis department
ΣF	other field operations
R	a summation of controlling 'regulators', constraining the operational response of system E in the light of 'context pressures' (intrinsic local factors and working conditions)

practical, to reorientate digging and recording methods in order to enhance the detection and understanding of 'significant' characteristics of archaeological sediments.

Central to this point is the claim that the maximum information must be extracted from the terrain immediately at the digging stage, or at least in the course of site operations. Some types of information may be captured exclusively at either the 'site', 'excavation', or 'digging' point (*cf.* Figs. 2 and 3). As Binford (1964) stated in discussing artifactual evidence, 'many of the formal observations must be made while the feature is being excavated'. Sampling problems become quite relevant in this context. Sampling seems to be an inherent necessity in the universe of our 'ecofactual' evidence (Binford, 1964), for sediments are today the only class of material evidence which cannot be recovered in full from a site.[4] Sediments must therefore be examined as far as possible in the field and suitable devices have to be prepared for this purpose (*in situ* observation, cascade sieving of *all* the earth or of random samples, desk computation and measuring, geological recording; see section 2.2).

As in geology, sampling is also crucial in archaeology. Heavy reliance on samples inevitably results in loss of information, unless the general theoretical background is borne in mind. The nature of samples and the proper procedure for selecting a sampling plan congruent to the variability of the problem must not be neglected (Cochran, 1963; Rootenberg, 1964; Binford, 1964; Clarke, 1968, p. 549). Sampling operators must be aware of the whole range of problems which are associated with samples (Coles, 1972) and must act within an appropriate operational context (see section 1.3). As an excavation has to be a set of experimental procedures to inspect testable hypotheses and to gain calibrated answers to selected problems (Atkinson, 1960; Coles, 1972; Piggott, 1973; Hole *et al.*, 1973; Fedele, 1973b), a fundamental feed-back circuit should be established between 'excavators' and 'sediment specialists'. A model for archaeological operations is required to place this in perspective (Figs. 2 to 4).

The place of archaeological fieldwork is illustrated in Fig. 2 within a basic and condensed operational model for archaeological research activities (*cf.* Struever, 1968; Clarke, 1968 p. 36; Alexander, 1970; Coles, 1972); the elements in the model are explained in the figure captions. The excavation segment of such a model is expanded as a flowsheet in Fig. 3. This diagram is intended to convey the relationships between the 'excavation' system and the broader archaeological-procedural system in terms of input-output. It also emphasises the role of controlling 'regulators' (R). R combines the 'context pressures' affecting the system's behaviour and the constraints on any operational outcome of the system itself. Context pressures are represented by intrinsic factors of the site, such as local geomorphological and topographical conditions, and by working conditions associated with the project. A thorough understanding of the *local*, 'microsetting' conditions (*cf.* Butzer, 1972) is essential; a pertinent technique will be given in the next section. The system 'input' (Fig. 3) is by itself a source of regulating information – questions to be posed, hypotheses to be proved, and demands and constraints to be met. All these regulators act upon the 'excavation' system in modulating strategy decisions and actual application of methods (for instance, the strategy to be adopted to tackle specific and/or new geological problems, and the selection of what analyses to perform at a single site; *cf.* Cornwall, 1960).

A detailed diagram (Fig. 4) illustrates the proposed model for the excavation-phase operations in archaeological fieldwork.[5] The activities and staff in excavation may be divided into three departments, at least in simple situations: D, digging department; P processing department; and A, analysis department. Feed-back circuits are manifest in such an excavation structure. With geological analyses, department A receives samples and cooperates in the control of sampling, both in digging and in processing; it may give timely directions or predictions for the excavation as well as suggestions for the recording of the evidence.

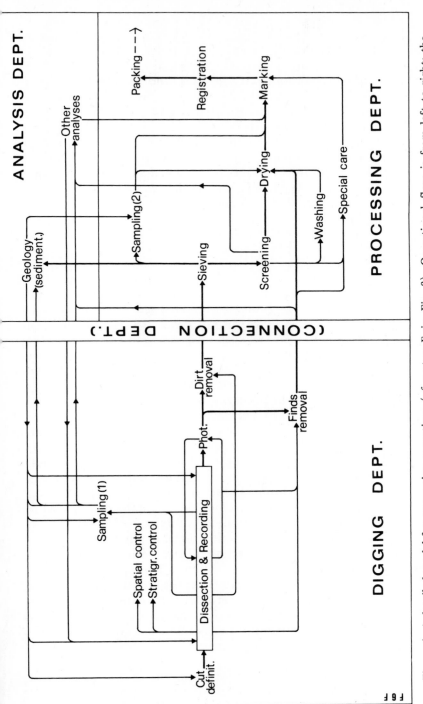

Figure 4. A detailed model for excavation operations (*cf.* system E in Fig. 3). Operational flow is from left to right; the existence of feed-back circuits must nevertheless be noted. The heavy arrows in the flowsheet indicate 'normal' channels. (Adapted from Fedele, 1973b.)

2.2 Some technical suggestions

It is anticipated that the excavation-phase study of sediments will give especially valuable information about, for instance:

    (i)   depositional environments, agencies, and regimens, allowing the detailed reconstruction of event sequences in a deposit;

    (ii) sequential transformations of a sediment through time ('sediment/deposit evolution');

    (iii) patterned ground phenomena and identification of natural in contrast to man-made objects or features in the field, in otherwise doubtful situations (e.g. bone assemblages in caves, Palaeolithic cobble pavings, and natural elements displaced or used by man, termed 'manuports' by M.D. Leakey, 1971).

A few trends and procedures which seem to be relevant to the approach advocated here are briefly discussed.

1. Digging: Strata definitions and the search for 'elemental' events. In the theory of archaeological deposits (Fedele, in press, a), the term 'elemental sediment unit' (ESU) might be suggested for a unit constituting the smallest geologically homogeneous entity as perceived in excavation. An ESU is, in other words, the smallest sedimentary body contained between two consecutive recognisable discontinuities. This definition is based on a criterion of relative 'elementalness' (*cf.* 'sedimentation unit' in *Am. Geol. Inst.*, 1962). An ESU may well be a stratigraphic division, a lateral (facies) differentiation, or even a pedological horizon. The nature of the sedimentary body and of the interfaces separating units must be determined in order to define an ESU (see section 1.3). Upon recognition in digging each ESU must be given a code (e.g. PGL3) and must be formally recorded on standard sheets; its description will be completed when it is examined in detail. Formal digging techiques must accordingly be brought into operation (Fedele, 1973b). Allowing for minor adjustments, this flexible method is applicable to a wide spectrum of field situations. The comments and the models set forth in section 1.3 for unspecified sediment units can be applied to ESUs.

The rationale of the ESU's method is the desire formally and unambiguously to isolate constituent elements. This approach may be extended to the search for 'elemental events' in the sedimentary evidence: i.e. multiple sets of clearly seriated and unambiguous events which can provide the basis for the reconstruction of changing palaeo-environments. The 'figure attribute' analysis which is later discussed is relevant to this topic. As a project progresses, floating small-scale sequences of simple events can easily be extracted from particular components of a deposit; then a more comprehensive sequence can tentatively be recognised. This can be updated as more information becomes available (Fig. 5). Examples from the MF Project have not yet

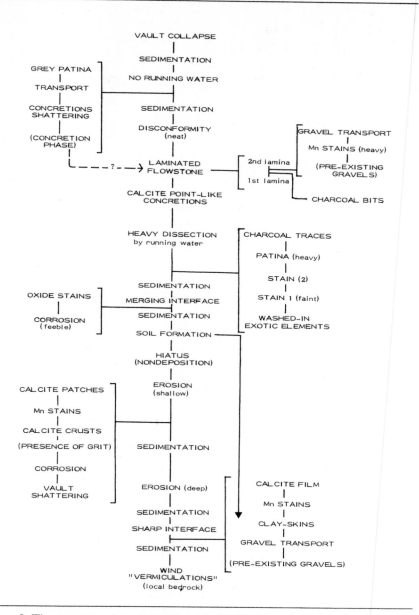

Figure 5. The reconstruction of the 'history' of a deposit using an 'elemental events' approach. An example.

been reported in full (Fedele, 1973c, and on files; multiple-coated elements in Neolithic stakeholes, etc.); see some intuitive approximations in H. De Lumley (1972b) and analogs in palaeontology ('taphonomy'-oriented fieldwork; for example, Voorhies, 1969) and in a rock-art context (Anati, 1973).

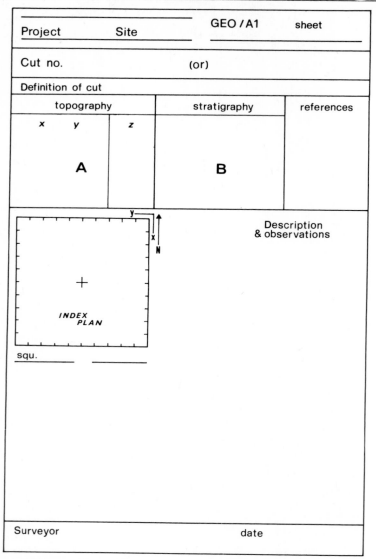

Figure 6. MF Project Papers. Record card for the control of dissection units ('cuts' or 'spits'). This form is designed to record how the deposit was dug, what was dug, and the overall relationships of the excavated material. (A) Spatial boundaries of the cut; *x, y* and *z* according to the rectangular coordinates (or Laplace-Méroc) method, 1954 (see Laplace, 1971; Fedele, 1973b). (B) 'Elementary sedimentary unit' or explicit stratigraphical designation.

2. Recording: Spatial relationships and data recording in excavation. Sediment studies in archaeology benefit from refinements in the methods of data recording. Detailed observation and recording of spatial relationships *in situ* before destruction in excavation is a well known procedure, yet progress towards a formalisation is very slow. It

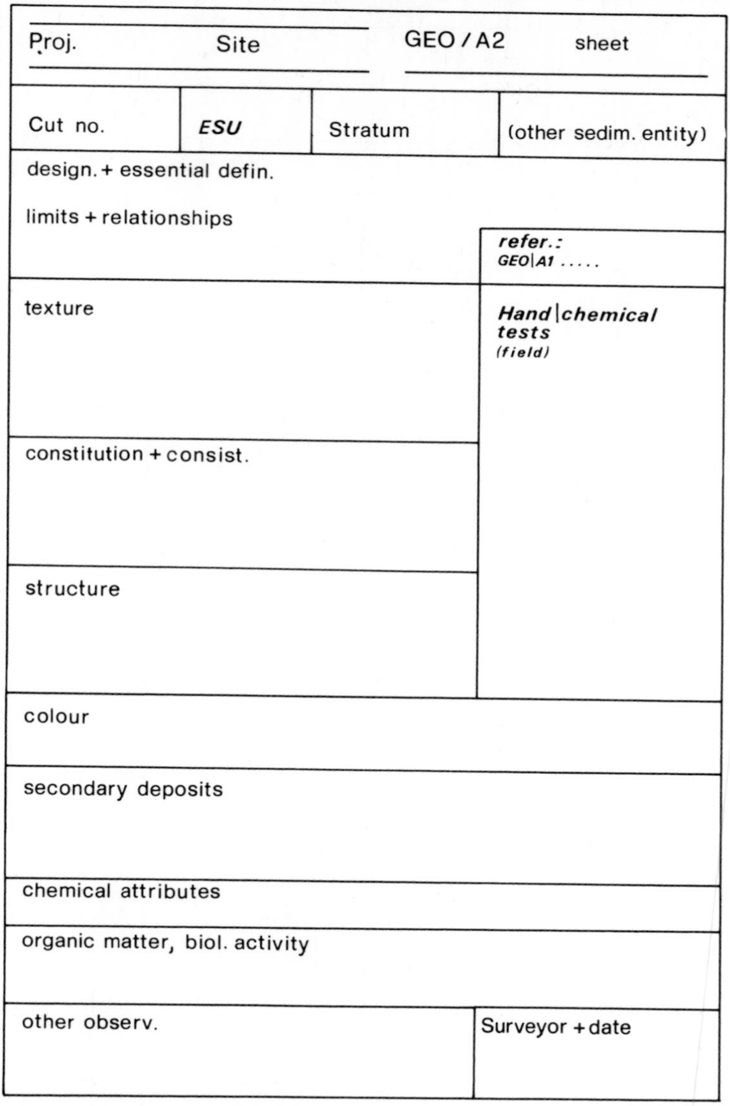

Proj.		Site		GEO / A2 sheet
Cut no.	**ESU**	Stratum		(other sedim. entity)

design. + essential defin.

limits + relationships

refer.:
GEO|A1 . . . . .

texture

Hand|chemical tests
(field)

constitution + consist.

structure

colour

secondary deposits

chemical attributes

organic matter, biol. activity

other observ.

Surveyor + date

Figure 7. MF Project Papers. Sedimentological record card.

must be applied to both sedimentary and non-sedimentary, or biogenic bodies. The spatial structure of a population of data encompasses the patterns of space states and values.* Formal mapping of spatial relationships is a basic device for data storage. Its value can be greatly enhanced by drawing scaled plans of each excavated square and level on standard cards or sheets according to

---

* Above/below, side/side or in/out relations; imbricate structures; position and metric values in the gridded space; strike and dip; magnetic bearing; relative position to natural or man-built features of any scale.

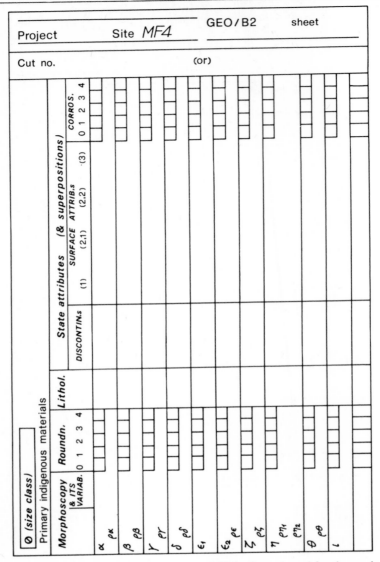

Figure 8. MF Project Papers (Belvedere Site, MF4). Record card for the analysis of the coarse fraction of sediments (I). See also Fig. 9.

standard rules. Only recently have explicitly stated methods been presented with more or less cursory accounts of general recording in excavations (Laplace, 1971; De Lumley, 1972a; Sauter *et al.*, 1972; Fedele, 1973b). In the MF Project a manual of field instructions and a comprehensive set of standard cards are used for data recording and storage (Figs. 6 to 11; see captions for details).

As to mapping, excavation plans for each square and level regularly include stones greater than or equal to 10 cm in length, surface

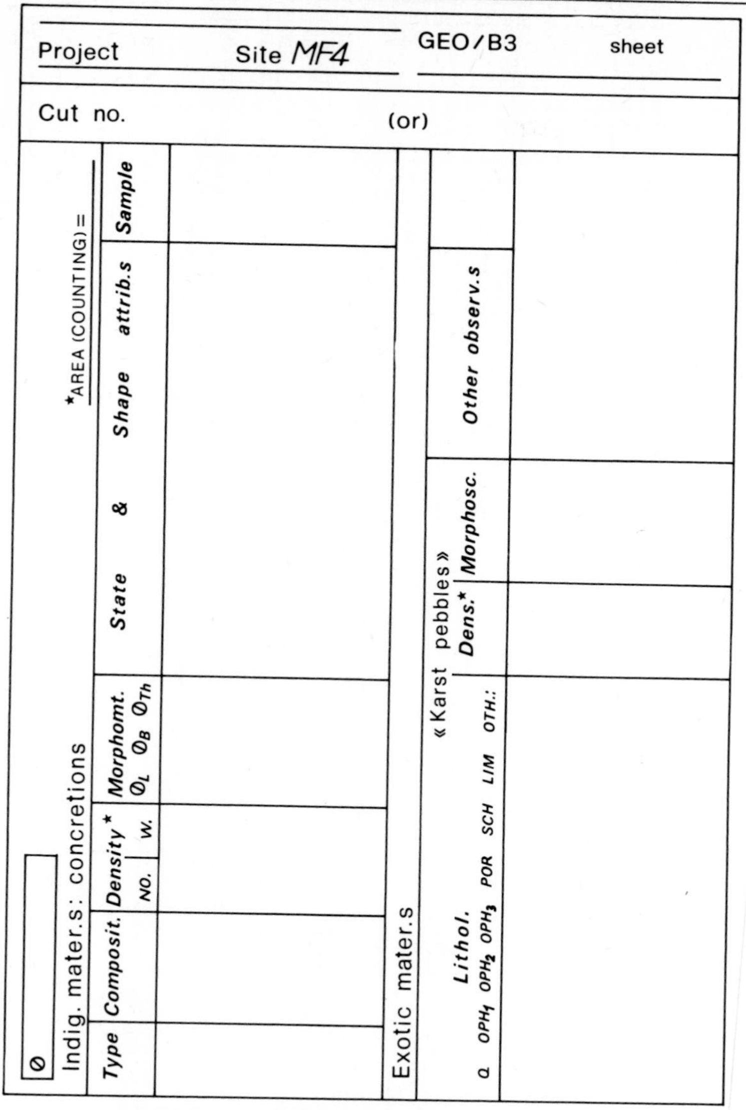

| Project | Site *MF4* | GEO/B3 | sheet |

Cut no.   (or)

AREA (COUNTING) =

Indig. mater.s: concretions

| Type | Composit. | Density* | | Morphomt. | State | & | Shape | attrib.s | Sample |

Density* NO. | W.

Morphomt. O_L O_B O_{Th}

Exotic mater.s

«Karst pebbles»

Lithol.
α OPH_1 OPH_2 OPH_3 POR SCH LIM OTH.:

Dens.* Morphosc.

Other observ.s

Figure 9. MF Project Papers (Belvedere Site, MF4). Record card for the analysis of the coarse fraction of sediments (II). In Figs. 8 and 9, it is important to note that each site, or *locus* within a site, may have its specific form.

contour lines and micro-morphologic traits, biogenic elements and features; all the spatial attributes and relationships of these components are expressed by means of a suitable symbolic notation. Selective mapping may be used to stress the distribution of special items (Fig. 12). The time of mapping as well as of sampling in routine digging is exactly specified. Dig plans come to resemble tesserae in a mosaic in being coordinated as a whole. Detailed sections may be

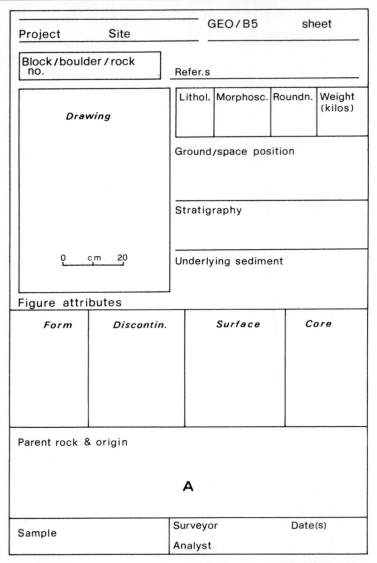

| Project | Site | GEO/B5 | sheet |
| Block/boulder/rock no. | | Refer.s | |

| Drawing | Lithol. | Morphosc. | Roundn. | Weight (kilos) |

Ground/space position

Stratigraphy

0   cm   20

Underlying sediment

Figure attributes

| Form | Discontin. | Surface | Core |

Parent rock & origin

A

| Sample | Surveyor   Date(s) |
| | Analyst |

Figure 10. MF Project Papers. Record card for the examination of blocks and rocks. (A) Parent rock and origin; for example, walls, ceilings, slopes, and outcrops; parent rock surface and its nature, pre-depositional events, etc.

drawn from such seriated plans in a semi-automatic way. Since the information is in a standardised form, it can be used for a variety of purposes in the laboratory. The information is also amenable to automatic data processing.[6]

3. The role of local conditions: 'Vault study' in caves. The relevance of considering local conditions whenever interpretations or generalisations are drawn from single point evidence (see sections 1.3

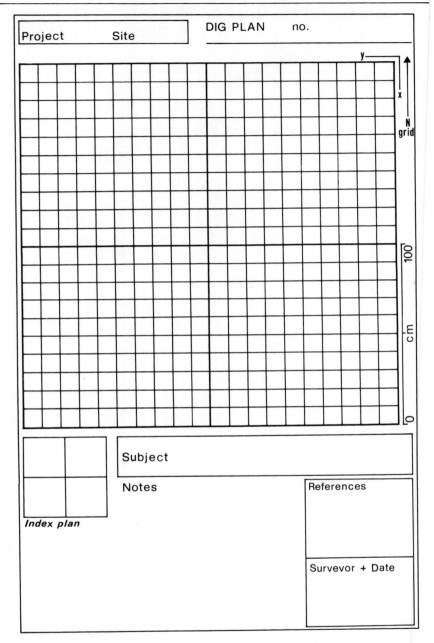

Figure 11. MF Project Papers. Record card for scaled 'dig plans' (four-square variant of the form). Scale of the actual sheet: 1:10.

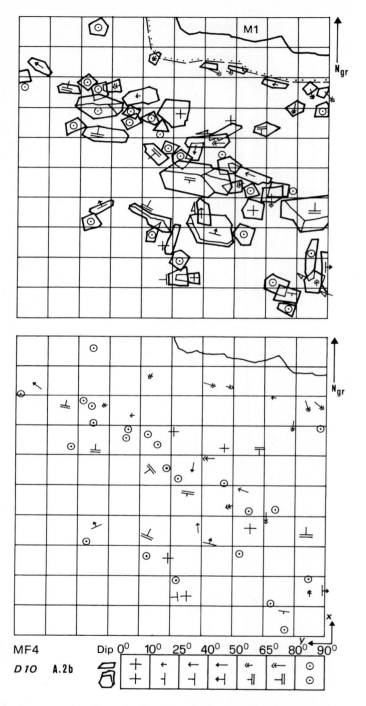

Figure 12. An example of a selective 'dig plan': mapping is used to detect the distribution of strike and dip values in a patterned sedimentary body – an alignment of angular stones connected by frost-weathering activity to a vault fissure and perhaps modified by soil-frost dynamics.

and 2.1) is prominent in the study of cave sediments. A technique will be outlined as an example. The 'cave vault study' approach is rather new in that it provides for a formal survey and mapping of a cave setting, as distinct from the mere topography. The close linkages between the fill and the ceiling and walls of a cave are well known (Fedele, 1966; Schmid, 1969; Butzer, 1972). This causal interdependence needs elucidation by studying the vault of a cave to the same degree of detail and accuracy as its deposits. There is need for the survey of complete cave micro-environments.

Survey and mapping of vault macro-morphology and of rock surface micro-morphology can be made in analogy with the standard procedures of open-air karst and limestone relief studies (e.g. Sparks, 1971), provided that cave vault maps are anchored to the same excavation grid. Special problems of sedimentological interpretation must be taken into account during mapping decisions. These vault maps match symmetrically the dig and site plans recorded in excavation. In addition to morphological attributes, other characteristics of vaults may be selected for mapping, such as water or moisture phenomena (water retention, condensation, dripping, influences by local rain), proneness to rock shattering and predetermination of detritus shapes, and thermal attributes. Legge (1972) has recently shown how productive parallel studies can be of cave atmospheres. On the interpretive side, cave vault studies promise to afford a key to understanding morphoscopic characteristics of detritus, correlations between vault facies and floor (sedimentary) facies, displacements of blocks, concretionary activity, incidence of weathering attacks in contrast to seismism, or obstruction of feeding chimneys.

4. Observation and analysis: A plea for scrutinising the coarse fraction of sediments. Visual observation and counting of the coarse sediment fraction have enjoyed considerable success, mainly to the credit of the French schools (Escalon *et al.*, 1957; De Lumley, 1957, 1972b; Laville, 1964, 1970; Miskovsky, 1969, 1972), after the original statement by E. Bonifay (1956) that they had been unduly neglected. Such re-evaluation still deserves agreement, and indeed further improvements could now be urged. The coarse fraction of a sediment corresponds to the angular and rounded materials coarser than sand; for a synopsis of definitions and classical methods of study, see Sparks (1971) and Butzer (1972). Visual analyses and computations apply to both individual elements and gross 'sediment patterning' (i.e. *in situ* texture and structure of coarse-sediment bodies). Within the operational model proposed in section 2.1, these methods are practised in digging and to a larger extent in analysis. Visual observation may focus in digging for example on collapse features, cryoturbate pockets, or on the orientation of ventifacts; in analysis, upon the type and state of concretions, or on the 'figure attributes' illustrated below. Counts in their turn may involve densities and ratios of indigenous/exotic elements and of all or most of the 'figured' elements already observed. In connection with instrumental measurements such as complete cascade sieving for grain-size analysis and selected morphometry, simple counting affords a powerful method to cope with deposits in their entirety with little sampling, during the short span of a site operation.

'Figure attribute' analysis describes a complex set of objectives and methods which are concerned with the surface, core, and continuity attributes of sedimentary elements. Desk analysis of 'figured' elements in the field should be supplemented by immediate consideration of the contexts of their occurrence and by subsequent laboratory analyses. A conspectus of the figure attributes is presented in Table 1. Useful suggestions for their visual examination may be sought in recent manuals as for example Maignien's (1969) and Clarke's (Clarke *et al.*, 1971) (*cf.* Icole, 1971; Butzer, 1972; Miskovsky, 1972); formal instructions are still unpublished (MF Project Papers; Fedele, in press, a). As already mentioned, this multilateral analysis seems to be an appropriate procedure to tackle the problems of postdepositional changes in deposits and of multigenerational suites, in the light of an 'elemental events' approach. The same type of analysis can be operated on non-sedimentary particles, and the results to date on faunal remains and artifactual elements appear promising.

Properties	*Origin*
1. State attributes	(a) by
1. Core attributes	ground processes
1. Primary structure	(b) by
2. Superimposed structure (chemical	biological
weathering/corrosion, impregnation)	activity (plants
2. Surface attributes	and animals)
1. Primary surface	(c) by
2. Coatings	cultural activity
1. Stains, patinae, flecks, skins	(trampling,
2. Crusts, aggregates	recreation, firing etc.)
3. Polish	
3. Discontinuity pattern attributes	
1. Fragmentation	
1. Fractures (craquelures, fissures)	
2. Split (fragments)	
2. Cavities and perforations	
1. Pits, grooves, etc.	
2. Holes	
3. Relief (crests, etc.)	
2. Shape attributes	
1. Solid geometry (original form)	(a) primary (natural)
[morphoscopy, morphometry]	(b) transportational
2. Superimposed shape (total, partial)	& depositional
[roundness, flattening, sphericity,	(c) postdepositional
faceting, and other shape attribute indices]	

Table 1. 'Figure attributes' of sedimentary elements. (Selected analytical techniques are shown in square brackets.)

Acknowledgments

Some aspects of the text benefited from discussion with Myra L. Shackley (University of Southampton), Susan Laflin (University of Birmingham), Dr. George J. Miller (El Centro, California), Professor Amos Rapoport (University of Wisconsin-Milwaukee), and Mr. R. Nisbet, my co-operator, in the Monfenera Project.

NOTES

1. Publication of the Monfenera Project (Prehistoric and palaeoecological research on the Quaternary deposits in the Monfenera area, Northern Piedmont), funded by the *Consiglio Nazionale delle Ricerche* (Rome), grant CT73.00037.05.
2. This systems theory approach (e.g. Struever, 1968; Flannery, 1968; Harris, 1969; Doran, 1970; Davidson, 1972; *cf.* human ecology of Vayda, 1964) has substantial advantages over the former empirical models or simple claims of man-environment interdependence (e.g. Clark, 1952; Mysterud, 1965; Alcock *et al.*, 1963; Godwin, 1965; Butzer, 1972).
3. A problem hardly touched on here is the definition of the units to be adopted in practice. Delimitation of a S_{tn} in particular is a matter of decision congruent to the specific parameters of each given instance. See section 2.2.
4. Some attempts at a 'full' transfer of the sediments of small-scale sites have been made, but they have remained exceptional measures inspired by local objectives and feasibility. (For example, the Tyrrhenian III beach at the Casino Site, Balzi Rossi area, West Liguria, by G. Vicino with the support of the Laboratoire de Paléontologie humaine et de Préhistoire of Marseille, 1968-9.)
5. This operational model is founded on the author's ongoing experiments within the Monfenera Project (Northwestern Italian Alps; Fedele, 1966, 1972, 1973b, in press, b). This research project is referred to as the 'MF Project'.
6. So far, there have not been any attempts at making automatic runs while the excavation is still in progress (see Hole *et al.*, 1967, for a seriation experiment). Myra L. Shackley (this volume) is trying to record field data on standard sheets already organised for direct transfer to computer input.

REFERENCES

Achinstein, P. (1971) *Concepts of science: A Philosophical Analysis*. Baltimore and London.

Alcock, L. and I.Ll. Foster, eds. (1963) *Culture and Environment*. London.

Alexander, J. (1970) *The Directing of Archaeological Excavations*. London.

American Geological Institute (1960) *Glossary of Geology and Related Sciences*. 2nd ed. Washington.

American Geological Institute (1962) *Dictionary of Geological Terms*. Garden City, New York.

Anati, E. (1973) 'La stele di Triora (Liguria)', *Bollettino del Centro Camuno di Studi Preistorici* 10, 101-26.

Ashby, W.R. (1965) *An Introduction to Cybernetics*. London.

Atkinson, R.J.C. (1960) *Stonehenge*. Harmondsworth.

Binford, L.R. (1964) 'A consideration of archaeological research design', *Am. Antiq.* 29, 425-41.

Binford, L.R. (1968) 'Archaeological perspectives', *in* S.R. Binford and L.R. Binford, eds., *New Perspectives in Archaeology*, Chicago, 5-32.

Bonifay, E. (1956) 'Les sédiments détritiques grossiers dans les remplissages des grottes. Méthode d'étude morphologique et statistique', *L'Anthropologie* 60, 447-61.

Braidwood, R.J. (1960) 'Levels in prehistory: a model for the consideration of the evidence', *in* S. Tax, ed., *Evolution after Darwin*, Chicago, vol. 2, 143-51.

Butzer, K.W. (1972) *Environment and Archaeology: An Ecological Approach to Prehistory*, 2nd ed. London.

Childe, V.G. (1956) *Piecing Together the Past: The Interpretation of Archaeological Data.* London.

Chorley, R.J. and P. Haggett, eds. (1967) *Models in Geography.* London.

Clark, J.G.D. (1952) *Prehistoric Europe: the Economic Basis.* London.

Clarke, D.L. (1968) *Analytical Archaeology.* London.

Clarke, G.R. and P. Beckett (1971) *The Study of Soil in the Field*, 5th ed. Oxford.

Clarke, W.C. (1971) *Place and People: An Ecology of a New Guinean Community.* Berkeley and London.

Cochran, W.G. (1963) *Sampling Techniques*, 2nd ed. New York.

Coles, J. (1972) *Field Archaeology in Britain.* London.

Cornwall, I.W. (1960) 'Soil investigations in the service of archaeology', *in* R.F. Heizer and R.S. Cook, eds., *The Application of Quantitative Methods in Archaeology*, New York, 265-99.

Davidson, D.A. (1972) 'Terrain adjustment and prehistoric communities', in P.J. Ucko, R. Tringham and G.W. Dimbleby, eds., *Man, Settlement and Urbanism*, London, 17-22.

De Lumley, H. (1957) 'La Baume des Peyrards (Vaucluse) – Campagne 1956', *Cahiers Ligures de Préhistoire et d'Archéologie* 6, 216-22.

De Lumley, H. (1972a) 'Méthodes de fouilles et d'études d'un gisement préhistorique', *in* H. De Lumley, ed., *La Grotte de l'Hortus (Valflaunès, Hérault).* Marseille, 17-21.

De Lumley, H. (1972b) 'La stratigraphie des dépôts quaternaires de la grotte de l'Hortus (Valflaunès, Hérault)', *in* H. De Lumley, ed., *La Grotte de l'Hortus (Valflaunès, Hérault).* Marseille, 57-99.

Doran, J. (1970) 'Systems theory, computer simulations and archaeology', *World Archaeol.* 1, 289-98.

Escalon de Fonton, M. and E. Bonifay (1957) 'Les niveaux solutréens de la grotte de la Salpêtrière (Gard)', *L'Anthropologie* 61, 207-38.

Fedele, F.G. (1966) "La stazione paleolitica del Monfenera in Valsesia. 2'', *Riv. St. liguri* 32, 25-78.

Fedele, F.G. (1972) 'Aperçu des recherches dans les gisements du Monfenera (Valsesia, Alpes Pennines)', *Bulletin d'études préhistoriques alpines* 4, 5-68.

Fedele, F.G. (1973a) Review of *Place and People: An Ecology of a New Guinean Community*, by W.C. Clarke (1971), *J. human Evol.* 2, 57-9.

Fedele, F.G. (1973b) 'Aspetti dell'archeologia preistorica moderna. Lo scavo', *Ad Quintum* 4.

Fedele, F.G. (1973c) 'Una stazione Vaso a bocca quadrata sul Monfenera, Valsesia (scavi 1969-72). Rapporto preliminare', *Preistoria alpina* 9, 151-222.

Fedele, F.G. (in press, a) *Paleoecologia umana e preistoria.* Torino.

Fedele, F.G. (in press, b) 'Stone Age discoveries on Monfenera, Northwestern Alps, and their bearing on human palaeoecology', *in* G. Novak *et al.*, eds., *Actes du VIIIe Congrès International de Sciences Préhistoriques et Protohistoriques, Beograd, 1971.* vol. 2. Beograd.

Fedele, F. and R. Nisbet (1974) 'Il problema dei ciottoletti esotici nei depositi pleistocenici del Monfenera (bassa Valsesia)', in *Atti dell'XI Congresso Nazionale di Speleologia, Genova, 1972,* vol. 1, 171-87. Como.

Flannery, K.V. (1968) 'Archeological systems theory and early Mesoamerica', *in* B. Meggers, ed., *Anthropological Archaeology in the Americas*, Washington, 67-87.

Godwin, H. (1965) 'The beginnings of agriculture in Western Europe', *in* J. Hutchinson, ed., *Essays on Crop Plant Evolution*. London.

Haggett, P. (1965) *Locational Analysis in Human Geography*. London.

Harris, D.R. (1969) 'Agricultural systems, ecosystems and the origins of agriculture', *in* P.J. Ucko and G.W. Dimbleby. eds., *The domestication and exploitation of plants and animals*, London, 3-15.

Hole, F. and M. Shaw (1967) *Computer Analysis of Chronological Seriation* (Rice Univ. Stud. 53, 3), Houston.

Hole, F. and R.F. Heizer (1973) *An Introduction to Prehistoric Archeology*, 3rd ed. New York.

Hyams, E. (1952) *Soil and Civilisation*. London.

Icole, M. (1967) 'Contribution à la connaissance des paléosols et à celle de l'altération des roches au cours du Quaternaire. Etude de galets à cortex du sommet des nappes d'alluvions du piémont nord-pyrénéen', *Science du sol* 2, 83-95.

Klir, G.J., ed. (1972) *Trends in General Systems Theory*. New York and London.

Laplace, G. (1971) 'De l'application des coordonnées cartésienna à la fouille stratigraphique', *Munibe* 23, 223-36

Laville, H. (1964) 'Recherches sédimentologiques sur la paléoclimatologie du Wurmien récent en Périgord', *L'Anthropologie* 68, 1-48 and 219-52.

Laville, H. (1970) 'L'abri magdalénien du Flageolet II (Bézenac – Dordogne). Etude géologique', *Bull. Soc. préhist. fr.* 67, 475-88.

Leakey, M.D. (1971) *Olduvai Gorge*, vol. 3: *Excavations in Beds I and II, 1960-1963*. London.

Legge, A.J. (1972) "Cave climates", *in* E.S. Higgs, ed., *Papers in Economic Prehistory*, London, 97-103.

Maignien, R. (1969) *Manuel de prospection pédologique*. Paris.

Miskovsky, J.C. (1969) 'Sédimentologie des couches supérieures de la grotte du Lazaret', *in* H. De Lumley, ed., *Une cabane acheuléenne dans la grotte du Lazaret*, Paris, 25-51.

Miskovsky, J.C. (1972) 'Etude sédimentologique du remplissage de la grotte de l'Hortus (Valflaunès, Hérault)', *in* H. De Lumley, ed., *La grotte de l'Hortus (Valflaunès, Hérault)*, Marseille, 101-53.

Mysterud, I. (1965) 'En kommentar till okologisk forskning', *Forskningsnytt* 1965, 5, 18 ff.

Oakley, K.P. (1969) 'Analytical methods of dating bones', *in* D. Brothwell and E. Higgs, eds., *Science in Archaeology: A Survey of Progress and Research*, 2nd ed., London, 35-45.

Piggott, S. (1965) *Approach to Archaeology*, 2nd ed. New York and Toronto.

Piggott, S. (1973) *Ancient Europe from the Beginnings of Agriculture to Classical Antiquity*. Edinburgh.

Rootenberg, S. (1964) 'Archaeological field sampling', *Am. Antiq.* 30, 111-88.

Ruhe, R.V. (1965) 'Quaternary paleopedology', *in* H.E. Wright Jr., and D.G. Frey, eds., *The Quaternary of the United States*, Princeton, 755-64.

Ruhe, R.V. (1967) 'Paleosol', *in* *McGraw-Hill Encyclopedia of Science and Technology*, revised ed., New York, 522.

Sauter, M.R., A. Gallay and L. Chaix (1972) 'Le Néolithique du niveau inférieur du Petit-Chasseur à Sion, Valais', *Jahrbuch der Schweizerischen Gesellscraft für Ur- und Frühgeschichte* 56, 1971, 17-76.

Schmid, E. (1969) 'Cave sediments and prehistory', *in* D. Brothwell and E. Higgs, eds., *Science in Archaeology: A Survey of Progress and Research*, 2nd ed., London, 151-66.

Sokal, R.R. and P.H.A. Sneath, (1963) *Principles of Numerical Taxonomy*. San Francisco and London.

Sparks, B.W. (1971) *Rocks and Relief*. London.

Struever, S. (1968) 'Problems, methods and organization: a disparity in the growth of archaeology', *in* B. Meggers, ed., *Anthropological Archaeology in the Americas*, Washington, 131-51.

Stjernquist, B. (1971) *Archaeological Analysis of Prehistoric Society* (Scripta minora, 1971-1972, 1). Lund.
Vayda, A.P. (1964) 'Anthropologists and ecological problems', *in* A. Leeds and A.P. Vayda, eds., *Man, Culture, and Animals*. Washington, 1-15.
Voorhies, M.R. (1969) *Taphonomy and Population Dynamics of an Early Pliocene Vertebrate Fauna, Knox County, Nebraska*. Laramie, Wyoming.

DISCUSSION

Mr. G. de G. Sieveking suggested that sampling was not always useful and quoted the excavations by Dr. Pales at Malarnaud-Soulabe where all the sediment from the cave had been kept. The speaker commented that sampling was not an exclusive means to record sediments and that it must be formally derived and supplemented by field observation. In certain cases total removal could be justified. Professor A.C. Renfrew then suggested that it might be simpler not to excavate the site!

I. Graham

The investigation of the magnetic properties of archaeological sediments

1. The magnetic properties of the soil

To the scientist interested solely in its magnetic properties the soil appears to consist of a mainly diamagnetic matrix, in which is dispersed a few per cent of ferrimagnetic minerals, typically iron oxides. Ferrimagnetic substances tend to concentrate the magnetic lines of force within them, while in contrast diamagnetic substances tend to repel the lines of force, but the diamagnetic effect is very much the smaller of the two. The magnetic properties of both ferri- and diamagnetic substances may at least partially be described by what is known as the susceptibility χ. This is defined as the ratio of the magnetic moment of a mineral in a given magnetic field, to that field; the magnetic moment is the force produced by a magnet that tends to align it with the magnetic field, as in a compass. For the volume susceptibility the magnetic moment is measured per unit volume, per unit mass for the mass susceptibility. χ is small and negative for diamagnetic minerals (a diamagnetic compass needle would point east-west); for ferrimagnetic minerals it is positive and very much larger, about 4×10^{-6} Si kg^{-1} for 1 per cent of γFe_2O_3, grain size 30 nm dispersed in a non-magnetic matrix.

The susceptibility is a rather inadequate description of ferrimagnetism. Diamagnetic susceptibility is reasonably constant with both grain size and dilution of grains in a non-magnetic matrix, but ferrimagnetic susceptibility is not. This arises from the fact that ferrimagnetic minerals are permanently magnetised. Interactions occur among the ions in the crystal lattice, which tend to line up the directions of magnetism of all the atoms. Normally a large ferrimagnetic grain splits itself up into a series of small volumes called domains. In certain directions relative to the crystal lattic magnetism is particularly easy; these are called the easy directions of magnetism of the mineral. Each domain will normally be oriented along one of these easy directions, the total magnetic moment of the crystal tending to be very small as the domains magnetised in different directions cancel each other out. It might be expected that in zero magnetic field the grain would contain a very large number of domains, giving a very good cancellation of the magnetic moments.

However the domains are surrounded by so-called 'walls', in which the direction of magnetism of the atoms changes slowly from the direction in one domain to that of the next. These walls themselves store considerable potential energy. The grain brings itself into an equilibrium state which minimises the total potential energy, sharing it out between the magnetic energy due to imperfect cancellation of the total magnetic moment, and the domain wall energy. A magnetic field applied to such a collection of domains causes some of those directed opposite to the field to reverse in direction, and some of those directed along the field to increase in size, until a new equilibrium is reached with a finite magnetic moment parallel to the magnetic field.

A large crystal will come into equilibrium containing very many magnetic domains with different direction of magnetism. However very small grains, with dimensions only a little larger than the thickness of the usual domain wall, will have their lowest potential energy when they contain but one domain. It is this size of grain that is responsible for most of the magnetism of the soil; their average size is thought to be about 30 nm. Such 'single domain' grains tend to have a rather larger susceptibility than multidomain grains.

The domain in a single domain grain cannot grow, as could a domain in a multidomain grain, but can only reverse to reduce its potential energy if a magnetic field is applied in the opposite direction to that of the domain magnetism. A certain amount of energy is

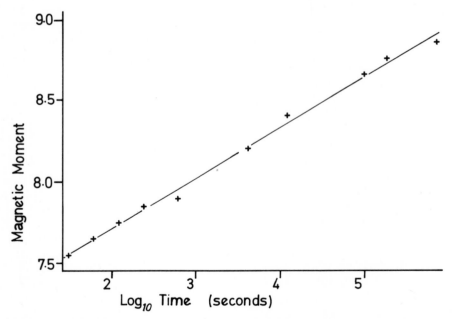

Figure 1. Plot of magnetic moment against time, which is shown on a logarithmic scale, for a loess sample. The crosses are the measured points; the straight line has a slope corresponding to a viscosity constant of 5.2 per cent.

required to surmount the energy barrier that prevents this reversal, but this energy can be provided by thermal energy coming from outside the grain.

Because of this the magnetic moment of soil samples increases slowly after they are put into a magnetic field, as from time to time a grain receives enough energy to enable it to reverse. This is called magnetic 'viscosity'. For a collection of grains of uniform size it can be shown that the increase in magnetic moment would be exponential, but a collection of grains with sizes uniformly distributed in the range 20-40 nm has a logarithmic increase in susceptibility with time. Fig. 1 shows an example of this, the increase in the magnetic moment of a soil sample (loess from a Bandkeramik site near Aachen); it fits the logarithmic law, shown by the straight line, very closely. Both Le Borgne (1960b) and the writer have measured samples for periods from a few seconds up to a few months and have found good agreement with a logarithmic law. It is to be expected that if the grain size distribution is smooth then the logarithmic dependence will carry on for the very long times which are archaeologically significant.

The increase in magnetic moment may be described by the equation

$$M_t = M_o (1 + V_c \log t/t_o) \qquad (1)$$

where M_t is the magnetic moment at experimental time t, M_o the magnetic moment at time t_o, and V_c the viscosity constant corresponding to time t_o. In our laboratory t_o is usually taken as 100 seconds. The increase in magnetic moment can be quite considerable over archaeological time. Typically $V_o \cong 5$ per cent, which leads to an

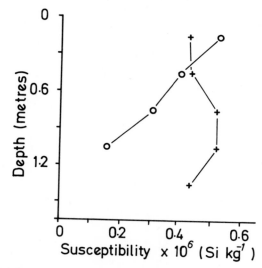

Figure 2. The change of magnetic susceptibility with depth for samples from part of the Dannewerke near Schleswig. One profile (circles) is from the undisturbed ground 20 m from the monument, the other (crosses) is from the ditch itself.

increase of magnetic moment of 45 per cent from 100 seconds to 3000 years (10^{11} seconds).

The main minerals responsible for the magnetism of the soil are haematite, αFe_2O_3, magnetite, Fe_3O_4 and maghemite, γFe_2O_3. Both magnetite and maghemite are strongly ferrimagnetic, whilst haematite is about a thousand times weaker. Haematite is however very important as it occurs, or may be produced by mild heating in very many soils, and as is explained below it may be converted into maghemite by strong heating.

2. *Variations of magnetic susceptibility within the soil*

The topsoil is generally found to have a rather higher susceptibility than the subsoil from which it is derived. (Le Borgne 1955). Fig. 2 shows the susceptibilities measured for samples from a ploughed out piece of the Dannewerke in Schleswig-Holstein. As well as the contrast between the topsoil and subsoil in the undisturbed profile it can be seen that the high susceptibility continues much deeper in the ditch section than in the undisturbed section. Where these samples were taken the bottom of the ditch is thought to lie about 2 m below the present ground surface. Thus this profile, taken with a short borer, does not quite reach the bottom of the ditch. Magnetometer surveys of sites can succeed because the mass of more magnetic material in a pit or ditch causes an anomaly in the earth's magnetic field above the feature. It will be measurable if the contrast between the susceptibilities of the ditch filling and undisturbed subsoil is great enough.

Fig. 3 shows a further example, a large ditch from the Roman town of Colonia Ulpia Triana, near Xanten in the Rhineland, part of which is clearly visible in the magnetometer map of Fig. 9. Samples were taken over a 1 metre grid, and they include some samples of the undisturbed subsoil; the outlines of the ditch and of a dark layer within it are also shown superimposed on the values of magnetic

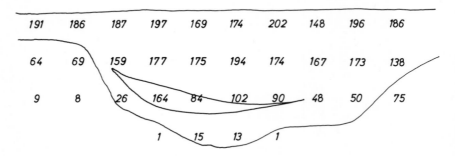

Figure 3. Susceptibility measurements from a section through a large ditch excavated in the Roman town of Colonia Ulpia Triana, near Xanten in the Rhineland. The susceptibility units are 10^{-7} Si kg^{-1}, the samples were taken over a 1 m grid.

susceptibility in Fig. 3. The susceptibility can again be seen to fall with depth, very much faster outside the ditch than within it. The dark silt layer, formed perhaps when the ditch lay open for some time, appears to coincide with a region of relatively high susceptibility, but more samples would be required to confirm this relationship.

3. Causes of the enhanced susceptibility of the topsoil

The susceptibility contrast between the topsoil and subsoil is thought to be caused by the effect of heat on the topsoil, during forest fires, slash and burn cultivation or any other burning associated with the occupation of the site.

Numerous experiments (among them Le Borgne, 1960a; Tite and Mullins, 1971) have shown that heating soil to a few hundred degrees Celsius in various atmospheres produces a tremendous increase in magnetic susceptibility. Fig. 4 shows the effect on the susceptibility of the same loess sample used in Fig. 1, heated to temperatures between 350 and 760°C. The sample was heated in nitrogen and maintained at peak temperature for 40 minutes and the atmosphere was then changed over to air. After 2 hours in air the sample was allowed to cool slowly, and the susceptibility then measured. The main increase in susceptibility occurs between about 450 and 550°C, the highest susceptibility being produced by heating to 600°C. Temperatures

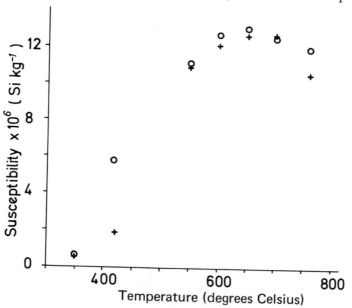

Figure 4. The susceptibilities of loess samples after heating, mixed with 5 per cent flour, to various temperatures. Two samples were heated in each experiment, in two boats one on top of the other. The circles are the results for the upper sample, the crosses those for the lower sample.

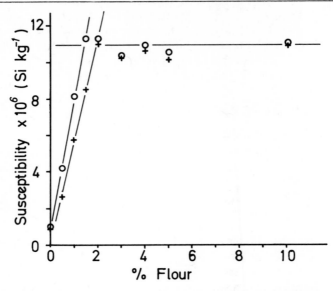

Figure 5. Susceptibilities of loess samples after heating to 550°C with various amounts of flour. The symbols are as for Fig. 4.

above this tend to produce susceptibility values rather lower than the maximum.

It is thought that carbon monoxide from the burning of organic matter in the soil reduces the aFe_2O_3 to Fe_3O_4, which is then re-oxidised to γFe_2O_3. The latter is normally unstable at such high temperatures when pure, but may be stabilised by the presence of sodium ions in the soil (Le Borgne, 1960a).

The dependence of this mechanism of susceptibility increase on the amount of organic matter may be investigated by heating soil, originally very low in organic matter, enriched with various amounts of some organic substance. Fig. 5 shows the results for a sample of the loess soil which has been enriched with household flour, and heated to 550°C. The susceptibility after heating rises very rapidly until about two per cent flour content is reached. After this the susceptibility is constant to within about one part in twenty, since there is sufficient flour for the reduction reaction to be complete.

Heating soils to 500°C in a hydrogen atmosphere also produces a large susceptibility increase and has been used by Tite (1972) to investigate the relative potential susceptibility of a large number of soils. However, X-ray diffraction investigations made here on soils heated in hydrogen have shown that the haematite is not converted to either magnetite or maghemite, but into metallic iron. This substance has magnetic properties very different from those of the ferrimagnetic iron oxides; in particular grains of iron of about 30 nm in size will contain many magnetic domains. Thus the susceptibility after heating in hydrogen is normally one half to one third of that produced by

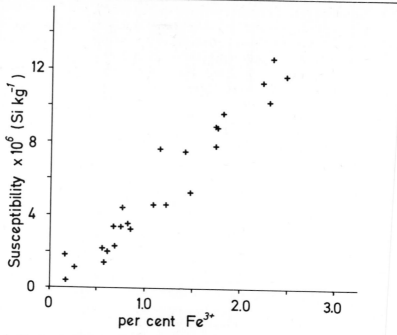

Figure 6. The variation of soil susceptibility after heating, with the iron content determined chemically.

heating the same soil with the nitrogen-air atmosphere cycle. The magnetic viscosity of samples heated in hydrogen is also rather less.

The nitrogen-air cycle is quite close to what might be thought to happen in nature. During the early phases of a fire there is too little air and plenty of fuel, giving a reducing atmosphere; as the fire burns itself out more air will enter, producing an oxidising atmosphere. The thermal conductivity of the soil is very low, and thus only the top few centimetres will be affected. With time, ploughing and the activities of worms, the newly formed maghemite will be distributed throughout the topsoil. A series of a few fires over the ages would be quite sufficient to produce the susceptibility contrasts typical today.

Comparison of the actual topsoil susceptibility on a site with what could be achieved with complete conversion to the strongly magnetic oxides provides a useful index of the length and intensity of the occupation of the site. Tite and Mullins have used the ratio χ_o / χ_H, where χ_o is the topsoil susceptibility and χ_H is the susceptibility of the topsoil after heating to 500°C in hydrogen. In this laboratory the value of χ_o / χ_N is calculated, where χ_N is the susceptibility produced by heating the soil, enriched with a suitable amount of flour to 550°C in nitrogen, followed by air. Tite and Mullins have found the mean value of χ_o / χ_H to be about 6 per cent for normal topsoil, with values up to 60 per cent on archaeological sites. The values of χ_o / χ_N are naturally two or three times lower as χ_N is always that much larger than χ_H.

In future it may not be necessary to heat soils in order to estimate their potential susceptibility. The iron content of a large number of soils heated here has been measured by complexometric titration (West, 1969), after dissolving the iron oxides out of the soil with concentrated hydrochloric acid. Fig. 6 shows the iron content plotted against the susceptibility after heating. It is clear that the susceptibility is proportional to it with a scatter of less than about 20 per cent. If this holds true for most soils then complicated heating experiments may be replaced with the much simpler measurement of iron content with very little lòss of accuracy.

4.   Methods of making magnetic measurements on soils

In this laboratory three instruments are used for soil measurements: an AC bridge, a translation magnetometer and a pulsed induction meter (PIM).

The AC bridge (Scollar, 1968) operates by measuring the change of inductance of a coil when a soil sample is placed within it, the change being proportional to the alternating current susceptibility (χ_{ac}) of the sample. The inductance measurements are made at a frequency of 1555 Hz. As the magnetic field applied to the sample varies sinusoidally at this frequency it is very difficult to know to what value of t in equation (1) these measurements correspond, but calibration has shown it to be about 10^{-4} seconds. Thus for a typical soil sample,

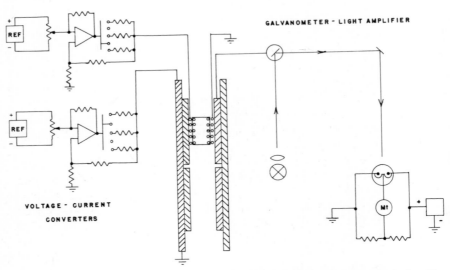

Figure 7. Schematic diagram of the translation magnetometer used in the Labor für Feldarchäologie. In the centre is the sample holder, surrounded by the bifilar coil and the pick up and field coils. To the left is the light amplifier (optical lever) system that is connected to the pick up coils; to the right the constant current power supplies for the bifilar compensation coil and the field coil.

Figure 8. The translation magnetometer installed in the magnetic laboratory. The four very large coils form the system that cancels out the Earth's magnetic field. The pickup coils and field coil are in the rectangular wooden box, within which the sample is suspended by a wire from the ceiling. This runs over pulleys to the operating position from where the sample may be moved by pulling the wire.

the susceptibility as measured in an AC bridge is about half the 2000-year susceptibility. The AC bridge is used mostly when very quick measurements are required of a large set of samples, such as when making susceptibility maps of large area, since only a few seconds are required for each reading.

With the translation magnetometer (Nagata, 1961, 41; Scollar and Graham, 1974), Figs. 7 and 8 it is quite clear what time constants are involved. Here the sample, in a constant magnetic field, is moved between two centres of two large pickup coils connected in series opposition. This induces a charge proportional to magnetic moment, which may be measured with a sensitive galvanometer. In our instrument the sample is surrounded by a bifilar coil and a small current through this is used to cancel out the sample's magnetic moment so that no deflection of the galvanometer is detectable. This changes the problem from that of measuring a small charge accurately into the much easier one of measuring the small steady current in the bifilar coil. For the greatest sensitivity the galvanometer deflection is magnified by a differential photocell-amplifier and displayed on the output meter, Mt in Fig. 7.

The pickup coils are surrounded by a large solenoid, used if higher fields than that of the Earth are required for measurements, for example for very weakly magnetic samples. The whole set of coils is within a large system which annuls the Earth's magnetic field so that soil samples may be shaken in zero field to remove their remnant magnetic moment before measurements are made. Fig. 7 shows the schematic diagram of the instrument, with the exception of the zero field coils. These may be seen in Fig. 8, which shows the instrument installed in our magnetic research laboratory, in an abandoned opencast mine, well away from all disturbances.

It is clear that with this instrument measurements of the viscosity constant are easily made. The sample is put into a special holder, shaken in zero field and then compacted so that the grains cannot move during measurements. The sample in its holder is placed in the instrument and the zero field coil switched off. Now measurements may be made for any desired period of time, to produce results such as Fig. 1.

The pulsed induction meter (Colani, 1966; Mullins, 1974) subjects the soil sample to an intense pulse of magnetic field, and then measures the rate of decay of the magnetic moment of the sample when the field is removed. This measurement is proportional to the product of the viscosity and the susceptibility. Thus the instrument can only be used to provide values of the viscosity constant for samples of known susceptibility when this proportionality constant has been found by calibration against the translation magnetometer. Our instrument has a special coil system which allows measurements on samples up to one kilogram in weight.

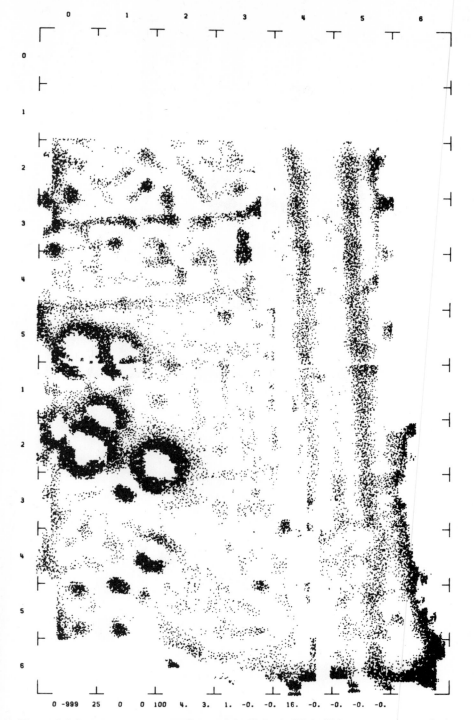

Figure 9. Magnetometer survey of part of the Colonia Ulpia Triana, near the modern town of Xanten, in the Rhineland. The area surveyed is about 140 x 200 m. On the right are the two large ditches; the circles are bomb craters and above and to the right of these are visible the remains of buildings.

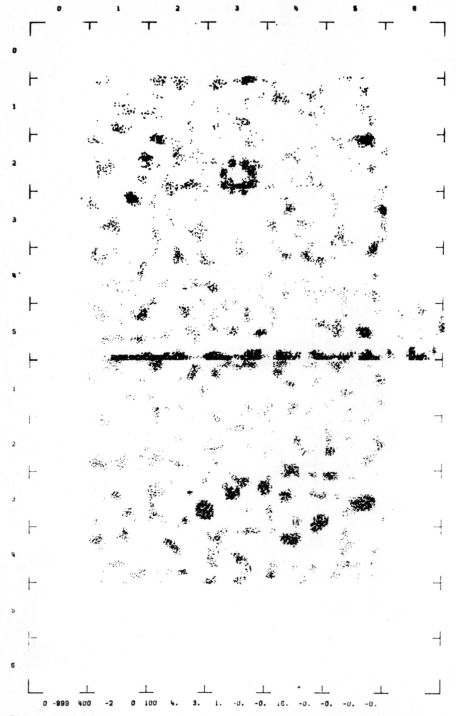

Figure 10. Magnetometer survey at Meil-Givvenkoven, Rhineland, showing a survey area of about 100 x 180 m. A road (modern) crosses the centre of the site, there is a bomb crater in the upper half but otherwise little is visible.

5.   Sample collection and preparation

On unexcavated sites samples are collected by boring, normally one every 0.3 metre's depth, down to about 1.5 m. Where clear horizon differentiation may be seen samples are taken from each horizon as well. If there is an open section the samples are taken with a brass trowel; this avoids the risk of the samples being contaminated by steel fragments from the tool. The samples are then air dried, until they reach humidity equilibrium with the laboratory atmosphere, a process that takes from three weeks to a month. Usually it is found that the fraction greater than 2 mm is almost non-magnetic, being almost unaffected by soil heating, but this can be quickly checked with the AC bridge. The less than 2 mm fraction is weighed into plastic sample holders, which fit into all of the measuring devices; thus the same sample may be used for all measurements.

6.   The archaeological uses of magnetic soil measurements

The most obvious use of magnetic soil measurements is prior to magnetometer prospecting. As has been mentioned above, the results depend very much on the soil on which the site is built. Surveys such as that at Xanten (Fig. 9) carried out on a soil with a susceptibility contrast of about 1.7×10^{-6} Si kg^{-1} have a much greater chance of success than those carried out on a soil with a contrast of 0.12×10^{-6} Si kg^{-1} or less (Fig. 10). The Xanten survey shows quite clearly the line of two large ditches (a section through one is shown in Fig. 5), the remains of buildings, and even bomb craters from the second world war. The second survey (Miel-Givvekoven, near Bonn) shows little except one bomb crater and some random disturbance.

It is certainly useful to be able to predict the success or failure of prospecting so that time and money are not wasted making surveys on sites where there is no hope of discovering anything except tin cans. As it is tedious to have to take soil samples from every site where survey is proposed, attempts are now being made to collect enough samples to make rough maps giving the probability of success of survey, or more specifically the typical magnetic anomaly strength to be expected given an average occupation density. Tite (1972) has done this for parts of England, and is continuing this work for Italy. At this laboratory a study is being made of German soils, concentrating on those common in the Rhineland, and also of some specimens from other European and Near-eastern countries. It is to be expected that this work will lead to far less disappointment in the use of magnetometer survey. It also enables us to predict what sensitivity is required of a magnetometer, so that successful survey could be made over a still larger number of sites in the Rhineland, and tells us what to strive for in instrumental sensitivity.

6.1 *Topsoil susceptibility variations*

As has been explained above, the fraction of the maximum susceptibility that is achieved by the topsoil depends on the length and density of occupation of the site. It is therefore in some cases possible to outline the occupation area of the site with a grid of topsoil measurements. These can be made two ways: on the soil *in situ* with an instrument such as the 'soil conductivity meter', SCM (Howell, 1966), which in fact measures the AC susceptibility, or by taking a grid of samples for measurement in the laboratory.

To obtain good results the value of the fractional conversion must in any case be sufficient to provide sufficient contrast to the normal agricultural topsoil. Scollar has used the grid method to good effect at Froitzheim in the Rhineland (Scollar, 1965, p50). Mullins (1974) used the SCM at Dragonby and showed very convincingly that the highest values of susceptibility coincide with the centres of occupation. On both these sites the fractional contrast was quite high;· for Dragonby $\chi_o / \chi_H > 50$ per cent, for Froitzheim $\chi_o / \chi_N \approx 10$ per cent.

Grids of samples have been taken recently from two other sites in the Rhineland, both with $\chi_o / \chi_N < 5$ per cent. Both showed no significant variation of the topsoil susceptibility with the known archaeological features. One site did, however show a steady decrease in topsoil susceptibility towards the badly drained part of the site. This lends support to the idea (Le Borgne, 1955) that maghemite is converted to more weakly magnetic oxides by gleying of the soil.

6.2 *Stratigraphy*

Under certain conditions it is possible to distinguish strata within the soil, or in the fillings of pits and ditches, by means of their magnetic properties. An example has already been given for the Xanten ditch; a further example is provided by the loess deposit at Rheindahlen, Rhineland. In this section the modern soil and two buried interglacial soils may clearly be seen, and indeed the susceptibility of the B-horizon of each soil is higher than that of the surrounding loess, whilst that of the A-horizon is lower. In general those strata which are clearly differentiable by magnetic properties are usually clearly visible to the eye. The application of this to archaeological practice is worthy of further study.

REFERENCES

Colani, C. (1966) 'A new type of locating device, I: The instrument', *Archaeometry* 9, 3-8.

Howell, M. (1966) 'A soil conductivity meter', *Archaeometry* 9, 20-3.

Le Borgne, E. (1955) 'Susceptibilité magnétique anomale du sol superficial', *Annls. Géophys.* 11, 399-419.

Le Borgne, E. (1960a) 'Influence du feu sur les propriétés magnétiques du sol et sur celles du schiste et du granit', *Annls. Géophys.* 16, 159-95.

Le Borgne, E. (1960b) 'Etude expérimentale du trainage magnétique dans le cas d'un ensemble de grains magnétiques très fins dispersés dans une substance non-magnétique', *Annls. Géophys.* 16, 445-93.

Mullins, C. (1974) 'The magnetic properties of the soil and their application to archaeological prospecting', *Archaeo-Physika* 5, 143-7.

Nagata, T. (1961) *Rock Magnetism.* Tokyo.

Scollar, I. (1965) 'A contribution to magnetic prospecting in archaeology', *Archaeo-Physika* 1, 21-92.

Scollar, I. and I. Graham (1974) 'A method of determination of the total magnetic moment of soil samples in a constant field', *Prospezioni Archaeologiche* 7, 85-92.

Tite, M.S. (1972) 'The influence of geology on the magnetic susceptibility of soils on archaeological sites', *Archaeometry* 14, 229-36.

Tite, M.S. and C. Mullins (1971) 'Enhancement of the magnetic susceptibility of soils on archaeological sites', *Archaeometry* 13, 209-19.

West, T.S. (1969) *Complexometry with EDTA and Related Reagents.* BDH Chemicals Ltd, Poole, England.

DISCUSSION

Dr. A.H. Weir began the discussion by enquiring whether the speaker had compared in the laboratory the effects of firing samples of geothite and lepidocrocite. The speaker said that he had not separated these out and had only analysed haematite by itself, but he thought that the experiment was worth trying. Dr. A.H. Weir noted that silt-sized particles containing maghemite tended to disappear in time from sites of heath fires on freely drained soils. Dr. D. Jenkins enquired whether the speaker had identified a component with the X-ray diffraction properties of δ FeO.OH which he had encountered when studying the magnetic concentrates from the clay fraction of soils. Dr. I. Graham commented that difficulties were caused by the presence of small particle-sizes resulting in broad diffraction peaks. He did not recognise any compound other than maghemite and magnetite.

Mr. B. Kerr raised the question of the effect of certain pedogenic conditions on magnetic conditions. The speaker noted that gleying destroyed magnetic minerals in the lower parts of the soil, and that in ditches with high gleying the magnetic content was negligible. He had not examined podzolic soils since the Rhineland agricultural soils being discussed were non-podzolic. In answer to a point made by Mr. K. Edwards concerning the relative merits of aerial photography and magnetic prospecting, the speaker concluded that the resolution of the two techniques was both about 1 m. However, aerial photographs were useless with loess and excellent for gravelly sediments, whilst the reverse was true for magnetic susceptibility.

J.A. Catt and A.H. Weir

The study of archaeologically important sediments by petrographic techniques

Archaeologists have never been slow to utilise techniques from other scientific disciplines; indeed, their own science would never have advanced this century in the way it has without multidisciplinary help, because, as in geology and other historical sciences, each rare clue to the past needs the fullest exploitation and the most careful evaluation. Various aspects of the study of sediments, principally stratigraphy, palaeontology, petrography and structural geology, have formed the core of geological science since the mid-nineteenth century, and current research still expands the store of information obtainable from sediments. Archaeologists have realised the importance of studying sediments and reconstructing past sedimentary environments, but have usually relied almost entirely on field observations. These are of course still fundamentally important, but geologists have found that the fullest understanding of sediments depends also upon a range of laboratory techniques, many of which demand specialist knowledge or apparatus. Some of these were developed in response to economic needs during and after World War I, but many new methods have arisen since then, and it is clear that the type of information obtainable from both old and new methods can greatly help the archaeologist in his task of tracing the environmental changes against which the development of human life and culture has been enacted.

Some of the techniques for studying sediments, such as the evaluation of the various types of palaeontological evidence, are dealt with both generally and specifically in other papers in this volume. We are concerned almost exclusively with the evidence provided by inorganic parts of the sediment, in particular by application of petrographic methods. Some petrographic methods, such as the use of thin sections to find the origin of rocks used for implements or megaliths, have long been used by archaeologists, but many others applicable to soft, unconsolidated sediments and to soils seem to have been almost completely ignored. Geologists have found the petrographic study of sediments useful in many ways. It can help to find the source rocks from which the sediment was derived, and trace changes in provenance both laterally and through time; it may indicate the transportation processes which brought material to the site, and the environmental conditions at and near the site; it can suggest the post-depositional processes that have changed the sediment during and after burial or when subsequently exposed to weathering; it can also be used to help compare detailed stratigraphic successions from place to place, which in unfossiliferous deposits is often the only means of indirect dating. Further, in helping to trace the course of many of these geological processes, petrography can and should become part of palaeogeographical or palaeoenvironmental reconstructions, which show and date, for example, changes of land and sea distribution, sea level, climate, or sub-aerial drainage patterns.

In reconstructing human cultural development, the archaeologist is dealing mainly with the Quaternary period, the last 1-2 million years of earth history, though in Africa and possibly other parts of the world later Tertiary time (the Neogene) may also be involved. Sediments of this age range are usually unconsolidated, though a few, such as limestones and volcanic deposits, may be hard rocks. Most of the petrographic techniques we discuss are consequently those applicable to unconsolidated sediments and incoherent, weathered materials in soil profiles. However, the necessity of recognising in Quaternary deposits clastic material or weathered residues derived perhaps from local bedrock often means that at least a partial petrographic study of much older rocks is needed. Methods particularly applicable to hard, consolidated sediments and even to igneous and metamorphic rocks cannot therefore be ignored, and the definition of which rocks are important archaeologically must vary according to the particular requirements of individual problems at each site.

The importance of posing specific questions to which petrographic work may provide or help provide an answer is emphasised also by the need to choose first as few samples as possible for laboratory analysis, and second the analytical methods best suited to each problem. There is often a bewildering array of possible methods for obtaining even only one expression of a sediment's composition (e.g. particle size distribution); some are better suited to particular types of sediment than others, and some can provide the evidence to solve particular problems better than others. Also, even the size of the sample and the way it is taken in the field should be determined by the type of analysis for which it is needed, and the uses to which the analytical results will be put. The amount of time needed for each technique is estimated later; many are very time-consuming, and the number of samples that can be analysed quickly and cheaply is often severely limited. Consequently, it is important to plan well ahead and pose the main questions early if reliable answers are eventually to be obtained. It is no longer justifiable to regard the petrographic study of sediments as a 'last hope' of solving problems presented by archaeological sections; nor is it fair to expect the petrographer to work efficiently with samples other than his own.

With fossiliferous Quaternary deposits, petrographic work often provides useful complementary evidence to palaeontological studies. However, many of the sediments encountered in Quaternary successions are barren of indigenous fossils (e.g. glacial and aeolian deposits), and petrographic work may then afford the only positive evidence from which the environmental significance, etc. of the deposits can be deduced. It may seem hardly worthwhile investigating sediments that contain no evidence of contemporary life, human or otherwise, but as evidence for the type of period separating for example distinct phases of human habitation they should never be ignored. Also, with the evidence provided by the fossiliferous horizons, occupation levels, etc., they can help the careful stratigrapher to trace a succession of events which, by comparison with the details of known successions elsewhere, may help date the deposits.

The petrographic techniques available for the study of sediments can be broadly divided into those concerning the sediment as a whole, and those in which the sediment is separated into various fractions on the basis of size, density or composition of particles. In subsequent parts of our paper, we give under these two headings some details of the main laboratory petrographic techniques useful in archaeology, which we hope will enable the reader to decide if he can muster the time, equipment and technical knowledge to do such work himself, or if he should seek specialist advice. The preliminary techniques of correct field sampling, laboratory pre-treatment and sub-sampling, etc., which are important to the acquisition of reliable analytical results, are outside the scope of this paper, but are adequately described by Milner (1962, Chapter II *et seq.*), Avery (1974) and others.

1. *Techniques applicable to the sediment as a whole*

1.1 *Thin sections and the study of microfabric*
Microscopic examination of thin sections is an especially valuable technique, which, following the introduction of modern synthetic resins for impregnation and hardening, can now be applied to unconsolidated sediments, clays, soils and organic deposits almost as easily as to hard rocks. Samples for thin sections of unconsolidated materials should be removed undisturbed by easing a rectangular box into the deposit, and carefully cutting away the surrounding material. The details of this technique, and of subsequently drying the sample, impregnating it with epoxy resins, and of cutting large (100 x 50 mm) thin sections are given by FitzPatrick (1970). An impregnation technique using polyethylene glycols was developed by Greene-Kelly *et al.*, (1970) for use with wet samples, to avoid the distortion of pores and other features in fine-textured materials during drying.

Thin sections afford the best means of characterising and recognising the rock types of artefacts, glacial erratics, water-transported cobbles, etc. With soils and unconsolidated deposits, they have been used increasingly in recent years for micromorphological studies, such as the determination of glacial transport directions from tills (Sitler and Chapman, 1955), the classification of soil microfabric (Brewer, 1964), the investigation of pedogenically important features, such as clay cutans (Buol and Hole, 1961), pores (Kubiena, 1967) and faecal pellets (Babel, 1968), and for microchemical analysis using staining techniques (Babel, 1964; Jenkins, 1970).

The archaeological significance of some of these applications will probably be apparent in specific problems, and require no further comment, but it is perhaps important to add a few remarks on soil micromorphology. The development of a soil profile at a given site implies a period of relative stability in the land surface, with little erosion to remove weathered materials, and little deposition of new sediment. However, subsequent erosion may remove all or part of the evidence for soil development, and renewed deposition may bury it. The type of soil profile that develops, as expressed by the nature and thickness of successive soil horizons, is determined by the type and extent of physical and chemical weathering, the humus form, and the redistribution of soluble or very fine soil components. These in turn depend on many factors, notably the composition of the soil parent material, climate, vegetation, relief, local drainage conditions, and various types of human interference with natural processes. The soil type changes laterally in response to one or more of these factors, but within a relatively small area the range of soil type is usually limited, and changes occur in characteristic ways, for example in response to simple changes in drainage conditions dependent upon local topography. The resultant catena of characteristic local soil types can provide much more information about the conditions of soil development than does a single profile type. Classification of soil types is based largley on a field assessment of profile characteristics and their genetic significance, but in some schemes (Kubiena, 1953; Soil Survey Staff, 1960) micromorphology is an important supplementary basis, as it provides considerable evidence for past and present soil forming processes, and sometimes may be the only means of detecting their incipient effects.

The use of micromorphology, particularly of buried soils, in reconstructing past environmental conditions has been discussed by Dalrymple (1958), Kubiena (1959, 1963, 1970) and Cornwall (1958, 1963), all of whom have as suggested that it does provide reliable palaeoenvironmental evidence. However, as in any argument invoking the law of uniformitarianism, one must be certain that the features attributed to present conditions are not themselves inherited wholly or even partly from past periods. Yet this and other fundamental research needed to establish the exact origin of the important microfabric types, and to distinguish climatic effects from those due to other factors determining soil type, still needs to be done. A useful start was made by Dalrymple (1972), who studied experimentally the fabrics that can be formed in clay/iron oxide mixtures, and suggested that *lehm* type fabrics, which Kubiena and others relate to tropical pedogenesis, reflect only the drying mechanism in the soil. Further problems of interpretation can arise through relic soil material from earlier profiles being incorporated in colluvial or solifluction deposits, and the possible diagenetic modification of weathered materials after burial. Nevertheless, the micromorphology of buried soils has been used with apparent success as a palaeoenvironmental indicator and stratigraphic tool in France (Federoff, 1967, 1968, 1969; Jamagne, 1972; Hetier *et al.*, 1972), Germany (Bronger, 1966, 1970, 1972) and elsewhere, mainly on loess-derived soils. Buried soils seem to occur much more commonly in aeolian sediments than other materials, but often present the additional difficulty of distinguishing soil forming processes that occur during deposition (e.g. gleying and mobilisation of carbonate) from those of later periods.

One of the most important morphological features of soils from an environmental point of view is rubification, the development of red colours (hues 5YR or redder) in mottles and cutans of translocated clay by partial recrystallisation of amorphous iron oxides and hydroxides. A humid and relatively warm climate is necessary for rubification, but its occurrence also depends on soil texture and the nature of the iron-bearing minerals in the soil parent material. Holocene rubification has occurred only in areas wetter and warmer than Britain, such as parts of Mediterranean and Atlantic France, so that any occurrence in Britain, for example on the North Yorkshire Moors (Bullock *et al.*, 1973) must be as a palaeosol feature inherited from an Ispwichian or earlier interglacial period of soil development.

The time needed for preparation and interpretation of thin sections can vary from less than a day to 6-8 weeks, depending mainly on that necessary for impregnation and hardening of unconsolidated deposits. Slow resin infiltration of fine sediments and soils is best, particularly if distortion of pores and other features is to be avoided. However, techniques and apparatus are available for simultaneous treatment of many samples.

1.2 Bulk density and porosity

Two kinds of density are recognised in the study of sediments, namely grain density and bulk density. Grain density (GD) is the average density of the constituent particles, weighted according to their relative proportions in the sediment, whereas bulk density (BD) is the weight per unit volume of the material in its natural state. The difference between the two is due to the proportion of voids, or spaces between particles, which is expressed by the porosity (P), calculated as $P = \dfrac{GD - BD}{GD}$, so that for sediments with the same grain density, bulk density is inversely proportional to porosity. Changes in bulk density and porosity result from differential compaction or cementation; the latter is often not applicable to archaeological sites, but changes caused by differential compaction can be of archaeological interest as evidence for former soil profiles, habitation horizons, pits, ditches, etc., and in distinguishing some types of sediments (e.g. those deposited beneath a heavy overburden of ice). Compaction can be judged qualitatively in the field, but quantitative estimates of bulk density and porosity require careful laboratory measurements of weight and volume.

The standard technique for measuring the bulk density of soils or soft sediments involves removing an undisturbed core of known volume (Dagg and Hosegood, 1962), and determining its oven dry weight. Porosity may also be determined on the same sample by weighing it at field moisture content, and by finding the grain density from the weight and volume of a fully disaggregated, representative sub-sample. From these measurements the volumes of solid, air and water in the original moist sample can be calculated. The difficulty of removing a core from a stony soil or sediment is overcome by measuring the volume of an extracted sample with sand from a graduated container or with plastic balls of uniform known diameter (Smith and Thomasson, in Avery, 1974).

Pore space may also be measured fairly accurately in thin sections made from samples impregnated with resin containing a fluorescent dye. These can be examined either with a microscope using an ultraviolet light source and a point counter to measure the ratio of voids to mineral grains (Chayes and Fairbairn, 1951), or with a Quantimet image analyser, which can measure pore size distribution and pore shape as well as total pore space. However, in samples with large pores a single thin section may not give an adequate representation of the pore size distribution. Also pore space measured in thin section is affected by shrinkage during drying and impregnation, so that this technique is best restricted to horizons which are too thin to give the large samples needed for the standard method.

1.3 Chemical analyses of bulk samples

Various methods of chemical analysis have been used by archaeologists to help date materials, such as bone, and to understand the origin of sediments and soils. Methods useful as part of a petrographic study include analyses for phosphate (Fogg and Wilkinson, 1958), calcium carbonate (Bascomb, 1961), organic carbon (Tinsley, 1950; Shaw, 1959), nitrogen (Bremner, 1965), and 'free' iron and aluminium, and the laboratory determination of pH. Most of these are fairly rapid methods; Avery (1974) gives many of the necessary experimental details.

The amounts of organic carbon and nitrogen reflect organic matter content. In agricultural soils the C/N ratio is approximately 10, and considerably larger ratios, for example in Quaternary sediments and buried soils, usually indicate the presence of charcoal. Also, highmoor peats have larger C/N ratios than the less acid lowmoor types (Waksman, 1936, 265). Ratios of carbon or clay content to 'free' iron and aluminium, as determined by extraction with various reagents (Kilmer, 1960; Mehra and Jackson, 1960; Schwertmann, 1959; McKeague, 1967; McKeague *et al.*, 1971) are used in soil classification, principally as a basis for the recognition of podsolised soils. The Soil Survey of England and Wales use pyrophosphate extraction for laboratory characterisation of such profiles (Bascomb, 1968), as it seems to provide the best correlation with conclusions based on field and micromorphological evidence.

Phosphate analyses are useful for showing horizons or areas enriched in bone material, such as occupation layers, burial sites, some fossiliferous horizons, etc. The 'background' phosphate content of some sediments (e.g. sands) is extremely small, but many clays and limestones contain larger quantities, principally as mineralised faecal pellets or collophane from fossil remains (Brown and Ollier, 1956). The technique is therefore likely to be most useful in sandy deposits, though phosphate is leached slowly from acid sandy soils. Further complications may arise because most surface soil layers in agricultural areas are enriched in phosphate from fertilisers, the remains of which are fixed in a relatively insoluble form near the surface by association with iron and aluminium oxides (Cooke, 1967, Chapter 2). Phosphate may also be increased in some sub-surface horizons enriched in the natural insoluble residues of limestones, for example Clay-with-flints *sensu stricto* (Loveday, 1962). Results should therefore be interpreted with some care, but the method is rapid and could be used more widely.

Silicate analyses of bulk sediment or soil samples are very time-consuming, and provide little useful information to the archaeologist. However, in some circumstances trace element analyses by spectographic or other methods may help characterise sediments or artefacts and hence facilitate correlation. They may also help in identifying small amounts of the metals used by early man, such as gold, silver, copper and tin. Spectrographic analysis is fairly rapid; 20-25 samples can be analysed quantitatively for up to about 18 elements per week.

1.4 Particle size analyses
Particle size analysis is discussed under the general heading of techniques applicable to the sediment as a whole, but in it one meets an important basic problem in the subdivision of a sediment or soil sample into different fractions, namely disaggregation. The individual particles in almost any sample, except possibly the purest silver sands, are bound together more or less firmly in various ways, and the particle size distribution obtained on analysis depends on the extent

to which the bonding is broken. Also, once the particles have been dispersed, they must be kept so until the particle size analysis or separation into fractions is complete; in particular, the flocculation of clay particles must be avoided.

Dispersion is achieved by various physical and chemical treatments of the sample (Elonen, 1971). The physical treatments applicable to soils and unconsolidated sediments include gentle crushing and grinding (e.g. Waters and Sweetman, 1955), extended shaking, and ultrasonic vibration (Genrich and Bremner, 1972). Chemical treatments include removal of organic matter with hydrogen peroxide or sodium hypochlorite (Anderson, 1963), removal of carbonates with hydrochloric or acetic acid, and removal of iron oxides with dithionite-citrate (Mehra and Jackson, 1960) or ammonium oxalate under ultraviolet irradiation (Le Riche and Weir, 1963). A weak solution of sodium hexametaphosphate (Calgon), made alkaline by addition of a little sodium carbonate or hydroxide, is also widely used as a dispersing agent in soil analysis (Kilmer and Alexander, 1949), and seems especially helpful with calcareous materials and those weakly cemented with hydrated iron oxides, but part of the phosphate can become fixed in the clay fraction.

No single dispersion method or combination of methods can be recommended universally, as the technique needs to be modified according to the peculiarities of individual samples. For routine use on soil samples, the Soil Survey of England and Wales first remove stones greater than 20 mm, then crush the remainder in a machine similar to that described by Waters and Sweeting (1955), to give fine earth (<2 mm) and fine gravel (2-20 mm). A sub-sample of the fine earth is then peroxidised, centrifuged, and shaken overnight in dilute sodium hexametaphosphate solution prior to particle size analysis (Avery, 1974). Ultrasonic treatment is more rapid than grinding and overnight shaking, especially if a high intensity probe-type vibrator is used; it also avoids the use of chemical reagents, many effects of which are still imperfectly understood, but only small amounts can be treated, which is a severe limitation with stony soils or sediments.

Decalcification is often necessary, for example to allow comparison of naturally decalcified horizons with calcareous layers or zones of secondary carbonate accumulation. Treatment with acetic acid buffered at pH 5 is preferred to hydrochloric acid, as the latter dissolves phosphates and other minerals in addition to carbonate. To prevent flocculation, the acetate should be removed from the suspension by centrifugation and two washes with distilled water. Another decalcification procedure, which is closer to the natural process and does not add unwanted anions, involves shaking an aqueous suspension of the sample with solid carbon dioxide in a bottle with a screw cap to allow periodic release of pressure; subsequent centrifugation is necessary to prevent reprecipitation of carbonate.

The use of hydrogen peroxide should be avoided with pyritic

sediments and with samples containing free manganese oxides (e.g. Clay-with-flints). The rapid oxidation of pyrites produces sulphuric acid, which may dissolve or modify some soil minerals and cause clay flocculation. Manganese oxides catalyse the decomposition of peroxide, often with the violent evolution of oxygen. Organic acids formed by peroxidation of organic matter can also cause clay flocculation, and should be removed by centrifugation before the final dispersion prior to particle size analysis.

Disaggregation of harder, consolidated sediments presents further special problems, some of which are insuperable with present techniques. Calcareous cements are easily removed by treatment with acetic acid, but siliceous and some ferruginous cements demand stronger chemical treatments, which are likely to destroy detrital constituents as well. Prolonged ultrasonic vibration often helps disaggregation of compacted or weakly cemented sediments, especially if they are first crushed into fairly small pieces, but has little or no effect on harder rocks. The particle size analysis of these is then best determined from thin sections by measuring the size of a large number of particles with a microscope fitted with a suitable graticule or micrometer eyepiece (Krumbein, 1935).

The range of particle size in many sediments and soils is extremely large on an arithmetic scale: the finest clay particles are often no more than 0.1 μm across and are usually considerably thinner, whereas a moderately large stone would have an equivalent spherical diameter at least a million times larger than this, and a volume approximately 10^{19} times greater. As a result, particle size divisions are based on the logarithmic ϕ scale, in which $\phi = -\log_2$ grain diameter in mm (Krumbein, 1934). The result can be expressed as a cumulative graph or histogram, the form of which may be summarised in terms of four moments, the mean grain size, standard deviation, skewness and kurtosis, all of which are quoted in terms of ϕ units. Greatest precision is obtained from small subdivisions, at half or preferably quarter ϕ intervals, throughout the entire range of particle size, but most well established methods of particle size analysis fall short of this ideal. We developed the method outlined below to give moderately detailed analyses fairly rapidly; 12-18 samples can be analysed at whole ϕ intervals per week, but analyses at half and quarter ϕ intervals are considerably slower.

The sand fraction (62-2000 μm; 4ϕ to -1ϕ) is removed from 150-200g of air dry sample by ultrasonic dispersion and wet sieving through a 62 μm sieve. This is oven-dried and subdivided in a tower of sieves at whole, half or quarter ϕ intervals (Table 1), which are shaken on a standard sieve shaker for ten minutes. The weight percentage of each sand fraction is then calculated on an oven dry basis, after correcting the weight of the 150-200 g sample according to the percentage weight lost from a small air dry sub-sample left for 24 hours in a 105°C oven.

The silt fraction (2-62 μm; 9-4ϕ) can be subdivided at whole ϕ

Table 1. Aperture sizes of sieves used for subdivision of sand fractions at quarter ϕ intervals.

Sieve mesh number	Nominal Aperture μm	ϕ
8	2000	-1.0
10	1680	-0.75
12	1400	-0.5
14	1200	-0.25
16	1000	0.0
18	850	0.25
22	710	0.5
25	600	0.75
30	500	1.0
36	420	1.25
44	355	1.5
52	300	1.75
60	250	2.0
72	210	2.25
85	180	2.5
100	150	2.75
120	125	3.0
150	105	3.25
170	90	3.5
200	75	3.75
240	63	4.0

intervals using the pipette sampling technique; more detailed analyses at half or quarter ϕ intervals (e.g. Weir *et al.*, 1971) can be achieved with the apparatus described by Stairmand (1950). Alternatively, a cumulative curve over this size range can be obtained with a sedimentation balance, such as that described by Bostock (1952). For the pipette sampling technique, a 10-15 g air dry sub-sample is treated as necessary to ensure prolonged dispersion, and stirred homogeneously in exactly 1000 ml distilled water contained in a tall vessel, such as a measuring cylinder; the water temperature should be stabilised within 1-2°C throughout the analysis. Samples taken with a 20 ml pipette from a depth of 200 mm below the water surface at the times given in Table 2 are then oven dried in crucibles and weighed. The insertion and removal of the pipette should stir the solution as little as possible, and is best done with a pipette stand (Elonen, 1971). The sedimentation times are based on Stokes' law (Tanner and Jackson, 1947), and for a given particle size decrease with increasing temperature. The amounts of each fraction are then obtained by multiplying the weight of oven dry sediment in each pipette sample by $\frac{1000}{20} = 50$. As before, the air dry weight of the 10-15 g sample should be corrected to an oven dry weight, and if sodium hexametaphosphate is used as a dispersing agent, a correction for the weight of this in each

Table 2. Pipette sampling times for particle size analysis at whole φ intervals of six samples simultaneously, started at nine minute intervals; all pipette samples are taken at 200 mm depth, except those for <9φ (<2 μm), which are taken at 80 mm depth. (h = hours; m = minutes; s = seconds).

Water Temp.	Particle Size	1st Sample Start 0m	2nd Sample Start 9m	3rd Sample Start 18m	4th Sample Start 27m	5th Sample Start 36m	6th Sample Start 45m
20°C	<4φ, 63 μm	57s	9m 57s	18m 57s	27m 57s	36m 57s	45m 57s
	<5φ, 31 μm	3m 48s	12m 48s	21m 48s	30m 48s	39m 48s	48m 48s
	<6φ, 16 μm	15m 12s	24m 12s	33m 12s	42m 12s	51m 12s	1h 0m 12s
	<7φ, 8 μm	1h 0m 48s	1h 9m 48s	1h 18m 48s	1h 27m 48s	1h 36m 48s	1h 45m 48s
	<8φ, 4 μm	4h 3m 12s	4h 12m 12s	4h 21m 12s	4h 30m 12s	4h 39m 12s	4h 48m 12s
	<9φ, 2 μm	6h 29m 7s	6h 38m 7s	6h 47m 7s	6h 56m 7s	7h 5m 7s	7h 14m 7s
21°C	<4φ, 63 μm	55s	9m 55s	18m 55s	27m 55s	36m 55s	45m 55s
	<5φ, 31 μm	3m 42s	12m 42s	21m 42s	30m 42s	39m 42s	48m 42s
	<6φ, 16 μm	14m 48s	23m 48s	32m 48s	41m 48s	50m 48s	59m 48s
	<7φ, 8 μm	59m 12s	1h 8m 12s	1h 17m 12s	1h 26m 12s	1h 35m 12s	1h 44m 12s
	<8φ, 4 μm	3h 56m 48s	4h 5m 48s	4h 14m 48s	4h 23m 48s	4h 32m 48s	4h 41m 48s
	<9φ, 2 μm	6h 18m 53s	6h 27m 53s	6h 36m 53s	6h 45m 53s	6h 54m 53s	7h 3m 53s
22°C	<4φ, 63 μm	53s	9m 53s	18m 53s	27m 53s	36m 53s	45m 53s
	<5φ, 31 μm	3m 36s	12m 36s	21m 36s	30m 36s	39m 36s	48m 36s
	<6φ, 16 μm	14m 24s	23m 24s	32m 24s	41m 24s	50m 24s	59m 24s
	<7φ, 8 μm	57m 36s	1h 6m 36s	1h 15m 36s	1h 24m 36s	1h 33m 36s	1h 42m 36s
	<8φ, 4 μm	3h 50m 24s	3h 59m 24s	4h 8m 24s	4h 17m 24s	4h 26m 24s	4h 35m 24s
	<9φ, 2 μm	6h 8m 38s	6h 17m 38s	6h 26m 38s	6h 35m 38s	6h 44m 38s	6h 53m 38s
23°C	<4φ, 63 μm	52s	9m 52s	18m 52s	27m 52s	36m 52s	45m 52s
	<5φ, 31 μm	3m 30s	12m 30s	21m 30s	30m 30s	39m 30s	48m 30s
	<6φ, 16 μm	14m 0s	23m 0s	32m 0s	41m 0s	50m 0s	59m 0s
	<7φ, 8 μm	56m 36s	1h 5m 36s	1h 14m 36s	1h 23m 36s	1h 32m 36s	1h 41m 36s
	<8φ, 4 μm	3h 46m 42s	3h 55m 42s	4h 4m 42s	4h 13m 42s	4h 22m 42s	4h 31m 42s
	<9φ, 2 μm	6h 2m 42s	6h 11m 42s	6h 20m 42s	6h 29m 42s	6h 38m 42s	6h 47m 42s

pipette sample should also be introduced. From experience it has been found convenient to analyse six samples simultaneously per day.

In the modification of Stairmand's method which we use, a suspension of 2 g sample in 350 ml distilled water contained in a sedimentation tube is homogenised by bubbling air through it for several minutes. Particles are then sedimented under free fall conditions, and as they accumulate in the tapered base of the tube are flushed out at times corresponding to half ϕ intervals by small aliquots of clear water from the side tube connected to a reservoir. Each fraction is then oven dried and weighed. The upper particle size limit of material that can be analysed by this method is approximately 3.5 ϕ, which is slightly coarser than the effective upper limit with the pipette sampling technique (4.0 ϕ), and allows a small overlap with the finest sieves available. Also the amount of suspension removed with each sample is smaller than with the pipette, and more subdivisions of particle size can be made. Two staggered sets of measurements at half ϕ intervals are used to obtain quarter ϕ data, but the apparatus is then limited to one sample per day.

The sedimentation time for particles of 2μm (9 ϕ) equivalent spherical diameter is so long that in the pipette sampling technique the fraction less than 2 μm (i.e. clay) is best sampled at only 80 mm depth (Table 2); otherwise the analysis cannot be completed easily the same day. Yet finer fractions would take much longer, and are not usually attempted. However, some centrifugation techniques (q.v.) allow the separation of clays into various size fractions, and the figures obtained on this basis can be used if necessary as part of the particle size distribution.

The size of sample required for particle size analysis of stones (>2 mm) is very large (Milner, 1962, Vol. 1, p. 171), so that with deposits containing large stones, the analysis is best done at least partly in the field, for example by sieving stones from the finer earth. The weights of soil or sediment and of separated stones can be determined accurately enough with a spring balance, and the water content can be found later in the laboratory from a small sample sealed in a polythene bag. However, the use of a 2 mm sieve in the field is difficult with wet or clayey samples, in which case a coarser sieve (e.g. 16 mm) can be used to separate the larger stones from a big sample, and the 2-16 mm fractions can be determined by a laboratory analysis of a somewhat smaller sample. The size of large stones in terms of ϕ units is perhaps best determined by weighing individual stones and calculating their equivalent spherical diameters assuming their specific gravity to be that of the rock type from which they are derived (e.g. 2.60 for flint).

The main use of particle size data is in descriptive petrography, especially as a means of characterising a sediment to allow comparison with other deposits. In favourable circumstances, the nature of mixed deposits containing sediment from two different sources, each with a characteristic particle size and distribution, can be determined (e.g. Mitchell *et al.*, 1973; Perrin *et al.*, 1974), but it is important to confirm this type of conclusion with mineralogical or other evidence. Considerable effort has gone into studying the use of particle size distribution in differentiating depositional environments (Folk and Ward, 1957; Friedman, 1961, 1962; Sevon, 1966; Buller and McManus, 1972, 1974), but its use in this is still only qualitative and empirical. Aeolian and beach sands are well sorted, whereas river and glacial sands are poorly sorted, and glacial till and gravitational deposits (e.g. colluvial and solifluction deposits, slide, slip and fall deposits) are very poorly sorted. Aeolian silt (loess) is also fairly well sorted (Russell, 1944), and is thus fairly easily recognised in mixed deposits (Harrod *et al.*, 1974).

2. *Techniques applicable to separated fractions of sediments*

2.1 *Separation into size fractions*

After adequate dispersion, a soil or sediment sample can be separated into size fractions by sieving, and by sedimentation in water under the influence of gravity or centrifugal acceleration. The optical microscope can conveniently be used to study the shape and composition of particles only in the range 16-250 μm (6-2 ϕ), which can be divided into fine sand (4-2 ϕ) and coarse silt (6-4 ϕ). Sand grains coarser than 2 ϕ are too large for easy examination with a research petrological microscope, and rarely offer any more information than can be obtained from the fine sand; however, if a sample contains a large amount of coarser sand and little or no fine sand, then a fraction near the mean grain size should be separated for analysis. Fractions too fine for optical examination are usually analysed by X-ray techniques, but these are most easily applied to particles less than 2 μm equivalent spherical diameter (i.e. clay). The fractions most easily analysed mineralogically are therefore the clay, coarse silt and fine sand; often these are also the most important on a weight percentage basis.

To separate the clay and coarse silt fractions, the material finer than 4 ϕ passing the 62 μm sieve is collected in a bucket or large beaker, and allowed to stand undisturbed overnight. The highest 200 mm of the suspension, which by then will contain only clay particles, is transferred to another large container without disturbing the sediment below 200 mm depth; this is best done with a siphon with a hooked end, which when inserted to a depth of 200 mm sucks in material only from above. The remaining sediment is transferred with distilled water to a tall container, such as a litre measuring cylinder, and stirred; after a further sedimentation period, the portion of the column free of particles less than 2 μm is again siphoned off. This procedure is repeated until a clear supernatant is obtained at the end of the sedimentation time. A few drops of dilute alkaline sodium hexametaphosphate solution should be added before re-stirring the suspension each time, to prevent flocculation. The sedimentation times for clay separation are temperature-dependent, 51 minutes/cm fall at 16°C, 49 at 18°C, 47 at 20°C, 45 at 22°C, etc. The volume of the separated clay suspension is decreased by flocculating the clay with approximately 100 ml of decinormal calcium chloride or magnesium chloride solution, and decanting the clear supernatant. The flocculated clay is then saturated with calcium or magnesium by prolonged shaking in further decinormal chloride solution, then centrifuged so that the excess solution can be poured off, and finally shaken with two aliquots of distilled water to remove most of the chloride, from which the clay is again separated by centrifugation. The coarse silt fraction is separated from the fine silt by further repeated sedimentation in water, the sedimentation times being 51 second/cm fall at 16°C, 48 at 18°C, 46 at 20°C, 43 at 22°C, 41 at 24°C, 39 at 26°C, etc. Like sand, it can be dried from water, but if the fine silt is to be kept for analysis, that should be freeze-dried to prevent caking. An alternative method of drying clays and fine silts is to wash the wet but chloride-free material twice with acetone, twice with petroleum ether, and finally evaporate the remaining ether at room temperature or just above.

Further subdivision of the material finer than 2 μm (clay) into less than 0.04 μm, 0.04-0.2 μm and 0.2-2.0 μm fraction is done by centrifugation. The less than 2 μm clay is converted to the sodium saturated form by shaking with normal sodium chloride solution, then dispersed in water adjusted to pH 8 with sodium hydroxide. The centrifuging times and speeds depend partly on the dimensions of the centrifuge used; Tanner and Jackson (1947) showed the method of calculating these, and gave corrections for different temperatures, specific gravities of clay particles, etc. Some allowance should be made for increases in temperature during long centrifuging periods. The finest clay (<0.4 μm) is removed first by repeated centrifuging for 60-75 minutes at 10,000 r.p.m. or 25-30 minutes at 15,000 r.p.m.; normally the sequence of dispersion, centrifugation and decantation needs repeating at least six times before this fraction is completely separated. The medium and coarse clays are then separated at 0.2 μm by repeated centrifuging for approximately 35 minutes at 2,500 r.p.m. or approximately 18 minutes at 3,500 r.p.m.

2.2 Fine sand and coarse silt mineralogy
In most deposits these fractions are composed mainly of quartz, with variable but generally quite small amounts of felspar, mica, iron ore minerals (magnetite, ilmenite, haematite, etc.), and various non-opaque heavy minerals, and sometimes also of glauconite, calcite and gypsum. Minor quantities of flint fragments are also extremely widespread in British Tertiary and Quaternary deposits. The quickest

and most reliable means of recognising all these minerals is with a research petrological microscope, identifications being based on measurements of optical and crystallographic properties of individual grains. The various techniques involved are described by Smith (1956), Hartshorne and Stuart (1960), Bloss (1961), Kerr (1959) and in other standard texts; useful tables to help identification are by Winchell (1939, 1965) and Larsen and Berman (1934). Temporary mounts on glass slides are preferable to permanent canada balsam mounts, as they allow manipulation of individual grains on the microscope stage, which is often necessary in identification.

Quantitative mineral analysis with the petrological microscope is, however, difficult and fairly time-consuming, partly because many of the minerals helpful in characterising a sediment or finding its origin are present in only very small amounts, and partly because counting grains, even with the aid of a point counter, is a tedious and rather unreliable means of estimating weight percentages. An important preliminary stage in quantitative analysis is the subdivision of the fine sand or coarse silt into sub-fractions on the basis of density, or of magnetic or chemical properties. Density separations in bromoform (specific gravity 2.9) or other heavy liquids (Milner, 1962, Vol. 1, p. 119) are widely used, but many of the liquids are toxic and need careful handling to prevent inhalation and skin contact (O'Connell, 1963). Sand fractions can be adequately separated in a separating funnel, providing any heavy particles adhering to the side walls are gently tapped to the bottom, but silt separations require centrifugation in special tubes which allow recovery of the heavy and light fractions. Three main methods of silt separation are used (Griffiths, 1967, p. 208); we prefer the freezing technique, in which the mixture is centrifuged in pointed glass tubes, and the heavy fraction then frozen in the pointed base of the tube, so that the light fraction can be poured off and flushed out of the top. Pure bromoform freezes at 8°C, so that the lower part of the tube rapidly solidifies in an ice-salt mixture. The heavy fraction rarely exceeds 2-3 per cent of the fine sand or coarse silt in most samples, and is often much less. Despite this, heavy mineral analysis is more widely used than light mineral analysis as a means of characterising a sediment, because the heavy fraction usually contains a much greater range of mineral species, and includes many which can indicate the source of the sediment more precisely than the almost ubiquitous quartz, felspar, mica, etc., contained in the light fraction.

Further subdivision of the light or heavy fraction by magnetic separation or chemical differential dissolution techniques may also help quantify mineral analysis of fine sands and coarse silts. Strongly magnetic minerals (magnetite, maghemite, pyrrhotite) can be removed from the heavy fraction with a small hand magnet, but a powerful electromagnetic mineral separator is necessary to separate moderately or weakly magnetic species from the non-magnetic.

Messrs C.W. Cook & Sons, 97 Walsall Road, Birmingham, manufacture a suitable electromagnetic mineral separator, which can be used with particles in the 40-250 μm size range. Removal of layer silicates, principally mica, from the light fraction by dissolution in fused sodium bisulphate (Kiely and Jackson, 1964) or hydrochloric acid (Reynolds and Lessing, 1962) leaves a quartz-felspar residue, which can be analysed chemically to give the proportions of K, Na and Ca felspar end-members and, by difference, quartz. However, both reagents have some effect on felspars, especially plagioclases, and the results are not completely reliable.

Completion of the quantitative mineralogical analysis of fine sands and coarse silts depends ultimately on the use of the microscope, both to check the efficiency of density, magnetic or chemical separations, and to determine the relative proportions of different minerals within the various subfractions. Some method of grain counting must then be adopted; the precision and accuracy of the result clearly depend on the number counted, the standard deviation for all but the rarest minerals being approximately the square root of the number counted. It is consequently impossible to state a minimum count applicable to all samples; this varies according to the composition of the sample, and the precision aimed at. As many of the more significant minerals are often present in such small quantities, it is usually possible to estimate the proportions of only the more common species, and to quote rare constituents as '<1%', '<1°/oo', or merely 'present'. This has the merit of speed; the proportions of major constituents can be expressed on the basis of counting perhaps 500-1000 grains, and a much larger though unknown number can subsequently be scanned for rarer minerals. In samples containing the usual small amounts of heavy minerals, this procedure can be adopted for both the light and heavy fractions, but the significance levels are different. The composition of the light fraction can be expressed in percentages and amounts less than 1 per cent of the total fine sand or coarse silt, to give the approximate bulk composition, but the proportions of heavy minerals are better expressed as percentages or parts per thousand (and amounts <1°/oo) of the heavy fraction itself. Working on this basis, two fine sand or coarse silt fractions can be separated and analysed per day, but more detailed subdivisions involving magnetic or other separations can take considerably longer.

Some of the sand and silt minerals have special palaeo-environmental significance; four examples are worth mentioning. Glauconite forms almost exclusively in marine conditions, though its exact origin is still debated (McRae, 1972). Pyrites also needs seawater to provide the necessary sulphate ions, which are bacterially reduced to sulphide in anaerobic conditions just below the sediment surface. Gypsum originates in various ways: in clayey soils under moderate rainfall, it forms by reaction between calcium carbonate and the sulphuric acid produced by oxidative weathering of pyrites;

however, in arid conditions it forms large aggregates of crystals ("desert roses") in sands. Grasses and some other plants produce characteristic forms of opal (Smithson, 1958), the presence of which in a buried horizon would indicate an old land surface with a fairly open vegetation. However, it is important to remember that all these minerals may be transported from the site of their formation, and incorporated in younger deposits, sometimes without appreciable alteration if transport and redeposition occur in cold conditions.

2.3 Clay mineralogy

Clay mineraology is often less useful in differentiating deposits than sand or coarse silt mineralogy, because fewer clay mineral species can be recognised by present techniques, quantitative analysis is more difficult than with coarser fractions, and clay fractions can be altered more easily by weathering, diagenesis, etc. Nevertheless, a considerable amount of archaeologically useful information can be obtained from careful analysis and study of clay fractions of soils and sediments, especially when it is interpreted in the light of data obtained from other studies.

Clay minerals are mainly sheet silicates, which are most readily identified from their basal spacings – the distances between the successive alumino-silicate layers that comprise the sheets. These are measured by X-ray diffractometry of specially treated oriented clay aggregates. Small sub-samples of the moist Ca or Mg saturated less than 2 μm fractions are slurried with water and then dried on to glass slips; the surface tension during drying orients the platy particles parallel to the glass surface, thus enhancing the intensity of low-angle X-ray reflections from the basal planes. Further details of specimen preparation, and the recording and interpretation of reflection patterns obtained are given by Bullock and Loveland (in Avery, 1974). Replicate specimens on glass slips are air dried at room temperature, treated with ethylene glycol vapour, or heated to 300 or 500°C. Sorption of glycol helps distinguish the basal spacings of swelling minerals, such as smectite (17 Å), vermiculite (14.4 Å) and complex interstratified minerals from non-swelling chlorite (14.2 Å), illite (10 Å), kaolinite (7.1 Å) and interstratified mixtures of these. Heating to 300°C collapses the expanding minerals and distinguishes them from chlorite, and heating to 500°C destroys kaolinite, so that its basal reflection can be distinguished from the nearly coincident reflection of chlorite.

The diffraction intensities of basal reflections from specimens treated in these ways may be used to obtain an approximate estimate of the relative proportions of minerals present in a mixture, and in some circumstances the differences between clays calculated in this way can help distinguish two or more deposits. However, figures obtained for mica, chlorite and expanding minerals should be used with care in such work, because these minerals are easily altered by weathering processes.

In addition to the major sheet silicate components, clay fractions often also contain small amounts of quartz, felspar, goethite, lepidocrocite, calcite, haematite, apatite and some other minerals. These may help distinguish different clays or indicate the source of the clay, the conditions of deposition, or the weathering history, but their recognition in small quantities and quantitative determination are not easy. When they are known to be present, the amounts of calcite can be estimated by calcimeter (Bascomb, 1961), of apatite by phosphate determination, and of total free iron by dithionite extraction (Mehra and Jackson, 1960); the amount of quartz can be determined on the residue from fusion with bisulphate (Kiely and Jackson, 1964), but felspar determination by this method is unreliable with clays. Amorphous silica and alumina, as estimated by sodium hydroxide extraction (Hashimoto and Jackson, 1960), can help indicate the extent of chemical weathering of the silicate clays.

Other methods used to identify the minerals in clay fractions include electron diffraction (Gard, 1971), differential thermal analysis (Mackenzie, 1957) and infrared spectroscopy (Farmer, 1968), but these are not as widely used as X-ray diffractometry, either because they are not capable of such detailed discrimination between mineral species, or because the techniques are not yet sufficiently well developed. Electron diffraction is potentially a very sensitive technique, because selected area diffraction with a transmission electron microscope enables single clay particles to be examined. However, platy clay particles settle with their basal planes normal to the electron beam, so that their basal spacings cannot easily be measured. Eberhardt and Triki (1972) overcome this by electron diffraction of ultramicrotomed sections of resin-impregnated clay aggregates cut perpendicular to the sedimentation plane. Differential thermal analysis suffers from the difficulty of discriminating between many clay mineral species, and infrared spectroscopy from its present lack of development to the stage where it can be used by non-specialist workers for routine clay mineral analysis, though it is obviously an extremely useful tool in clay research.

2.4 Fine silt mineralogy

With samples containing fairly large quantities of fine silt (2-16 μm; 9-6 ϕ), which are fortunately fairly rare, it may be necessary to analyse this fraction mineralogically. This presents the immediate difficulty that fine silt particles are generally too small to study with an optical microscope, yet too large to give good X-ray diffraction patterns. However, the fine silt of many sediments contains interesting and often highly significant minerals, such as jarosite (Weir and Catt, 1969) and clinoptilolite (Brown *et al.*, 1969), so that it is often important to study this fraction. It also contains up to 20 per cent layer silicate minerals, which are usually slightly different from those in the clay fraction and give sharper diffraction peaks, indicating that the crystal size is larger and that the grains are not simply undispersed aggregates of clay size material. The fraction is thus best analysed by a combination of density separations, chemical techniques and X-ray diffraction. To prevent flocculation of fine silt in polar heavy liquids, such as bromoform or mixtures of this with acetone or alcohol, it should be treated with a special suspending agent, such as polyvinylpyrrolidone (Francis, 1973). Many of the layer silicates found in fine silt fractions have densities in the range 2.7-2.9, and can thus be separated from quartz, felspar and the heavy minerals to give a concentrate that will orient well for X-ray diffractometry.

2.5 Particle shape

The shape of sand and larger particles has long been used as evidence for conditions of deposition, but is one of the most difficult properties of particles to describe adequately. Wadell (1933) used measurements of sphericity and roundness to define particle shape, and Krumbein and Pettijohn (1938) added the important concept of surface texture. Sphericity and roundness can be assessed, though somewhat subjectively (Griffiths, 1967, 114-120), by simple visual comparison of particles with standard charts, such as those given by Krumbein (1941), Rittenhouse (1943) and Powers (1953). Two-dimensional shape in thin section can also be expressed by measuring axial ratios (Griffiths and Rosenfeld, 1950; Bokman, 1952, 1957).

As shape depends partly on crystallographic properties, particularly on characteristic cleavage directions, measurements of sphericity and roundness of sand grains are normally confined to a single mineral species, usually quartz. The most angular particles are from sediments which have suffered diagenetic chemical dissolution, or physical weathering processes involving crushing and fracturing (e.g. by glacial action) rather than abrasion. Abrasion by fluvial, aeolian or marine transportation initially increases roundness and sphericity quite rapidly, but subsequent changes in shape occur more slowly, so that long periods of stable conditions are necessary to give a large proportion of well-rounded grains in a sediment. However, it is not clear to what extent roundness and sphericity are dependent on particle size. For example, aeolian silt particles are usually much more angular than aeolian sands, but this could reflect (a) an original difference in angularity before the aeolian phase, the silt being largely derived from glacial sources, or (b) a difference in the mode of transport, the silt being subjected to fewer collisions in the higher atmosphere than the saltating sand particles near ground level, or (c) chemical attack of silt particles similar to that noticed by Cegla *et al.* (1971) in some loesses. The surface textures of sand grains have been studied with the scanning electron microscope by Krinsley and

Donahue (1968) and Krinsley and Margolis (1969), who concluded that particles from different depositional environments are characterised by differences in their surface markings. However, Brown (1973) suggested that the physical conditions producing certain types of mark may be replicated in different environments, so that more fundamental research may be needed to establish the exact origin of most surface types.

Clay particles sometimes have characteristic shapes, which can be recognised in transmission or scanning electron micrographs, though the finer clay particles are too small for the scanning electron microscope, as they do not back-scatter enough secondary electrons. Apart from palygorskite, which usually forms long thin laths, no clay mineral species can be recognised from shape alone, but individual minerals from certain sources often have characteristic shapes. For example, the clay mica and smectite in the Upper Chalk and other Cretaceous strata often form six-rayed star-like aggregates or have authigenic overgrowths with the same 60° orientation (Weir and Catt, 1965). This type of feature can help locate the origin of clay in some deposits. Clay samples for electron microscopy should be separated from a freshly prepared, neutral or slightly alkaline, aqueous suspension, as there is some evidence that prolonged exposure to even the slight acidity of distilled water can cause minor breakdown of some clay minerals, and the breakdown debris may obscure the outlines of other particles.

The shape and surface texture of stones in Quaternary and other deposits can often help determine their transport history. Their shape is partly dependent on the original form of the rock fragment; this would be determined in some instances by the bedding, jointing or cleavage of the source rock, and in others by the original form and structure of nodules or concretions. Subsequent physical weathering and abrasion modifies this original shape to a variable extent, according to the hardness of the rock, the type of abrasion it undergoes, and the length of time abrasion lasts. As several factors are involved, it is difficult to formulate rigid rules by which to interpret stone shape. Nevertheless, stones are affected more rapidly than finer particles by many transportation processes and physical weathering environments, and because of their size can be studied more readily. Consequently, several distinctive shapes and surface markings are known; these include the wind-faceted ein- and dreikanter, the "flat-iron" shape, striations and scars produced by glacial transport, and the crescentic impact scars or percussion marks on hard siliceous stones, which result from beach battering or high velocity stream flow. Glacial striae are usually restricted to softer rocks, such as limestones. The roundness and sphericity of softer rocks are increased rapidly by stream transport or wave action; an example of special environmental significance, discussed by Bell (1940), is the armoured mud-ball. The stones present in periglacial deposits may be angular, as in many head deposits, or wind-facetted or rounded, as in the aeolian sediments. The lines of stones in coversands or loess deposits are commonly interpreted as fossil desert pavements, but as Cooke (1970) showed, modern desert pavements can have various origins.

3. Discussion

Many different methods of study and analytical techniques can be employed in the petrographic examination of sediments and soil materials, but the limitations imposed by time, money, expertise, interests, apparatus, suitable samples, etc. usually combine to force the petrographer to use only a few of the methods available. Some of those we have mentioned (e.g. particle size analysis) could easily be learnt and practised by an archaeologist with little or no petrographic training, but ultimately he would either have to seek the advice of a specialist petrographer with respect to methods or interpretation of results, or become such a specialist himself. This is mainly because the petrographic study of sediments is, like many other branches of natural science, still at the stage where much more fundamental research is needed before processes and their effects are fully understood. The archaeologist who ventures into this field will soon find himself at the frontiers of knowledge, probably asking questions to which there is at present no answer, and feeling tempted to disgress into the fundamental petrological research which could provide answers. Such disgressions should be encouraged, as they can only benefit both archaeology and sedimentary petrography. However, for the present petrography should not be regarded as an analytical tool providing simple, direct answers to important archaeological questions, but should be used, together with other methods for studying sediments, to help provide the evidence on which tentative answers to specific questions may be based.

The importance of good fieldwork as a basis for petrographic and other laboratory methods of studying sediments cannot be over-emphasised. This involves not only the adequate excavation, investigation and collection of artefacts, etc. from the "interesting" horizons, such as occupation levels, but also a careful examination of the apparently less interesting deposits, possibly containing no artefacts or fossils, which may occur stratigraphically above or below, or extend laterally beyond the boundaries of the excavation or natural exposure. Tracing the deposits laterally is especially important where old land surfaces are involved, as much more palaeoenvironmental evidence can be obtained when the configuration of the land surface over a large area and the soil types in various situations are known. It is also useful to know the lateral and altitudinal distribution of certain deposits when considering their mode of origin and age; for example, solifluction or aeolian deposits may mantle large parts of an old land surface, whereas marine sediments occur only up to a certain level, and fluvial deposits only within valley systems.

Petrographic investigation of unfossiliferous deposits is especially important at Palaeolithic sites, where many such deposits (till, loess, solifluction deposits) are often the only evidence of cold periods, and relative dating may be facilitated by establishing sequences of climatic change. Many cold deposits are immediately recognisable in the field, but when strongly weathered, extensively disturbed, or mixed with other materials, their true nature and origin may be revealed only be laboratory petrographic studies.

Despite the advice about incorporating petrographic aspects of sediment study early in a site investigation, a late-stage situation may still arise in which it becomes important to assemble at least some petrographic information from inadequate samples, possibly to check theories based on non-petrographic data. If the choice of petrographic methods remains open, it is important to select techniques which need only small amounts of material, and are as non-destructive as possible, also to arrange the analyses chosen in a logical order so that the material used for one method remains suitable for subsequent studies. The petrological microscope is extremely useful equipment at this stage, as it can be adapted to a range of analytical methods, all providing useful information yet comparatively non-destructive. For example, the bulk mineral composition of a very small sample (<0.01g) can often be rapidly estimated, giving also an approximation of bulk chemical composition, including carbonate content; at the same time some diagnostically important minor mineral constituents may be noted, the approximate particle size distribution calculated from measurements with a micrometer eyepiece, and the shapes of sand and silt grains noted. Further, the sample can be recovered virtually unaltered from a temporary mount in clove oil or similar medium by washing with alcohol or acetone, centrifuging and drying from petroleum ether. In view of its usefulness in these ways and in examining and interpreting thin sections, emphasis is given to mastery of petrological microscope techniques as perhaps the most useful technical knowledge in sedimentary petrography, but the writers agree with Griffiths (1967, 204) that it is "a special skill developed by long practice, and . . . requires a psychophysical conditioning to develop a mental habit".

One of the main difficulties encountered in petrographic studies of Quaternary sediments results from the relatively rapid changes of climate and sea level, and hence of conditions of deposition, that have occurred since the late Tertiary. Many of the characteristics of sediments useful as palaeo-environmental indicators may be only poorly expressed in Quaternary sediments because they have not had time to develop fully. This probably applies to particle size distribution, shape and surface texture of sand grains as indicators of mode of transport, and to soil micromorphology and clay mineralogy as indicators of weathering regimes. Further, the most recent period of relative stability, up to approximately the last 10,000 years at the most, may not in itself have been long enough to characterise effectively the sediments and soils we often use as standards by which to judge older environments. Has the sedimentary petrologist been limited in the past by his failure to view results against the background of changes in Quaternary environments? It is suggested that in the future he will need the archaeologist, along with other Quaternary scientists, as much as they need him.

REFERENCES

Anderson, J.U. (1963) 'An improved pre-treatment for mineralogical analysis of samples containing organic matter', *Clays Clay Miner., Proc. 10th Conf.*, Texas, 1961, 380-8.

Avery, B.W. (1974) 'Soil Survey Laboratory Methods', *Soil Survey Technical Monograph 6*, Harpenden.

Babel, U. (1964) 'Chemische Reaktionen an Bodendünnschliffen', *Leitz-Mitt. Wiss. Tech.* 3, 12-14.

Babel, U. (1968) 'Enchytraeen-Losungsgefuge in Löss', *Geoderma* 2, 57-63.

Bascomb, C.L. (1961) 'A calcimeter for routine use on soil samples', *Chemy Ind.*, 1826-7.

Bascomb, C.L. (1968) 'Distribution of pyrophosphate-extractable iron and organic carbon in soils of various groups', *J. Soil Sci.* 19, 251-68.

Bell, H.S. (1940) 'Armoured mud balls – their origin, properties and role in sedimentation', *J. Geol.* 48, 1-31.

Bloss, F.D. (1961) *An Introduction to the Methods of Optical Crystallography.* New York.

Bokman, J. (1952) 'Clastic quartz particles as indices of provenance', *J. sedim. Petrol.* 22, 17-24.

Bokman, J. (1957) 'Comparison of two and three dimensional sphericity of sand grains', *Bull. geol. Soc. Am*, 68, 1689-92.

Bostock, W. (1952) 'A sedimentation balance for particle size analysis in the sub-sieve range', *J. scient. Instrum.* 29, 209-11.

Bremner, J.M. (1965) 'Total nitrogen', in C.A. Black, ed., *Methods of Soil Analysis, Part 2: Chemical and Microbiological Properties.* Madison.

Brewer, R. (1964) *Fabric and Mineral Analysis of Soils.* London.

Bronger, A. (1966) 'Lösse, ihre Verbraunungszonen und fossilen Boden. Ein Beitrag zur Stratigraphie des oberen Pleistozäns in Südbaden', *Schr. geogr. Inst. Univ. Kiel* 14.

Bronger, A. (1970) 'Zur Mikromorphogenese und zum Tonmineralbestand Quartärer Lössboden in Südbaden', *Geoderma* 3, 281-320.

Bronger, A. (1972) 'Zur Mikromorphologie und Genese von Paläoboden aus Löss im Karpatenbecken', *Soil Micromorphology, Proc. 3rd Int. Working Meeting, Wroclaw, Poland, 1969*, 607-15.

Brown, J.E. (1973) 'Depositional histories of sand grains from surface textures', *Nature* 242, 396-8.

Brown, G. and C.D. Ollier (1956) 'Collophane from the Chalk', *Mineralog. Mag.* 31, 339-43.

Brown, G., J.A. Catt and A.H. Weir (1969) 'Zeolites of the clinoptilolite-heulandite type in sediments of south-east England', *Mineralog. Mag.* 37, 480-8.

Buller, A.T. and J. McManus (1972) 'Simple metric sedimentary statistics used to recognize different environments', *Sedimentology* 18, 1-21.

Buller, A.T. and J. McManus (1974) 'The application of quartile deviation-median diameter curves to the interpretation of sedimentary rocks', *J. geol. Soc. Lond.* 130, 79-83.

Bullock, P., D.M. Carroll and R.A. Jarvis (1973) 'Palaeosol features in northern England', *Nature Phys. Sci.* 242, 53-4.

Buol, S.W. and F.D. Hole (1961) 'Clay skin genesis in Wisconsin soils', *Proc. Soil Sci. Soc. Am.* 25, 377-9.

Cegla, J., T. Buckley and I.J. Smalley (1971) 'Microtextures of particles from some European loess deposits', *Sedimentology* 17, 129-34.

Chayes, F. and H.W. Fairbairn (1951) 'A test of the precision of thin section analysis by point counter', *Am. Miner.* 36, 704-12.

Cooke, G.W. (1967) *The Control of Soil Fertility*. London.

Cooke, R.U. (1970) 'Stone pavements in deserts', *Ann. Ass. Am. Geogr.* 60, 560-77.

Cornwall, I.W. (1958) *Soils for the Archaeologist*. London.

Cornwall, I.W. (1963) 'Soil micromorphology and the study of prehistoric environment', *Microscope* 13, 342-5.

Dagg, M. and P.H. Hosegood (1962) 'Details of a hand sampling tool for taking undisturbed soil cores', *E. Afr. agric. For. J.*, supplement to special issue, 129-31.

Dalrymple, J.B. (1958) 'The application of soil micromorphology to fossil soils and other deposits from archaeological sites', *J. Soil Sci.* 9, 199-209.

Dalrymple, J.B. (1972) 'Experimental micropedological investigations of iron oxide-clay complexes and their interpretation with respect to the soil fabrics of paleosols', *Soil Micromorphology, Proc. 3rd Int. Working Meeting, Wroclaw, Poland, 1969*, 583-94.

Eberhardt, J.P. and R. Triki (1972) 'Description d'une technique permettant d'obtenir des coupes minces de minéraux argileux par ultramicrotime. Application à l'étude de minéraux argileux interstratifiés', *J. Microscopie* 15, 111-20.

Elonen, P. (1971) 'Particle-size analysis of soil', *Suom. Maatal. Seur, Julk.* 122.

Farmer, V.C. (1968) 'Infrared spectroscopy in clay mineral studies', *Clay Miner.* 7, 373-87.

Federoff, N. (1967) 'Un exemple d'application de la micromorphologie à l'étude des paléosols', *Bull. Ass. fr. Etude Quat.* 3, 193-209.

Federoff, N. (1968) 'Genèse et morphologie de sols à horizon B textural en France Atlantique', *Sci. Sol.* 1, 29-65.

Federoff, N. (1969) 'Caractères micromorphologiques des pédogenèses quaternaires en France', *in* M. Ters, ed., *Etudes sur le quaternaire dans le monde*, 8th INQUA Congress, Paris, 1969, 341-49.

Fitzpatrick, E.A. (1970) 'A technique for the preparation of large thin sections of soils and unconsolidated materials', *Micromorphological Techniques and Applications, Soil Survey Technical Monograph 2*, 3-13.

Fogg, D.N. and N.T. Wilkinson (1958) 'The colorimetric determination of phosphorus', *Analyst, Lond.* 83, 406-14.

Folk, R.L. and W.C. Ward (1957) 'Brazos River Bar: a study in the significance of grain size parameters', *J. sedim. Petrol.* 27, 3-26.

Friedman, G.M. (1961) 'Distinction between dune, beach and river sands from their textural characteristics', *J. sedim. Petrol.* 31, 514-29.

Friedman, G.M. (1962) 'On sorting coefficients and the lognormality of the grain size distribution of sandstones', *J. Geol.* 70, 737-53.

Francis, C.W. (1973) 'Adsorption of polyvinyl-pyrrolidone on reference clay minerals', *Soil Sci.* 115, 40-54.

Gard, J.A. (1971) *The Electron-optical Investigation of Clays*. Mineralogical Society, London.

Genrich, D.A. and J.M. Bremner (1972) 'A re-evaluation of the ultrasonic vibration method of dispering soils', *Proc. Soil Sci. Soc. Am.* 36, 944-7.

Greene-Kelly, R., S. Chapman and K. Pettifer (1970) 'The preparation of thin sections of soils using polyethylene glycols', *Micromorphological Techniques and Applications, Soil Survey Technical Monograph 2*, 15-24.

Griffiths, J.C. (1967) *Scientific Method in Analysis of Sediments*. New York.

Griffiths, J.C. and M.A. Rosenfeld (1950) 'Progress in measurement of orientation in Bradford Sand', *Bull. Miner. Inds Exp. Stn Penn. St. Univ.* 56, 202-36.

Harrod, T.R., J.A. Catt and A.H. Weir (1974) 'Loess in Devon', *Proc. Ussher Soc.* 2, 554-64.

Hartshorne, N.H. and A. Stuart (1960) *Crystals and the Polarising Microscope.* London.

Hashimoto, I. and M.L. Jackson (1960) 'Rapid dissolution of allophane and kaolinite-halloysite after dehydration', *Clays Clay Miner.* 7, 102-13.

Hetier, J.M., M. Rodrigues-Lapa and F. Le Tacon (1972) 'Etude micromorphologique de quelques sols de l'est de la France', *Bull. Ass. fr. Etude Sol* 1/2, 49-61.

Jamagne, M. (1972) 'Some micromorphological aspects of soils developed in loess deposits of northern France', *Soil Micromorphology, Proc. 3rd Int. Working Meeting, Wroclaw, Poland,* 1969, 559-82.

Jenkins, D.A. (1970) 'Some analytical techniques applicable in soil micromorphology', *Micromorphological Techniques and Applications, Soil Survey Technical Monograph* 2, 25-32.

Kerr, P.F. (1959) *Optical Mineralogy,* (3rd ed.). New York.

Kiely, P.V. and M.L. Jackson (1964) 'Selective dissolution of micas from potassium felspars by sodium pyrosulfate fusion of soils and sediments', *Am. Miner.* 49, 1648-59.

Kilmer, V.J. (1960) 'The estimation of free iron oxides in soils', *Proc. Soil Sci. Soc. Am* 24, 420-1.

Kilmer, V.J. and L.T. Alexander (1949) 'Methods of making mechanical analyses of soils', *Soil Sci.* 68, 15-24.

Krinsley, D.H. and J. Donahue (1968) 'Environmental interpretation of sand grain surface textures by electron microscopy', *Bull. geol. Soc. Am.* 79, 743-8.

Krinsley, D.H. and S. Margolis (1969) 'A study of quartz sand grain surface textures with the scanning electron microscope', *Trans. N.Y. Acad. Sci.* 31, 457-77.

Krumbein, W.C. (1934) 'Size frequency distributions of sediments', *J. sedim. Petrol.* 4, 65-77.

Krumbein, W.C. (1935) 'Thin section mechanical analysis of indurated sediments', *J. Geol.* 43, 482-96.

Krumbein, W.C. (1941) 'Measurement and geological significance of shape and roundness of sedimentary particles', *J. sedim. Petrol.* 11, 64-72.

Krumbein, W.C. and F.J. Pettijohn (1938) *Manual of Sedimentary Petrography.* New York.

Kubiena, W.L. (1953) *The Soils of Europe.* London.

Kubiena, W.L. (1959) 'Prinzipien und Methodik der paläopedologischen Forschung im Dienst der Stratigraphie', *Z.. dt. geol. Ges.* 111, 562-643.

Kubiena, W.L. (1963) 'Palaeosoils as indicators of paleoclimates', *Arid Zone Res.* 20, 207-9.

Kubiena, W.L. (ed.) (1967) *Die micromorphomorphometrische Bodenanalyse.* Stuttgart.

Kubiena, W.L. (1970) *Micromorphological Features of Soil Geography.* New Brunswick.

Larsen, E.S. and H. Berman (1934) 'The microscopic determination of the non-opaque minerals', *Bull. U.S. geol. Surv.* 848.

Le Riche, H.H. and A.H. Weir (1963) 'A method of studying trace elements in soil fractions', *J. Soil Sci.* 14, 225-35.

Loveday, J. (1962) 'Plateau deposits of the southern Chiltern Hills', *Proc. Geol. Ass.* 73, 83-102.

Mackenzie, R.C. (1957) *The Differential Thermal Investigation of Clays.* London.

McKeague, J.A. (1967) 'An evaluation of 0.1 m pyrophosphate and pyrophosphate-dithionite in comparison with oxalate as extractants of the accumulation products in podzols and some other soils', *Can. J. Soil Sci.* 47, 95-9.

McKeague, J.A., J.E. Brydon and N.M. Miles (1971) 'Differentiation of forms of extractable iron and aluminium in soils', *Proc. Soil Sci. Soc. Am.* 35, 33-8.

McRae, S.G. (1972) 'Glauconite', *Earth Sci. Rev.* 8, 397-440.

Mehra, O.P. and M.L. Jackson (1960) 'Iron oxide removal from soils and clays by a dithionite-citrate system buffered with sodium bicarbonate', *Clay Clay Miner.* 7, 317-27.

Milner, H.B. (1962) *'Sedimentary Petrography,* 2 vols. 4th ed. London.

Mitchell, G.F., J.A. Catt, A.H. Weir, N.F. McMillan, J.P. Margarel and R.C. Whatley (1973) 'The late Pliocene marine formation at St. Erth, Cornwall', *Phil. Trans. R. Soc.* B, 266, 1-37.

O'Connell, W.L. (1963) 'Properties of heavy liquids', *Trans. Am. Inst. Min. metall. Petrol. Engrs.* 226, 126-32.

Perrin, R.M.S., H. Davies and M.D. Fysh (1974) 'The distribution of Late Pleistocene aeolian deposits in eastern and southern England', *Nature* 248, 320-4.

Powers, M.C. (1953) 'A new roundness scale for sedimentary particles', *J. sedim. Petrol.* 23, 117-19.

Reynolds, R.C. and P. Lessing (1962) 'The determination of dioctahedral mica and potassium feldspar in submicroscopic grain sizes', *Am. Miner.* 47, 979-82.

Rittenhouse, G. (1943) 'A visual method of estimating two-dimensional sphericity', *J. sedim. Petrol.* 13, 79-81.

Russell, R.J. (1944) 'Lower Mississippi valley loess', *Bull. geol. Soc. Am.* 55, 1-40.

Schwertmann, U. (1959) 'Die fraktionierte extraktion der freien eisenoxide in Boden, ihre mineralogischen Formen und ihre Entstehungsweisen', *Z. Pfl-Ernähr. Dung. Bodenk.* 84, 194-204.

Sevon, W.D. (1966) 'Distinction of New Zealand beach, dune and river sands by their grain size distribution characteristics', *N.Z. Jl Geol. Geophys.* 9, 212-23.

Shaw, K. (1959) 'Determination of organic carbon in soil and plant material', *J. Soil Sci.* 10, 316-26.

Sitler, R.E. and C.A. Chapman (1955) 'Microfabrics of till from Ohio and Pennsylvania', *J. sedim. Petrol.* 25, 262-9.

Smith, H.G. (1956) *Minerals and the Microscope.* 4th ed. London.

Smithson, F. (1958) 'Grass opal in British soils', *J. Soil Sci.* 9, 148-54.

Stairmand, C.J. (1950) 'A new sedimentation apparatus for particle size analysis in the sub-sieve range', *Symposium Particle Size Analysis, Inst. Chem. Engrs. and Soc. Chem. Ind., London, 1947,* 128-34.

Tanner, C.B. and M.L. Jackson (1947) 'Nomographs of sedimentation times for soil particles under gravity or centrifugal acceleration', *Proc. Soil Sci. Soc. Am.* 12, 60-5.

Tinsley, J. (1950) 'The determination of organic carbon in soils by dichromate mixtures', *Trans. 4th Int. Congr. Soil Sci.* 1, 161-4.

Soil Survey Staff (1960) *Soil Classification: a Comprehensive System, 7th Approximation.* Washington.

Wadell, H. (1933) 'Sphericity and roundness of rock particles', *J. Geol.* 41, 310-31.

Waksman, S.A. (1936) *Humus: Origin, Chemical Composition, and Importance in Nature.* London.

Waters, D.F. and I.C. Sweetman (1955) 'The Rukuhia soil grinder', *Soil Sci.* 79, 411-13.

Weir, A.H. and J.A. Catt (1965) 'The mineralogy of some Upper Chalk samples from the Arundal area, Sussex', *Clay Miner.* 6, 97-109.

Weir, A.H. and J.A. Catt (1969) 'The mineralogy of Palaeogene sediments in northeast Kent (Great Britain)', *Sedim. Geol.* 3, 17-33.

Weir, A.H., J.A. Catt and P.A. Madgett (1971) 'Postglacial soil formation in the loess of Pegwell Bay, Kent (England)', *Geoderma* 5, 131-49.

Winchell, A.N. (1939) *Elements of Optical Mineralogy, Part III. Determinative Tables,* 2nd ed. New York.

Winchell, H. (1965) *Optical Properties of Minerals. A Determinative Table.* New York.

DISCUSSION

The discussion on this paper began with additional comments by Dr. J.A. Catt and Dr. A.H. Weir about the amount of time required for petrographic analysis of sediments. In answer to a question by Dr. D. Jenkins, the authors did not feel that at the present stage of knowledge soil micromorphological examination was sufficiently developed to make possible the use of thin-section evidence for the interpretation of palaeosols in archaeological contexts. However, they pointed out several examples of the use of such techniques to examine a soil fabric, notably the work of Dr. J. Dalrymple. Following a point made by Professor A.C. Renfrew, the speakers stated that in their opinion it was always necessary for a specialist to visit a site and take his own samples, and they emphasised the need for archaeologists to pose specific questions rather than generalisations. These views were supported by Professor V.B. Proudfoot and Professor G.W. Dimbleby who were of the opinion that specialists working on a site should always collaborate in sampling so that conclusions could be integrated.

Dr. M.S. Tite enquired about the role of the scanning electron microscope in mineralogical analysis of sediments, and Dr. A.H. Weir quoted an example of its use in the analysis of sand grains from the St. Erth Beds. The surface features of the grains were examined after a particle size analysis had shown that the assemblage was composed of two well-sorted but mixed sand populations of sizes 2.4 and 3ϕ. The coarser sands had surface characteristics suggesting high-energy beach conditions, and the finer grains were a dune sand deposited by aeolian processes. Dr. A.H. Weir suggested that this was to be interpreted as a mixture of beach and dune sand produced by a rise in sea level.

B. Proudfoot

The analysis and interpretation of soil phosphorus in archaeological contexts

1. Introduction

Phosphorus is present in all ecosystems. Some is involved in short term recycling from soil to plant to animals and back to soil again, and some is lost by leaching. The ultimate source of all such phosporus is the break down of phosphate minerals in the soil parent material. Human activity which disrupts such a cycle may lead either to loss or gain of phosphorus. For example, overgrazing may reduce the amount of primary production in a grassland system, leading to the loss by erosion of the organic and phosphorus-rich top-soil. Conversely, the concentration of livestock on part of an inhabited site, or the day-to-day occupation of a restricted area by a group of humans, may increase the accumulation of phosphorus-rich wastes which will be incorporated within an ensuing soil-plant-animal system. Phosphorus gains of this sort tend to be long persistent because of the low solubility and limited movement of soil phosphorus. Increased amounts of soil phosphorus may, therefore, be used as indicators of early habitation sites, and on individual sites as indicators of different intensities or types of use.

The high stability, or low solubility, of phosphorus in soils, which allows such conclusions to be drawn, is also the immediate cause of deficiences of soil phosphorus for plants. Although the total amount of phosphorus in the soil may be considerable, only a very small fraction may be available for plant growth at any one time. Similarly, if manures or fertilisers are applied to crops, only a small fraction of the total phosphorus may be available for plant growth, the rest being converted almost immediately into forms which are unavailable to the plant. The problems of tracing the fate of phosphorus added to soils, and of distinguishing between soils not only in terms of their total phosphorus content, but also of their available phosphorus content, have proven amongst the most intractable in soil science. The identification of different forms of phosphorus in soils has also proven to be analytically extremely difficult. Some understanding of the nature of these problems is essential to the interpretation of phosphate analyses from archaeological sites.

1.1 Sources of phosphorus on archaeological sites
Major sources of phosphorus in archaeological contexts are domestic refuse, food wastes, plant and animal remains, excreta, bodies, and, in

early field systems, deliberate applications of manure or fertiliser. Cook and Heizer (1965) have summarised the amounts of phosphorus likely to be excreted by human or animal populations. Their standard living group of 100 people would excrete a total of some 62 kg of phosphorus annually. They suggest that a similar amount might be added through wastes or garbage – food residues, especially bones, and plant and vegetable materials used domestically, for example, in clothing construction and burning. There are of course likely to be considerable variations between populations because of different life styles and the different phosphorus contents of their residues. The amount of phosphorus in plants, for example, depends on species, soil conditions and stage of growth (Black, 1968). Within the plant the phosphorus is unequally distributed so that use of different parts of plants would add different quantities of phosphorus to food supply and waste material. The amount and distribution of phosphorus in animal – and human – bodies depends on such factors as age, sex, and food supply or nutrition level (Just Nielsen, 1972), while the composition of solid and liquid excrement varies between animals (Thompson, 1952; and for a recent bibliography McQuitty and Barber, 1972).

However, even allowing for such variations as these, a living group of 100 people occupying a 0.81 ha site as postulated by Cook and Heizer (1965) is likely to contribute a total of some 146 kg of phosphorus annually per hectare. Such an annual increment might be as much as 10 per cent or as little as 0.5 per cent of the phosphorus already present in the soil. If accumulation were concentrated within the occupied area, or a site were occupied over some time, then the phosphorus resulting from human activity would comprise a considerable addition to that originally present. Under many soil conditions much of this added phosphorus is likely to be retained in the soil and hence provide indications of human occupance of sites.

The concentration of phosphorus likely to be deposited as a result of human interment has also been considered by Cook and Heizer (1965) and Johnson (1954-6). The body of a 68 kg man contains some 630 gm of phosphorus, 86 per cent of which is in the skeleton. In a burial this amount of phosphorus would be concentrated in a relatively small area, and could result in locally high phosphate values.

1.2 Soil phosphorus analysis

Chemical laboratory methods to determine the amount of phosphorus present in the soil have been aimed, firstly, at establishing the amount of phosphorus available for plant growth, that is, at finding some index of phosphate fertility; and, secondly, at separating the phosphorus present into its various components.

Early attempts to derive indices of phosphate fertility were based on total phosphate content, but these were unsatisfactory, since not all the phosphate present in a soil is available for plant growth. Next, extractions of the soil with various chemical mixtures or dilute acids to recover phosphorus from the soil were tried. These extractants were designed initially to duplicate the acids thought to be given out by plant roots, or otherwise present in the soil. Although the theoretical basis for this approach is weak, it has proved useful for assessing relative fertiliser requirements among soils of broadly similar characteristics, but is of lesser use in assessing relative fertility between soils of different types. For the same soils different extractants may give widely different results. In Northern Ireland, for example, where practically all soils are acid – only 6 per cent having pH values above 6.4 – results obtained using 0.02N calcium lactate and 0.02N hydrochloric acid (modified Egnér extractant) agree fairly closely with those obtained using 0.5N acetic acid, but correlate poorly with those using citric acid (McConaghy and McAllister, 1952). Presumably the differences between the amounts of phosphorus recorded by different extractants is due not so much to differences in the amounts originally dissolved by the acids but to chemical precipitation and adsorption during the extracting process. To counteract these reversionary processes Ghani and Islam (1957) have described the use of 8(OH)quinoline and selenious acid as a deactivating agent. The problems associated with the use of different methods for evaluating phosphate fertility of soils within Britain and Ireland were discussed at a conference held in 1962 (Great Britain, Min. of Agriculture, 1965), the result of which was that the soil chemists in the National Agricultural Advisory Service decided to use 0.5M acetic acid plus 0.5M ammonium acetate as a standard routine extractant for phosphorus in soils, a ratio of soil to extractant of 1 to 5, and a shaking time of 15 minutes. More recently, the Dutch Agricultural Extension Service has decided to use distilled water as an extractant for determining available phosphorus in arable soils, again under closely specified laboratory conditions (van der Paauw, 1971; Sissingh, 1971). Under these conditions there is a high correlation between water extracted phosphorus (Pw) and crop response. Changes of Pw effected by additions to the soil of different kinds of fertilisers, by liming and by mixing soil with peat, correspond to the response of the plant to these treatments.

The various components present in the total soil phosphorus may be divided into organic and inorganic. Several methods have been utilised in determining the organic phosphorus. In the first of these the organic matter is extracted from the soil by acid and alkali extractions and the organic phosphorus is determined as the difference between the inorganic and total phosphorus. The second method involves measuring the inorganic phosphorus extracted by a suitable acid solution before and after the soil organic matter has been destroyed either by ignition or by oxidising agents such as peroxide. McConaghy (1960) has indicated some of the difficulties associated with each of these methods and discussed the forms of organic phosphorus which may be present and their availability to plants. Black (1968) has claimed that with care most of the organic phosphorus can be extracted from a soil as organic phosphorus, and using suitable fractionation methods most of the organic phosphorus can be separated from most of the nonphosphatic organic matter that has been extracted with it. Methods for the determination of organic phosphorus in soils are generally time-consuming, but recently Steward and Oades (1972) have described a method based on a pretreatment with 1M hydrochloric acid followed by ultrasonic vibration in 0.5M sodium hydroxide which is both more efficient and speedier than previous procedures. Several techniques have been used to isolate individual components of the organic phosphorus present, some of which may be altered during extraction (Omotoso and Wild, 1973). However, the nature of the major part of the organic phosphorus in most soils is still unknown.

Different extractants have been used either separately or consecutively in determining the constituents of the inorganic phosphorus present. Much uncertainty surrounds the identification of individual constituents determined by these methods (Black, 1968; Larsen, 1967). Walker (1965) has tabulated the conventional descriptions of three fractions separated by individual acid treatment and ignition as follows:

Pt = total phosphorus extracted by $HF\text{-}HNO_3$ digestion

Pa = inorganic phosphorus extracted by NH_2SO_4 without prior ignition

Pa+Po = phosphorus extracted by NH_2SO_4 after ignition

Po = organic phosphorus obtained by subtracting Pa from Pa+Po
Pf = inorganic phosphorus, insoluble in NH_2SO_4 after ignition and obtained by subtracting Pa+Po from Pt.

The precise forms of phosphorus represented by Pa and Pf are uncertain. In addition to Ca-bound phosphorus, Pa will also include some Fe- and Al-bound phosphorus. Pf may include phosphorus re-fixed during extraction with N sulphuric acid, or artifacts formed by ignition, in addition to very insoluble forms of Fe- and Al-bound phosphorus. Walker concludes that after fine grinding and ignition N sulphuric acid extracted virtually all the phosphorus from the soils he analysed.

Most of the methods of using several extractants consecutively derive from the method described by Chang and Jackson (1957), the results of which may be tabulated as follows, individual extractants being listed alongside the conventional descriptions of the constituents:

N ammonium chloride	water soluble phosphate
0.5N ammonium fluoride	aluminium phosphate
0.1N sodium hydroxide	iron phosphate
0.5N sulphuric acid	calcium phosphate
0.3M sodium citrate plus sodium dithionite	occluded iron phosphate
0.5N ammonium fluoride	occluded aluminium phosphate

The term occluded has been used since it was thought that the last two fractions were inside iron oxides that were not dissolved by earlier extractants.

Equilibration methods have also been used to determine the composition and form of inorganic phosphorus compounds present in the soil. Various soils and known phosphate minerals are equilibrated individually with solutions of different pH, then the phosphorus concentration in solution is plotted against the pH after equilibration, and the shapes of the curves for soils are compared with those for mineral standards. Alkaline soils usually display a solubility-pH curve similar to that of apatite, whereas curves for acid soils are similar to those of iron or aluminium phosphates (Black, 1968).

Equilibration between soil phosphate and introduced radioactive phosphorus has provided useful results on the form of the soil phosphates. Machold equilibrated samples of soils with a trace quantity of radioactive orthophosphate (cited in Black, 1968). This method can be used in conjunction with other methods of analysis, such as those of Chang and Jackson (1957), to compare the tendency of the different phosphate fractions to exchange with the radioactive phosphate. Since radioactive phosphates equilibrate only to a limited extent with crystalline phosphates, the tendency to equilibrate with different fractions is a measure of their crystallinity. Machold reported that much of the extractable phosphorus in soils was not present in crystalline form, or was in units so small that crystallinity did not make the phosphorus inaccessible to exchange (Black, 1968).

The final method for determining the types of inorganic phosphates present in soils compares the solubility of the soil phosphate with that of known phosphates (Russell, 1961; Black, 1968). When a compound goes into solution it dissociates into positively and negatively charged particles or ions. In the case of hydroxyl-apatite, the phosphatic constituent of bones and teeth, and one of the forms of calcium phosphate known to occur in soils, the solution of the crystalline solid may be represented as

$$Ca_5 (PO_4)_3 OH \xrightarrow[\text{solution}]{\text{on}} 5Ca^{++} + 3 PO_4^{---} + OH^-$$

If it is in equilibrium with the solution then the product of the ion activities is a constant. If activity is represent by a, then

$$\left(a_{Ca}^{++}\right)^5 \; \left(a_{PO_4}^{---}\right)^3 \; \left(a_{OH}^{-}\right) \;\; = k \text{ (constant)}$$

The calcium and other ion activities in a soil solution, and their product, can then be compared with those of known minerals in solution and the types of inorganic phosphate present in the soil determined, provided the phosphorus in the soil solution is at equilibrium with crystalline phosphate minerals. However, several investigators have found that the phosphorus concentration of the soil solution does not conform to simple solubility-product principles, and reasons for this have been discussed in detail by Larsen (1967). Importantly, discussion of soil phosphorus in terms of solubility-product principles not only provides a method for identifying the compounds present, but may help to elucidate the mechanisms by which phosphorus is transferred or held within the soil-plant system.

1.3 Soil phosphorus minerals
In soils most of the inorganic phosphorus occurs in the clay fraction from which it cannot be separated by physical methods, hence the use of the methods outlined in the preceding section. However, small crystals of some phosphates, especially apatite, have been found in the sand and silt fractions of a number of soils, and thin films of apatite have been identified on calcium carbonate nodules in soils treated with superphosphate fertiliser for a long period of years (Russell, 1961). Probably the commonest, and the most stable form of apatite is fluorapatite $(Ca_{10}(PO_4)_6F_2)$. Even in very dilute fluoride solutions the fluoride slowly displaces the hydroxyl (OH) in the apatite structure. Presumably as a result of long periods of weathering other displacements can occur to yield such minerals as gorceixite and florencite, containing barium and cerium respectively, both of which have been recorded in Australian soils. Ferrous phosphate, vivianite $(Fe_3(PO_4)_2.8H_2O)$ has commonly been identified under reducing conditions. Two aluminium phosphates have been found in soils developed under rather special circumstances: wavellite $(Al_3(PO_4)_2(OH)_3.5H_2O)$ in soils developed on a Florida phosphate deposit, and crandalite $(Ca\ Al_3(PO_4)_2(OH)_6)$ in soils on coral limestone islands in the south Pacific (Russell, 1961; Larsen, 1967; Black, 1968).

On the basis of the solubility-product principle the occurrence of phosphorus compounds of calcium, aluminium and iron have been discussed by Larsen (1967), and the following description is derived largely from this discussion.

Calcium compounds
Hydroxylapatite is certainly the most important calcium phosphate in soils and probably the only calcium phosphate permanently present in slightly acid, neutral and alkaline soils. Under laboratory

conditions di- and tri-calcium phosphates ($CaHPO_4$; $Ca_3(PO_4)_2$) may act as precursors for hydroxylapatite. It is probable that the formation of hydroxylapatite in soils occurs through chemical changes in the solid phase after an initial precipitation of some undefined calcium phosphate has taken place, the rate of formation of hydroxylapatite seeming to increase with pH. Again under laboratory conditions, it has been shown that the formation of hydroxylapatite is very sensitive to the presence of impurities (Russell, 1961). The presence of magnesium inhibits the growth of hydroxylapatite nuclei by being adsorbed onto their surfaces and blocking further lattice growth. In contrast strontium can replace calcium and a strontium hydroxylapatite is known (Larsen, 1967). The most important impurity in hydroxylapatite is carbonate, which is always found in hydroxylapatites in sediments and in many laboratory preparations. At least in some circumstances three carbonate (CO_3) ions seem to replace two phosphate (PO_4) ions. An effect of the carbonate in the hydroxylapatite lattice is to make it more chemically reactive than would otherwise be the case. There is evidence that other calcium phosphates may form a surface complex on hydroxylapatite crystals, the particular phosphate depending on pH. Apparent discrepancies between various results may be reconciled by assuming that different surface complexes are possible and that the boundary between them is diffuse. The nature of the surface complex is important since it is this, rather than the deeper layers of the hydroxylapatite crystal, which determines solubility.

Aluminium and iron compounds
The exact nature of the aluminium phosphates present in soils is uncertain. It has been suggested that the mineral variscite ($AlPO_4.2H_2O$) may be present, but in pure systems variscite only controls the phosphorus concentration in solution when the pH of the equilibrium solution is less than 3.1. At higher pH the ratio of phosphorus to aluminium dissolved differs from that in the solid state and a more basic hydroxyl phosphate may be formed as a surface complex which controls the phosphorus concentration in solution in acid soils. In soil solutions and extracts there is some evidence for aluminium hydroxyl and oxy-hydroxyl ions ($AlOH^{2+}$ and $Al(OH)_2^+$) and phosphorus may be present sorbed on the surface of these compounds.

As with variscite, the comparable iron mineral, strengite ($FePO_4.2H_2O$) has not been shown definitely to exist in the soil, although there is some evidence that surface complexes with this composition may exist in soils within the pH range 3.8 to 4.2. Similarly the occurrence of the intermediate mineral barrandite, a mixture of these iron and aluminium phosphates in almost any proportions, is uncertain.

The iron and aluminium phosphates in the soil most likely occur in the form of thin films no more than a few molecules thick (Russell, 1961). These films are probably held on the surface of hydrated films of iron and aluminium oxides, or on iron and aluminium ions forming part of the surface layer of clay minerals.

1.4 Phosphorus sorption

Adsorption, the concentration of liquid material on the surface of a solid, and absorption, the incorporation of liquid material into a solid, both seem to occur with phosphorus in soils, hence the use of the term sorption to describe the processes by which phosphorus is held in the soil. Phosphorus sorption is a complex process, associated especially with calcium ions, and with iron and aluminium ions. It is complicated by precipitation and the formation of phosphate compounds of uncertain composition as already discussed (Russell, 1961: Larsen 1967).

The calcium ions which hold phosphorus in a soil may be calcium ions in solution, exchangeable ions forming calcium phosphates on the surface of clays, or calcium ions held on the surface of calcium carbonate crystals. The processes involving phosphorus and calcium are important in neutral or calcareous soils with pH above about 5.5. The processes involving phosphorus and iron and aluminium ions are important at lower pH. The iron and aluminum ions which can hold phosphate may be present in films of hydrated oxides, or in the case of acid soils the aluminium may be present as exchangeable cations.

A third process whereby phosphorus is held on the edges of clay particles through hydrogen bonding between hydroxyls in the broken edges of the clay and oxygens in phosphorus tetrahedra is also possible (Russell, 1961).

Generally, a water soluble phosphate added to a soil in the field is rapidly converted into water insoluble forms. Similarly if a soil is shaken up in the laboratory with a water-soluble phosphate sorption will occur. Such phosphates would not be immediately available for plant growth. Hence the use of the term phosphorus-fixation to apply to phosphorus sorption, and the designation of particular fractions of the phosphorus present in soils as 'calcium-or-aluminium-bound' or 'fixed'. Similarly, that fraction of the phosphorus present in the soil which was assumed to be usable by plants was distinguished as 'available'.

Because of the imprecise and often rather different uses of these terms they have been avoided in recent discussions of soil phosphorus (e.g. Larsen, 1967). Increasing use is being made of the term 'labile', strictly defined as that fraction of the soil phosphorus which can enter the soil solution by isoionic exchange within a specified time span. Ultimately, over geologic time, all soil phosphorus may be able to enter the solution phase, but in a restricted time period only a part of the phosphorus is labile. Measurement of the labile phosphorus is by use of the radioactive isotope ^{32}P. In experiments using water-soluble phosphates a known proportion of the phosphate ions contain radioactive ^{32}P. This labelled phosphate will be sorbed with the added phosphate and will also undergo isotopic exchange with any ions which can dissociate from the soil surface. The greater the amount of this phosphate, the more the ratio of radioactive to non-radioactive phosphorus will be lowered from its initial value, and the amount of this phosphate can be calculated from this lowering. This isotopically exchangeable soil phosphate is the defined labile phosphate.

The behaviour of phosphorus in a soil can be likened to the behaviour of water in a well system (Thompson, 1952; Larsen, 1967; Wilson, 1968). Larsen (1967) suggests a central well immediately surrounded by highly porous material, beyond which there is an infinite extent of less porous material. The well and porous material are underlain by a trough in the bedrock which underlies the less porous material. Until the central trough is full all the water will be confined to the well and highly porous material. In this analogy phosphorus in the soil solution, readily usable by plants, is represented by mobile water in the highly porous material and non-labile phosphorus is equivalent to the almost non-mobile water in the outer zone of low porosity. As the plant removes the small amount of phosphorus present in the soil solution the equilibrium is disturbed and phosphorus in the labile fraction will be drawn upon. Non-labile phosphorus is unlikely to contribute to the supply over a period as short as one growing season since its rate of release is too slow. If phosphorus is added the concentration in solution will rise until an equilibrium level is reached, corresponding to the solubility-product of some phosphorus mineral. A crystalline phase will then precipitate and the phosphorus within the crystal lattice will no longer be labile. If further phosphorus is added the concentration of phosphate will rise above the equilibrium value, but in time the level will fall until equilibrium is re-established – the well and underlying trough are filled, the porous material fills from the base up while seepage continues into the low porosity material, until equilibrium is finally re-established.

1.5 The addition of phosphorus to soils
The effects of adding phosphorus to soils will depend on the type of material added, the plants growing in the soil and the nature of the soil itself. Gunary (1970) has suggested that a soil will adsorb a little phosphorus firmly, a slightly greater amount less firmly and so on, until a limiting value is reached when all the components of the phosphate adsorption system are saturated. In general, where the initial phosphorus status of a soil is very low, or the sorption capacity of a soil is high, added phosphorus will be readily sorbed, labile phosphorus will be formed, and there will be a gradual loss over time of labile phosphorus to a non-labile form. An exponential rate of loss

of labile phosphorus has been suggested by Larsen (1967). Rate of loss will depend also on soil type. In experiments within the pH range 5.5 to 7.5 Larsen (1965, 1967) found that more rapid loss was associated with soils of higher pH. Suggested mechanism for the loss of lability could be the slow formation of crystalline calcium phosphate, presumably hydroxylapatites. At lower pH and in non-calcareous soils presumably other, iron and aluminium phosphates would be formed in comparable fashion. From the results of other experiments Larsen and Widdowson (1970) have suggested that the upper limit of phosphate solubility in a calcareous soil is that of dicalcium phosphate ($CaHPO_4$) and the lower limit that of fluorapatite. Although dicalcium phosphate may be formed initially when soluble phosphate is added to a calcareous soil, it is not stable in the soil, and will be converted to less soluble calcium phosphates and ultimately to the least soluble form, fluorapatite.

When organic matter such as plant debris is added to a soil decomposition occurs and the conversion of organic phosphorus present in the organic matter into soil phosphate begins. The identity and composition of the organic phosphorus compounds present in the organic matter originally are not certainly known, nor have all the organic compounds formed in the soil been identified (Omotoso and Wild, 1972). Over time some of the organic phosphates present in the soil will ultimately be converted into mineral phosphates, but little is yet known of the mechanisms of conversion or the rates at which they occur.

Applications of farmyard manure to a neutral arable soil of pH 7.1-7.4 for a century at Rothamsted have resulted in an increase in organic phosphorus but the proportions of inositol phosphates have remained unchanged (Oniani *et al.*, 1973). Superphosphate alone slightly lessened the total organic phosphorus in the soil. However, in similar permanent grassland soils superphosphate applied for the same length of time increased organic phosphorus in both the surface and sub-surface layers. Organically combined phosphorus, whether present in added organic matter or in the soil organic matter, can only be utilised by plants after it has been converted into mineral forms. At least some of this conversion must occur during humification, that is, during the transformation of plant material and the breakdown of the products (Swift and Posner, 1972). Using a radioactive tracer method Larsen (1966) has shown that the rate of conversion of organic phosphorus compounds into inorganic phosphates, as represented by the rate of hydrolysis of pyrophosphate, depends on the biological activity and pH of the soil.

In conformity with the general view that organic phosphates will be converted ultimately into inorganic forms, Walker (1965) has illustrated an idealised chronosequence of soils in a humid climate. Through time organic phosphorus (Po) declines proportionately and insoluble phosphorus (Pf) becomes increasingly important. He suggests that the proportion of inorganic phosphorus extracted by N H_2SO_4 without prior ignition (Pa) would decrease exponentially, remaining at increasingly longer time periods as an even smaller proportion than the organic phosphorus (Po).

Evidence that organic phosphorus does remain in the soil for some time has been provided by Mattingly and Williams (1962) who found that perhaps one-third of the organic phosphorus originally present in the soil buried beneath the Roman amphitheatre at Winterslow had survived into modern times.

Other problems concerned with the re-distribution of added phosphorus must be examined in terms of the vertical distribution of phosphorus in soils under varying conditions.

1.6 Vertical distribution of phosphorus in soils

Plants growing in a soil will cause an upward movement of phosphorus to take place. All the labile phosphorus will be involved in this movement, the rate of movement depending upon the quantity of phosphorus taken up by the roots, transported through the plant and released into the upper soil horizons by subsequent decay. The ratio of organic to inorganic phosphorus will generally decrease down the profile (Williams and Saunders, 1956). However, this generalised pattern of phosphorus distribution resulting from plant growth will, of course, be modified by other soil processes. There may be some loss of phosphorus from the surface by leaching of soluble phosphates. Because of the close association of soil phosphorus with the organic matter and clay fractions of soils, re-distribution or concentration of these fractions within the soil may lead to similar re-distribution or concentration of phosphorus. Since the organic phosphorus in the soil is a part of the organic matter it tends to follow the pattern of distribution of organic matter as a whole, although there is no evidence for a fixed ratio of organic phosphorus to such other organic constituents as carbon, nitrogen and sulphur (Black, 1968).

Although clay rich horizons often exhibit high phosphorus contents there may be significant differences in the types of phosphates found in different size fractions of soils. Black (1968) reported that Scheffer and others in Germany found a tendency for iron and aluminium phosphates to accumulate in the finer fractions of soils, and calcium phosphates to accumulate in the coarser fractions, as would be expected if coarser apatite particles were weathering to iron and aluminium phosphates associated with the finer size fraction. Again, there is the common observation that iron and aluminium phosphates will be dominant at low pH, calcium phosphates at high pH. Where there is, therefore, vertical variation in texture and pH within a soil profile, variation in phosphate content would be expected. The association of phosphorus with iron and aluminium in acid soils will often lead to high phosphate levels in illuvial B horizons and low levels in eluviated A_2 horizons as clearly shown by Linnermark in his study of podsols and brown earth soils in southern Sweden (1960). Patterns of accumulation of different phosphates varied greatly. Within the soils examined by him lactate-soluble phosphorus was uniformly high in the litter layer F_0 and fell generally to a low level in the A_2 and deeper horizons. The acetic acid-soluble phosphorus was very variable in distribution but in the Strong Podsols tended to be high in the F layers and lower in the A and B horizons, sometimes rising to high values in the C horizon. The sulphuric acid-soluble phosphate, lower in the brown earths than in the podsols, was generally high in the B_2 horizons on the podsols and the B horizons of the brown earths, maxima often occurring at the same depths as the maxima of iron and aluminium in Tamm's extract of the soils.

This varied vertical distribution of phosphorus in soils emphasises that there is some mobility of phosphorus in soils. Not all the phosphorus added to the surface of a soil will necessarily remain at the surface. In sandy soils and peats that have little tendency to sorb phosphorus, added phosphorus may be leached rapidly and lost from the soil (Black, 1968). Presumably it is under such conditions that even the phosphorus derived from the decay of waste residues, animal remains, or bodies in archaeological contexts, is leached. On subsequent analysis there may be no significant difference between the phosphate content of the soil where the artifact is thought to have been, and the surrounding soil to which it is presumed no phosphorus had been added. Phosphate analysis (total phosphate) of a body stain in the Great Barrow at Bishop's Waltham, Hampshire failed to show any significant difference between phosphate levels of the stain and the presumed grave filling, although the method has been used successfully elsewhere (Bascomb, 1957; Johnson, 1954-6).

1.7 Variation and sampling
The variation of characteristics with depth in soils is universally recognised and, indeed, forms the basis for the study and definition of soil horizons. Much less attention has been paid to local areal or lateral variation in soil characteristics, and variability within single soil horizons. It is surprising that there are so few published discussions on local lateral variation in soils when the resultant differential growth in vegetation, both in terms of species and quantity of production, is visually so obvious (Bunting, 1961; Weeda, 1967). It should be clear from the above discussion of the various forms of phosphorus compounds present in soils that some variation in amount and form of phosphorus present may be expected laterally in the same horizon of undisturbed soils, even within quite short distances. Recognition of this is extremely important when assessing whether or not the amount of phosphorus present at a given location is the result of natural processes, or the result of human activities. The fundamental problem is to establish the expected range of natural variation so that samples with higher phosphate content can rightly

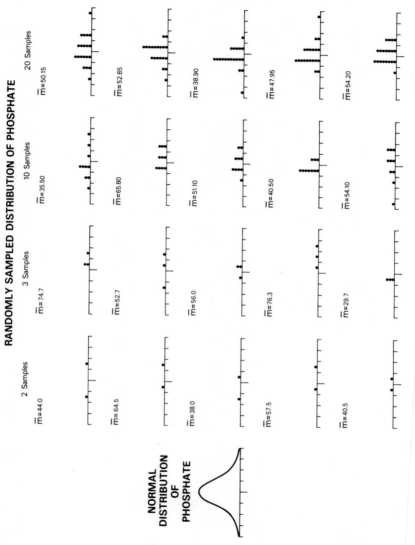

Figure 1. Effects of sample size in random sampling of a phosphate population of normal distribution.

be attributed to human interference. Some indication of the variation likely to be encountered in soils was given by Jenny (1941) who sampled along a transect in a virgin prairie and an adjoining cultivated field. Significantly, he was able to establish that cultivation had rendered the topsoil areally more uniform in terms of the organic matter content, and carbon-nitrogen ratios of the soils. A note drawing attention to some more recent work in the variation of both physical and chemical characteristics of soils has been published by Gerrard (1973).

In the present paper a simple graphical illustration is given of the difficulties of establishing the range of variation within a natural soil property at different levels of sampling (Fig. 1). Using the same numbers of samples, the difficulties of distinguishing whether or not additions have been made to the original soil are illustrated (Fig. 2). Data from the excavated site at Goodland, Northern Ireland are then considered in the light of these simple graphical models (Fig. 3).

For simplicity it is assumed that the amounts of phosphorus present in an undisturbed soil are normally distributed about the mean value (Fig. 1). It therefore follows, for example, that 68.3 per cent of the occurrences will lie between +1 standard deviation and -1 standard deviation, that is, there is approximately a 2 to 1 chance that a value will lie within these limits, and a 1 to 2 chance that it will not. Assuming that values lie between 0 and 100 the total population is sampled using randomly chosen values derived from a table of random numbers (e.g. Gregory, 1963). The effect of sample size is clearly shown, both in terms of the range of possible values and in terms of the mean values obtained from samples of different size. For example, the means of 5 sets of 2 samples range from 38.0 to 63.5; those of 5 sets of 20 samples range from 38.9 to 54.2. It is clear only in the sets of 20 samples that higher values are part of a single population. The evidence from smaller samples is quite ambiguous as to whether or not the higher values are part of a population distinct from a population of lower values. This is particularly the case with the first example of 2 samples drawn from the population where the values are 3 and 85, and with the last example of 10 samples drawn from the population. In the first instance there is a considerable gap between the values, in the second there is a heavy weighting of higher values: in both instances if such values were determined in an archaeological context it would be tempting to argue that the higher values represent additional phosphorus as a result of human activity.

In the second case, a higher value phosphate distribution is added to the same normal distribution of phosphate with values between 0 and 100 (Fig. 2). The higher value distribution has a mean of almost 100 and 28 values are arranged symmetrically in a narrow cluster about the mean. This new universe of two separate but overlapping populations is then randomly sampled in the same manner as before. The difficulties of interpreting the results are significantly greater

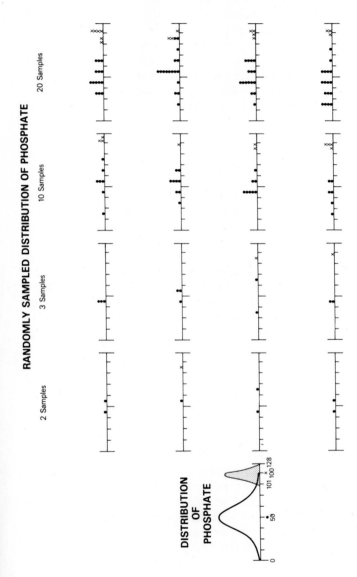

Figure 2. Effects of sample size in random sampling of a bi-modal phosphate population.

than in the simple case of a single population. In only one set of 2 samples is the added phosphate sampled, and then with a value within the range of the original single population. The added phosphate is recorded in 3 sets of 3 samples, but in only one instance with a value higher than that of the original single population. Only in the sets of 20 samples is there any obvious indication of the presence of a substantial phosphate population of higher value, and even in these sets the establishment of the form of the original population is uncertain. It should be stressed that the interpretations given in this second case depend on knowing the values within the two overlapping populations present. Without this knowledge it would be impossible to assign higher valued samples to the original population or to the added higher value population.

2.　*An archaeological site at Goodland*

Soil samples were collected from a number of locations on the archaeological site in Goodland Townland, County Antrim, Northern Ireland. Phosphate determinations were carried out using 3N H_2SO_4 as an extractant in a manner similar to the method description by Cornwall (1958). Part of the site is on rolling downland, and part is buried beneath peat in a shallow valley along the eastern edge of the downland. The whole area is underlain by chalk covered with a variable mantle of glacial till. The till contains a wide variety of pebbles, and is now largely decalcified, although its high flint content indicates that it must originally have had a considerable content of chalk. The features of immediate archaeological interest are, firstly, a series of houses and field systems on the downland dating to the medieval and late-medieval period, and, secondly, a complex of pits, ditches, post- and stake-holes of Neolithic date cut into the sub-soil beneath the peat (Case *et al.* 1969). It proved possible to link the various archaeological phases of human occupance of the area with the history of soil development in the area (Proudfoot, 1958). Prior to the use of the Neolithic site, leaching of the soil had lead to the formation of a zone of iron accumulation. The Neolithic pits and other features seem likely to have belonged to a ritual site. After this site fell into disuse the area was cultivated, truncating the tops of the pits; a thin iron pan podzol developed over most of the site, and then the blanket peat grew over the site. As the blanket peat deepened there was periodic cultivation of the downland, leaching of the downland soils, and finally the establishment of the field system and overlying huts.

　　It was hoped that the phosphate determinations would show whether or not the Neolithic pits and other features contained added phosphates which might have been derived from material put into the pits in Neolithic times. Similarly, phosphate values in the samples from the interiors of the downland huts which were different from

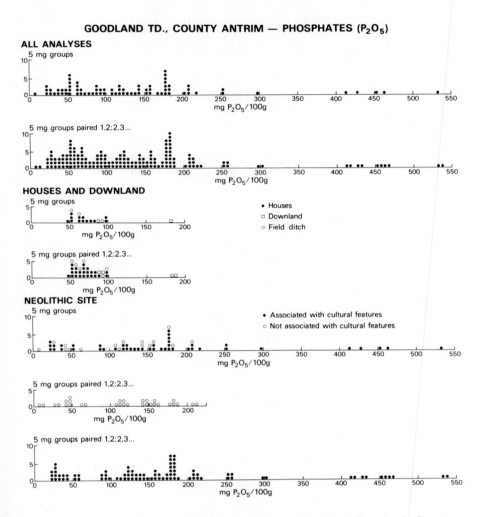

Figure 3. Phosphate content of soil samples from Goodland Townland, County Antrim, N. Ireland.

those in samples from outside the huts would indicate different patterns of use.

The phosphate content of relevant samples is shown in Fig. 3. There is clearly a wide range of values across the site. This would be expected even under natural conditions, taking into account the heterogeneity of the parent material. Some of the pebbles in the till were of relatively phosphate-rich igneous rocks, others were of rocks low in phosphorus. Presumably also there was likely to be minor variation in the chalk content of the till and in its phosphate content. To the initial variation should be added that induced by processes of soil development, including decalcification, differentiation of clay-rich and clay-poor horizons, and movement of iron, aluminium and some organic matter in the soil profiiles. Superimposed on this variation is perhaps that introduced by differential human activities at various times past.

It can be concluded from the data that there was no significant accumulation of phosphorus-rich debris in the houses sampled on the downland. Some at least of the Neolithic pits seem to have been filled not only with boulder clay similar to that in which they were dug but also with phosphate rich material. Detailed arguments to support these conclusions are beyond the limits of this paper and will be discussed elsewhere (Proudfoot, in prep.). Consideration of a site such as that at Goodland emphasises the difficulties of interpreting the analytical data in the light of what is currently known about phosphate distribution and analysis.

3.　*Conclusions*

At least since Hughes observations in Egypt in 1911 (Russell, 1957), it has been recognised that the phosphate content of soils on formerly occupied archaeological sites may be measurably higher than that of soils on adjacent areas that were never occupied. The higher phosphate levels on occupied sites derive from the accumulation of organic wastes, domestic refuse, and general occupation debris. Some of the phosphorus in these materials may be rendered insoluble by the normal processes which affect all phosphorus compounds in soils, and therefore be retained in the soil.

Systematic surveys of the phosphate content of the soils of extensive regions in order to locate achaeological sites were pioneered in Europe, by Arrhenius in Sweden, and by Lorch in Germany. Reviews of the use of phosphate analyses in archaeology have been published by Dauncey (1952), and by Cook and Heizer (1965). Cornwall (1958; 1960) has noted one of the methods available for analysis and commented on the role of phosphate analysis in relation to other types of information of archaeological significance to be gained from soils.

The present paper has summarised some of the recent developments in soil phosphate studies. In spite of the very considerable amount of research already carried out in this area, the composition of phosphorus compounds, and their mode and rate of formation in soils, are still uncertain. Traditional methods of extracting phosphorus from soils give discrepant results on different soils and, especially in the case of inorganic phosphates, various fractions isolated by such methods may be of little significance in studying the genesis of phosphorus compounds in the soil. Ratios between the various phosphorus fractions and other soil constituents generally fail to follow consistent patterns.

Natural variation both laterally and with depth in the amount and type of phosphorus present in soils would be expected. Different methods of analysis isolate different fractions of the phosphorus present, and different percentages of the same fraction, under different conditions. It is therefore, often difficult to determine unequivocally the effect of human occupance on a site. Different methods will yield different results as Johnson (1954-6) showed in comparing total and readily soluble phosphate content of soils in the burial chamber at Corrimony.

These difficulties need not deter continued use of phosphate analyses in helping to solve archaeological problems. The necessity is for an understanding of the complex nature of the problems involved. Research needs to be conducted much more systematically than has often been the case in the past. More rigorous sampling designs need to be developed so that there is adequate representation of the wide variation of natural and artificial features present both vertically and horizontally on sites. There is need for the systematic evaluation in archaeological contexts of the traditional methods of differentiating between the phosphorus fractions of soils. New methods of analysis, which distinguish, for example, between labile and non-labile phosphorus, should be similarly evaluated. Finally, archaeological sites provide the pedologist with unique opportunities for examining variations in the phosphorus content of soils within a chronological framework. Since many of the mechanisms of phosphorus interaction in the soil are time dependent, further work on the phosphorus content of soils in archaeological contexts could lead to considerable advances in our fundamental understanding of soil phosphorus.

REFERENCES

Bascomb, C.L. (1957) 'The silhouette in the great barrow at Bishop's Waltham, Hampshire', *Proc. prehist. Soc.* 23, 1623.

Black, C.A. (1968) *Soil-Plant Relationships*, 2nd ed. New York.

Bunting, B.T. (1961) 'The role of seepage moisture in soil formation, slope development and stream initiation', *Am. J. Sci.* 259, 503-18.

Case, H.J., G.W. Dimbleby, G.F. Mitchell, M.E.S. Morrison and V.B. Proudfoot (1969) 'Land use in Goodland Townland . . .', *J. Royal Soc. Antiquaries Ireland* 99, 39-53.

Chang, S.C. and M.L. Jackson (1957) 'Fractionation of soil phosphorus', *Soil Sci.* 84, 133-44.

Cook, S.F. and R.F. Heizer (1965) *Studies on the Chemical Analysis of Archaeological Sites*, University of California Publications in Anthropology 2. Berkeley.

Cornwall, I.W. (1958) *Soils for the Archaeologist*. London.

Cornwall, I.W. (1960) 'Soil investigations in the service of archaeology', *in* R.F. Heizer and S.F. Cook, eds., *The Application of Quantitative Methods in Archaeology*. Viking Fund Publications in Anthropology 28. Chicago, 265-84.

Dauncey, K.D.M. (1952) 'Phosphate content of soils on archaeological sites', *Advmt. Sci.* 9, No. 33, 33-7.

Gerrard, A.J. (1973) 'Soil particle densities', *Area* 5, 236-7.

Ghani, M.O. and Islam, A. (1957) 'Use of 8 (OH) quinoline and selenious acid in determining available phosphorus', *Soil Sci.* 84, 445-51.

Great Britain, Ministry of Agriculture (1965) *Soil Phosphorus*, Technical Bulletin 13, London.

Gregory, S. (1963) *Statistical Methods and the Geographer*. London.

Gunary, D. (1970) 'A new adsorption isotherm for phosphate in soil', *J. Soil Sci.* 21, 72-7.

Jenny, H. (1941) *Factors of Soil Formation*. New York.

Johnson, A.H. (1954-6) 'Examination of soil from Corrimony chambered cairn . . .', *Proc. Soc. Antiq. Scotl.* 88, 200-7.

Just Nielsen, A. (1972) 'Deposition of calcium and phosphorus in growing pigs determined by balance experiments and slaughter investigations', *Acta Agric. Scand.* 22, 223-37.

Larsen, S. (1966) 'A tracer method for determination of phosphate hydrolysis in soil', in *The Uses of Isotopes in Soil Organic Matter Studies*, Suppl. *J. appl. Radiat. Isotopes*, 491-4.

Larsen, S. (1967) 'Soil phosphorus', *Adv. Agron.* 19, 151-210.

Larsen, S., D. Gunary and C.D. Sutton (1965) 'The rate of immobilization of applied phosphate in relation to soil properties', *J. Soil Sci.* 16, 1418.

Larsen, S. and A.E. Widdowson (1970) 'Evidence for dicalcium phosphate precipitation in a calcareous soil', *J. Soil Sci.* 21, 364-7.

Linnermark, N. (1960) *Podsol och Brunjord*, Publications of Institutes of Mineralogy, Palaeontology and Quaternary Geology, University of Lund, 75, Lund.

Mattingly, G.E.G. and R.J.B. Williams (1962) 'A note on the chemical analysis of a soil buried since Roman times', *J. Soil Sci.* 13, 254-8.

McConaghy, S. (1960) 'Soil phosphates, with special reference to organic forms and their availability to plants', *Agric. Prog.* 35, 82-93.

McConaghy, S. and J.S.V. McAllister (1952) 'The evaluation of the available phosphate status of agricultural soils in Northern Ireland', *Proc. Int. Soc. of Soil Sci. Comm. 2 and 4*, 2, 354-63.

McQuitty, J.B. and E.M. Barber (1972) *An Annotated Bibliography of Farm Animal Wastes*, Water Pollution Control Directorate, Environmental Protection Service, Environment Canada, Ottawa.

Omotoso, T.I. and A. Wild (1970) 'Occurrence of Inositol phosphates and other organic phosphate components in an organic complex', *J. Soil Sci.* 21, 224-32.

Oniani, O.G., M. Chater and G.E.G. Mattingly (1973) 'Some effects of fertilizers and farmyard manure on the organic phosphorus in soils', *J. Soil Sci.* 24, 1-9.

Paauw, F. van der (1971) 'An effective water extraction method for the determination of plant-available soil phosphorus', *Pl. Soil* 34, 467481.

Proudfoot, V.B. (1958) 'Problems of soil history . . .', *J. Soil Sci.* 9, 186-98.

Proudfoot, V.B. (in prep.) 'Soils and soil history', *in* H.J. Case, ed., *Excavations at Goodland*. (Northern Ireland Government Archaeol. Report).

Russell, E.J. (1957) *The World of the Soil*. London.

Russell, E.W. (1961) *Soil Conditions and Plant Growth*, 9th ed. London.

Sissingh, H.A. (1971) 'Analytical technique of the Pw method used for the assessment of the phosphate status of arable soils in the Netherlands', *Pl. Soil* 34, 483-6.

Steward, J.H. and J.M. Oades (1972) 'The determination of organic phosphorus in soils', *J. Soil Sci.* 23, 38-49.

Swift, R.S. and A.M. Posner (1972) 'Nitrogen, phosphorus, and sulphur contents of humic acids fractionated with respect to molecular weight', *J. Soil Sci.* 23, 50-7.

Thompson, L.M. (1952) *Soils and Soil Fertility*. New York.

Walker, T.W. (1965) 'The significance of phosphorus in pedogenesis', in E.G. Hallsworth and D.V. Crawford, (ed.), *Experimental Pedology*, London. 295-316.

Weeda, W.C. (1967) 'The effect of cattle dung patches on pasture growth, botanical composition, and pasture utilisation', *N.Z. Jl. agric. Res.* 10, 150-9.

Williams, E.G. and W.M.H. Saunders (1956) 'Distribution of P in profiles and particle-size fractions of some Scottish soils', *J. Soil Sci.* 7, 90-108.

Wilson, A.T. (1968) 'The chemistry underlying the phosphate problem in agriculture', *Aust. J. Sci.* 31, 55-61.

DISCUSSION

Mr. J.L. Bintliff expressed some reservations on the fact that intense occupation could be correlated with waste disposal areas. The speaker replied that the matter was a question of relative intensity, and that if one site of 100 m² had 100 people living on it, the concentration would be higher that if only 10 people had been living there. An even spread of phosphate values across a site, compared with lower values outside the site, would indicate that the occupants were rather scruffy.

Mr. K. Edwards enquired whether any comparisons had been made with carbon or nitrogen values, but the speaker replied that this had not been done since the variations in these elements between different soils was so great. On a range of soils the results would be of doubtful value.

J. Ll. W. Williams and D. A. Jenkins

The use of petrographic, heavy mineral and arc spectographic techniques in assessing the provenance of sediments used in ceramics

Ceramic sherds are essentially transformed geological sediments and as such are therefore amenable to the same methods of analysis. The resultant data have two major and interrelated contributions to make to archaeological ceramic studies. Firstly, analysis offers the possibility of classifying pottery by composition, independently of the more formal typological classifications that form the cornerstones of modern archaeology; secondly, it can provide information on the provenance of some of the sedimentary materials used to make the pottery. But, if the study of ceramic compositions is effectively a problem of sedimentary analysis, the subsequent interpretation of the results must be governed by the fact that human intervention will have greatly modified the original natural sediments by such diverse actions as the admixture of clays,[1] the introduction of filler, or in the physical changes wrought by fabrication and firing. Consequently, there are limitations to the archaeological inferences that can be drawn from the compositional data in addition to those which originate in the particular analytical techniques used.

The aim of this paper is to assess in general terms the applicability and limitations of selected techniques, namely petrographic, mineralogical and trace element analysis, by reference to three separate ceramic studies set in contrasting geological environments. The choice of these three techniques from the increasing array of those now available (X.R.D., X.R.F., D.T.A., I.R.A., Neutron activation, Mössbauer, etc.,) reflects both their availability at Bangor and also the considerable return of information which they offer to the archaeologist.

1. Petrographic analysis

The procedure of petrographic analysis of rocks and more recently of unconsolidated sediments, for example soils, is well established and does not require detailed description here (Brewer, 1964). The

---

1. Throughout this paper the term 'clay' is used in its ceramic sense (i.e. fine grained plastic deposits used in the manufacture of pottery) rather than with its sedimentological definition (i.e. e.s.d. $<2\mu$).

technique has been successfully applied to the study of ceramics in a number of studies (Shepard, 1965; Peacock, 1970). At Bangor friable pottery is resin impregnated under vacuum using the same procedure as with soils; more durable pottery can be placed in hot Lakeside 70 to gain a protective coating of resin, (Cornwall, 1958), though this procedure hardly provides impregnation of the ceramic material. The former procedure, though more elaborate and time consuming, is much less wasteful of material and allows more than one section to be taken from a relatively small piece of pottery within a cylinder of polymerised resin. Thin sections are ground to the standard thickness of 25μ and examined visually by transmitted light in the polarising microscope. Mineral grains, rock fragments and artificial constituents, such as grog, are indentified, thus establishing the composition and relative frequencies of all the clastic components. Much of this data can be quantified and, with certain provisos, assessed statistically. In addition, quite important information can be obtained on the texture, shape and degree of sorting of the clastic components thus adding a further dimension to the study of fabric and composition.

The potential of such data for archaeology is immediately apparent. Many rock fragments or mineral suites are very distinctive, or of limited occurrence, and in some cases can be compared and related to specific outcrops or sedimentary deposits in the field. In theory polymineralic assemblages in the size range 100μ to 5 mm offer the most specific information, but inevitably their usefulness and sensitivity will be largely governed by the spatial distribution and unique nature of the parent materials. In practice much pottery contains undiagnostic assemblages composed of common mineral suites (quartz, feldspar, mica, calcite, etc.) or widespread rock types (flint, limestone, etc.). Areas offering a great diversity of rock types, such as parts of highland Britain, therefore provide greater potential than do areas of uniform and widespread geological outcrops such as occur in areas of lowland Britain.

Petrography has other less obvious limitations in its application to ceramics. The sampling reproducibility of a thin section of less than 5 cm^2 in area (that is 0.025 g of a 2-5 g subsample), from a vessel of between 1000 to 5000 cm^2 in area (1 to 5 kg) is inevitably limited and thus a specific diagnostic mineral or rock constituent could occur as sparingly as once per 10 thin sections. Furthermore, the subtle variations of composition within and between related rock types or sediments, which in themselves may be quite diagnostic of particular geological terrains, may, on the other hand, lead to ill-defined and diffuse group or unit boundaries. Frequently only extreme members can be clearly differentiated from the main body of intergrading variants; such data are often difficult to quantify and may not always be amenable to statistical treatment.

Finally, the study of the clay matrix with the petrographic

microscope is limited since clay mineral identification is not yet possible, and observations on the fabric, texture or optical properties of different clays are complicated because of modification by firing or incomplete oxidation of organic matter. In exceptional circumstances variations in the detrital mineralogy associated with the clay body may be used to distinguish petrographically between different clay types (Williams, 1975).

2.   Heavy mineral analysis

Heavy mineral studies of pottery follow the same routine as for sediments. A sherd fragment of between 10-20 g is carefully disaggregated and fractionated by sieving ($>60\mu$) and sedimentation ($<60\mu$) and the small fraction (usually <5 per cent) of the more diagnostic high specific gravity minerals (>2.95) separated from the less informative light minerals (quartz, feldspars) in appropriate heavy liquids by centrifugation. The resultant minerals, either in the more easily studied size range 200μ-60μ, or if necessary in the smaller size range 60μ-20μ, are identified under the polarising microscope. From such an analysis the mineralogical composition of the sample can be established, the data quantified (Griffiths, 1967) and the information used to classify the sherds into groups. The data can be usefully processed by such statistical techniques as Principal Component Analysis, and correlation matrices can help in recognising suites of associated minerals. So far the technique has not been so widely applied to the study of ceramics as petrographic or trace element analysis (Peacock, 1967, 1973; Peacock and Thomas 1967).

For the purpose of classification, units are not restricted to mineral species, but may also include distinctive varieties differentiated by colour, habit, presence of inclusions, strained extinction, etc., although the definition of such units may then be difficult. The technique is particularly relevant for pastes in which rock types are absent or undiagnostic, or which are composed of very fine textured sediments and therefore not amenable to petrographic study. By contrast, distinctive mineral species or varieties need only be present in minute quantities to identify the provenance of a sample. Compared to petrographic analysis, the technique requires a much larger initial sample (10-20 g) but the sampling error is reduced by about one thousandth since a representative fraction of the total sample is used and also a much smaller component (i.e. a mineral grain) is sought.

The technique is most effective when operating as an accessory to petrography, for it is the latter technique that can best define in thin section which pottery pastes merit further study by heavy mineral analysis. Furthermore, it is only by reference to a thin section that it may be possible to discriminate between the heterogeneous source materials used in pottery making and thus of establishing whether heavy minerals are derived from the clay body, a rock filler, or, indeed, from inclusions of grog. This is not possible in the disaggregated homogenised sample from which the heavy minerals are concentrated.

3.   Trace element geochemistry

The value of trace element analysis is based on the fact that the complement of trace elements varies between different types of rock and is, to varying degrees, characteristic of them. The theoretical background to trace element distribution is now reasonably well established, so that sediments may therefore be classified and, to a limited extent, their provenance deduced by reference to their trace element content: with certain reservations, the same applies to ceramic sherds. Such analysis has been successfully applied to a number of ceramic problems (e.g. Catling *et al.*, 1963; Asaro *et al.*, 1971).

One of the most convenient analytical techniques is that of arc spectrography. This enables the concentrations of a large number of elements to be estimated simultaneously at the μg/g level in samples of only a few mg in size. There are several modes of using this technique, according to the balance sought between accuracy and sensitivity. In Bangor, a semiquantitative procedure, adapted from that of Mitchell (1964) has been established for the routine analysis of

Table 1. Trace element contents in 19 potsherds from North Wales.

	Range μg/g		
Ba	18	—	1300
Be	<3.2	—	5.6
Co	4.2	—	32
Cr	42	—	1000
Cu	<10	—	180
Ga	7.5	—	32
La	3.2	—	42
Li	7.5	—	75
Mn	42	—	5600
Mo	<5.6	—	13
Ni	18	—	320
Pb	<10	—	180
Rb	180	—	4200
Sc	<10	—	32
Sr	3.2	—	750
Ti	2400	—	>1%
V	32	—	750
Y	13	—	130
Yb	1.0	—	7.5
Zn	<240	—	1300
Zr	130	—	2400

N.B. in the technique used, values are limited to the log. series $10^{n}/8$.

rocks and soils (Jenkins, 1964) and this has been used for the analysis of ceramic sherds. Although it has a relatively low order of accuracy (± 50 per cent) it has the advantage of sensitivity in that some 20 elements can normally be detected (Table 1). For this purpose a sample of 1 to 2 g is required. This is then subsampled after homogenisation by grinding in an automatic mechanical agate mortar, firing at 470°C overnight, and regrinding. Compared to the other two techniques, this procedure therefore results in a relatively low sampling error.

The data from trace element analysis differ fundamentally from those obtained by petrographic and mineralogical analysis in that they are the weighted averages of all the rocks and minerals contained in the ceramic paste body. The identity of the different sources that contributed to the sediment are thus not preserved intact, as in the form of discrete mineral grains or rock fragments. It is therefore more difficult to interpret the geochemical composition of a sediment of mixed origin in terms of the provenance of its individual component sources, a difficulty encountered in a number of spectrographic projects on ancient ceramics without a successful outcome. Exceptions may occur when the sediment happened to be dominated by some material of extreme geochemical composition (e.g. orthoquartzite, serpentine, etc.) or there is an 'indicator' element (e.g. Be,Sc, Mo, etc.). In rare cases it is even possible that geochemical analysis could discriminate between sediments (and thus sherds) composed of rock and mineral assemblages that were visually similar but derived from different geochemical provinces.

As in all trace element analysis, due care must be taken to avoid contamination during sampling and analysis, and the possibility of contamination prior to sampling (e.g. contemporaneous use of pots with metal implements; subsequent calcite deposition etc.) should also be borne in mind when interpreting the results. This problem of contamination is not so significant in mineralogical analysis and is virtually absent in petrographic analysis.

However, the data from trace element analysis are obtained in the form of continuous variables (concentrations) for discrete units (individual elements) whose definition poses no problem. The results are therefore more amenable to statistical handling in the classifying of the ceramic sherds, either by ordering (e.g. Principal Component Analysis) or by clustering (e.g. Average Link, and k-means Cluster Analysis) procedures, as discussed by Hodson (1970).

4.   Bronze Age cinerary urns from Anglesey

The first example quoted to illustrate the varied applicability of these techniques is concerned with a particularly rich collection of cinerary urns and sherds from a number of burial sites in Anglesey. The pottery comes mainly from Bronze Age multiple burial barrow sites,

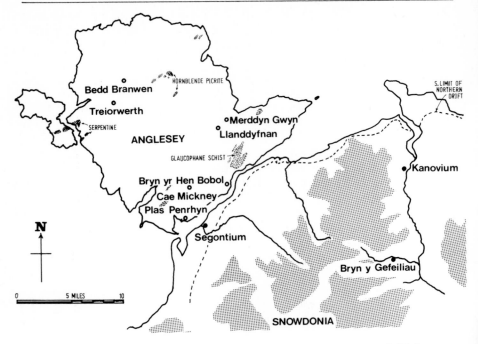

Figure 1. Bronze Age (o) and Roman (●) archaeological sites in north Wales.

such as Bedd Branwen, Treiorwerth and Llanddyfnan (Fig. 1), or from flat urnfield type sites, such as Plas Penrhyn and Cae Mickney, and, more rarely, as secondary associations in earlier Neolithic cairns as at Bryn yr Hen Bobl (Lynch, 1970). Typologically the best represented vessel is the collared urn and at several sites these contained the personal belongings of their inurned occupants in the form of such articles as small knives, daggers, awls, beads and buttons (Lynch, 1970). The basic function of the pottery as funerary receptacles might suggest that they are of local manufacture, but specific typological features reflect both British and Irish influences, which have been absorbed to produce a fairly typical local tradition native to Anglesey.

The problem was thus one of establishing the compositional characteristics of the sherds and, if possible, the provenance of their constituent materials. The geological setting for this archaeological problem is remarkably complex, and this determined the relative value of the different techniques. A diversity of rock types is exposed on the planated surface of Anglesey ranging from Precambrian to Tertiary in age and consisting of a variety of igneous, metamorphic and sedimentary rocks. These have been described fully by Greenly (1919) whose rock nomenclature is adopted in this section.

4.1 Petrographic studies

In view of the large rock fragments (>1 cm) visible in the pottery and

Table 2. Petrographic classification of 51 sherds from burial sites in Anglesey.

Site (No. samples)	Altered Dolerite		Olivine Dolerite		Hornblende Dolerite	Hornblende Picrite		Serpentine Gabbro	Schist
	Exl	Mxd	Exl	Mxd	Exl	Exl	Mxd	Exl	Mxd
Bryn yr Hen Bobl (1)	1	—	—	—	—	—	—	—	—
Plas Penrhyn (7)	5	—	—	1	—	—	—	—	1
Cae Mickney (11)	2	2	1	3	3	—	—	—	—
Merddyn Gwyn (5)	3	—	1	—	1	—	—	—	—
Llanddyfnan (5)	4	—	—	1	—	—	—	—	—
Treiorwerth (8)	1	—	—	—	—	6	1	—	—
Bedd Branwen (14)	8	1	—	—	—	3	—	2	—

Exl = exclusive assemblage; Mxd = mixed assemblage

the diversity of the local geology, petrography was considered the most suitable technique of analysis. The initial results, based on the analysis of 51 sherds from a maximum available sample of 68 vessels, are both surprising and very interesting. A primary classification of the pottery, based on the variety of rock inclusions present in the pastes (Table 2) shows how five principal igneous rock types of basic and ultra-basic composition dominate the whole sample.

The most consistently represented rock type is altered dolerite with highly saussuritised and chloritised feldspar and ophitic or granular pyroxene as the dominant minerals; other distinctive types include hornblende pictrite, olivine dolerite, hornblende dolerite and serpentine. Fragments of these rocks occur exclusively in the majority of pastes ('Exclusive Assemblages') but in others additional rock types, or inclusions of grog, are present ('Mixed Assemblages'). The most commonly occurring of these additional rocks are muscovite and biotite schists, feldspathic sandstones, quartzites and epidosites. In an exceptional paste from Plas Penrhyn, fragments of muscovite and glaucophane schist and quartzite dominate the assemblage to the practical exclusion of the basic igneous suite, although in all other 'Mixed Assemblages' pastes it is this latter suite which is most abundant. The distinction between 'Exclusive' and 'Mixed Assemblage' pastes raises the problem of possible sampling error discussed previously (p. 116) concerning the presence or absence in thin section of fragments of individual rock types, which is crucial to this type of classificatory scheme.

All the rock types identified in the pottery outcrop in Anglesey, though it does not necessarily follow that the pottery was manufactured there since other parts of Britain might have comparable rock associations. However, in terms of the geology of Anglesey, the exclusive and highly selective nature of the rock fragments reduces the potential petrological complexity of the island to a minimum, as illustrated in Table 3.

Furthermore, some of these rocks outcrop in only a few localities on the island – the hornblende picrite in the north-east, the serpentine/gabbro mainly in the extreme western quadrant, and the glaucophane schist in the eastern and southern sectors (Fig. 1). In contrast, the altered dolerite, and to a much lesser extent the olivine dolerite, are widely scattered throughout the greater part of the island in the form of small, insignificant dykes: hence the isolation of

Table 3. Percentage totals of mapped outcrops of specific rock types in Anglesey.

Rock Type	Percentage
Altered Dolerite (Palaeozoic)	0.7
Olivine Dolerite (Tertiary)	<0.2
Hornblende Picrite	<0.2
Serpentine/Gabbro	0.7
Glaucophane Schist	1.2

potential sources is more complex. Although it is highly unlikely that all the altered dolerite fragments are derived from a single outcrop, yet it is very difficult to prove petrographically how many different dolerites are involved, and indeed whether one particular variety recurs at more than one site. Variations of texture, fabric and degree of alteration abound in all the fragments, but these are highly unreliable criteria for differentiating between various dolerites; mineralogical composition is a much more effective index, although in the present context, the fragments appear to be remarkably uniform with the exception of those from Plas Penrhyn which are rich in apatite. This limitation will, therefore, restrict some of the more interesting archaeological inferences that could be drawn from the petrographic data since, at this level, it is difficult to quantify the latter.

The physical condition of the rock fragments and the discrete mineral grains of sand grade size is also informative. The rock fragments are large and angular and the mineral grains, particularly the ferromagnesian minerals derived from the rocks, are fresh and unweathered and often retain original prism faces or have developed jagged fracture surfaces as the result of physical comminution. Such features would not have survived attrition during transport in normal sedimentary processes. An initial study of the various rock components present in an assortment of fluviatile silt and gravel sediments and marine sand deposits collected from the environs of some of the main burial sites, shows that basic and ultra-basic rock fragments do not account for more than one per cent of the total constituents. The most abundantly represented are the granites, quartzites, schists, gneisses, limestones, sandstones and siltstones which form the commonest rock types in the island. From this evidence it would therefore appear that, since no natural deposits of the type encountered in the pottery have yet been identified in Anglesey, the makers of the urns consciously selected the comminuted fragments of basic and ultra-basic igneous rocks. This conclusion raises very interesting archaeological and socio-economic questions that cannot be pursued further here.

4.2 Heavy mineral studies

An exploratory heavy mineral study of eight selected sherds was conducted as a corollary to the petrographic work. The eight sherds provided a fairly good cross section of the variations existing within and between the different petrographic groups. The main aim was to discover whether specific non basic igneous minerals (e.g. staurolite, kyanite, andalusite, hypersthene from the 'Red Northern Drift') could be identified which might therefore provide some information on the origins of the clay deposits, as distinct from the rock fragments, used in the pottery.

Mineral counts were conducted on the 60μ-20μ fraction, a size range

Table 4. Percentage totals of heavy minerals in eight Anglesey Bronze Age urns.

Sample No.	1	2	3	4	5	6	7	8
SITE	BEDD BRANWEN	BEDD BRANWEN	PLAS PENRHYN	CAE MICKNEY	CAE MICKNEY	TREIORWERTH	BEDD BRANWEN	CAE MICKNEY
Zircon	Tr.	0.8	0.4	2.3	0.4	1.6	1.9	4.2
Rutile	—	0.4	—	—	—	—	—	0.4
Tourmaline	Tr.	0.8	—	2.7	—	—	—	0.4
Apatite	1.6	Tr.	33.7	1.9	—	0.4	—	1.9
Garnet	0.4	0.4	0.4	—	—	—	—	12.0
Augite	57.0	50.0	41.0	52.4	13.8	9.0	1.1	20.8
Amphibole: green/blue-green	—	Tr.	0.4	—	—	—	0.4	—
Amphibole: Olive-green/brown	1.8	—	1.5	—	46.5	63.8	68.5	1.1
Amphibole: orange	3.6	1.9	3.0	1.1	25.8	8.4	27.3	2.3
Glaucophane	—	—	—	—	—	—	—	0.8
Clinozoisite	2.0	8.8	3.7	1.5	4.6	6.6	—	10.8
Epidote	—	0.4	—	1.9	0.8	1.2	—	1.5
Chlorite	0.4	—	1.4	—	—	—	—	—
Unidentified	33.2*	36.5*	14.5	36.2*	8.1	9.0	0.8	43.8*
PETROGRAPHIC CLASSIFICATION	PALAEOZOIC ALTERED DOLERITE	PALAEOZOIC ALTERED DOLERITE	PALAEOZOIC ALTERED DOLERITE	TERTIARY OLIVINE DOLERITE	PALAEOZOIC HORNBLENDE DOLERITE	HORNBLENDE PICRITE	HORNBLENDE PICRITE	MIXED ASSEMBLAGE PALAEOZOIC DOLERITE/ FELDSPATHIC SANDSTONE

N.B. Tr : denotes trace amounts (<0.4%)

   * : these assemblages contain a relatively abundant and distinct mineral, as yet unidentified.

that was considered more likely to contain the clay heavy minerals than the coarser, but more easily studied, 200μ-60μ fraction. The results are summarised in Table 4 and still show very high concentrations of ferromagnesian minerals from the basic igneous rocks. This is a special situation in which the heavy mineral fraction of a presumed mixed assemblage is dominated by a source with a high heavy mineral content (i.e. basic igneous rocks >30 per cent). Under these conditions the likelihood of detecting a diagnostic mineral component from a source with a low heavy mineral content (e.g. 'Red Northern Drift', <2 per cent) is minimal since its concentration in the mixed assemblage can be shown to be hyperbolically, rather than linearly, related to the proportion of its source material.

It can be seen from Table 4 that only a few non basic igneous minerals assume significance, the most important being the zircon, rutile and tourmaline which probably belong to the clays, but the particular varieties present are not diagnostic as indicators of provenance. The highly rounded outlines of these minerals contrast directly with the angular and unetched condition of the more unstable ferromagnesian minerals, particularly the pyroxenes. The only distinctive mineral identified in the assemblages is glaucophane, a mineral with a very restricted distribution in Britain, and therefore, in all probability, confirming a local Anglesey source for the relevant vessel (No. 8).

In this instance the analysis largely confirms the validity of the petrographic classification, a conclusion which suggests that in this case the sampling error inherent in thin sections, referred to previously, is not a limiting factor. It is interesting that heavy mineral analysis does not differentiate between pastes containing Tertiary olivine dolerite and Palaeozoic altered dolerite as their principle rock constituents; neither does it allow new sub-divisions to be established within the existing framework, with the possible exceptions of the apatite-rich Plas Penrhyn assemblage (No. 3) and the garnet-rich suite from Cae Mickney (No. 8).

4.3 Geochemical studies

The same eight samples used for the heavy mineral studies were analysed for their trace element contents. These proved to be fairly similar, all reflecting the strong basic igneous influence (i.e. relatively high Co, Cu, Li, Mn, Sc, Ti and V and low Be, Pb, Rb and Zr). Consequently they afford little scope for classification, two groups only being distinguished by Principal Component Analysis (Fig. 2). These groups followed the petrographic divisions (Table 4), but it was not possible to discriminate between the altered dolerites (1, 2, 3 and 8) and hornblende dolerite (5), or between the olivine dolerite (4) and hornblende picrite (6 and 7).

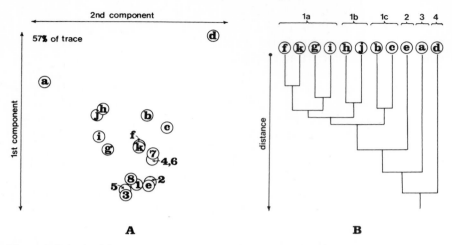

Figure 2. Principal Component (A) and Average Link Cluster (B) analysis of Roman potsherds from Bryn y Gefeiliau (a-k) and Bronze Age potsherds from Anglesey (1-8).

4.4 Summary

Anglesey has proved to be an unusual case by virtue of the very distinctive and localised rock types present in the island, rock types moreover which had been selectively utilised in the manufacture of the pottery so far analysed. In such a situation, petrographic analysis can realise its full potential and in this instance has not only suggested a detailed classification but also offered the possibility of precision in locating sources of the material used for the pottery. Conversely, heavy mineral analysis suffered from the high heavy fraction percentages so losing its sensitivity, whilst geochemical analysis added no new criteria for classification. More detailed analyses using the latter two techniques are planned in the hope of overcoming these limitations.

5.   *Roman pottery from Snowdonia*

The second investigation involved a small corpus of pottery from the strategic second century Roman fort of Bryn y Gefeiliau (Hall, 1923), in the heart of Snowdonia (Fig. 1). The pottery, which forms a selected rather than a random sample of eleven sherds, consisted of flanged and plain jars and cooking pots, the exact typological parallels of which cannot be found either within or outside northern Wales, with the single exception of a sherd of black-burnished I ware (Gillam, 1970; Farrar 1973).

The setting in terms of solid geology was again very complex. Snowdonia comprises Lower Palaeozoic strata encompassing a wide range of basic and acidic, extrusive and intrusive, igneous as well as sedimentary rocks, many of which have been modified by low grade regional metamorphism. The geology of the Capel Curig area has been comprehensively studied by Williams (1922) and Howells *et al.* (1973). However, in contrast with Anglesey, the superficial deposits are essentially of local derivation since Snowdonia was not penetrated by Irish Sea ice due to the development of its own ice cap. In this setting, a multi-analytical approach was again employed in an attempt to determine the provenance of the ceramic raw materials (Williams *et al.*, 1974).

5.1 Petrographic studies

The petrographic analysis of the pottery, based on the rock and mineral constituents in the pastes, produced a tripartite classification.

Group I contained fragments of rhyolites, welded tuffs and slates, rocks which can be matched with the local geology or with that of comparable areas like the Lake District. The group was represented by one sherd (a).

Group II contained similar rock fragments to Group I, but also included distinctive rounded quartz sand grains with overgrowths over Fe_2O_3 pellicles. The latter are associated with the Triassic sandstones or derived 'Red Northern Drift', rather than with the Lower Palaeozoic rocks of Snowdonia. Two sherds (b) and (c) fell within this group.

Group III is characterised by the undiagnostic nature of its coarser constituents, being composed mostly of detrital quartz and feldspar grains of various shapes and sizes. Such a limited and commonly occurring range of minerals restrict the potential of petrography since they provide poor criteria for subdivision and are insensitive as indicators of provenance. Three subdivisions were, however, tentatively suggested by details in the sorting, shape and size of the mineral grains; they were:

Sub-type (i): sorted angular quartz [sherd (d)]. This paste also contained a very interesting grog fragment to be referred to later.

Sub-type (ii): sorted rounded quartz [sherd (e)].

Sub-type (iii): unsorted rounded/subrounded quartz [sherds (f), (g), (h), (i), (j) and (k)].

5.2 Heavy mineral studies

A complementary heavy mineral study of eight out of the eleven sherds was resorted to with the general aim of testing the validity of the petrographic classification and, more specifically, of establishing whether sensitive indicator minerals were present in the undiagnostic Group III pastes.

In general the eight analyses confirm the petrographic subdivision, producing a range of minerals (zircon, actinolite, augite, clinozoisite/epidote) that are typically representative of indigenous Snowdonian suites (Jenkins 1964) for the Group I and Group II pastes: the latter contained in addition garnet and tourmalines. The

Group III sherds were, however, divisable into two categories, corresponding to sub-types (i) and (ii/iii) respectively of the petrographic classification. The two categories were characterised by distinctive mineral assemblages:

Sub-type (i) was practically a mono-minerallic assemblage composed of 85 per cent tourmaline, and therefore comparable with the unique tourmaline rich assemblages identified by Peacock in the black-burnished I wares, and traced to the Tertiary beds of the Poole area of Dorset (Peacock, 1973). Significantly, it is the black-burnished I sherd in the present corpus that has produced this tourmaline rich assemblage.

Sub type (ii/iii): in this group a distinctive hypersthene was consistently present, while topaz, andalusite and yanite were sparingly, though significantly, present. These minerals are not present in Snowdonian assemblages, but are considered to be fairly typical indicators of the Trias and of the 'Red Northern Drift' (Smithson, 1953), which occurs along the northern margins of Snowdonia (Fig. 1). The identification of a specific source within the territory covered by the source materials is not, however, feasible in the light of current knowledge.

5.3 Geochemical studies

The eleven-sherd sample displayed a wide range of trace element compositions, though one which falls within the normal span for rocks and sediments. These compositions can be compared with those of local Snowdonian rocks, although the ranges in the latter are also broad due to their diverse geochemical compositions, (Jenkins, 1968): for this reason it is not easy to interpret the geochemical data in terms of a Snowdonian or extraneous provenance for the raw materials. However, this diversity provides a good basis for classification, as illustrated in the results of Principal Component and Average Link Cluster analysis (the latter kindly carried out by Professor R. Hodson) shown in Fig. 2. From this it is evident that three of the sherds (a, d, and e) are isolated from the remainder, and that one in particular (d) is of extreme composition, comparable to that, for example, of a quartzite.

This division agrees with the classifications derived petrographically and mineralogically and further substantiates the tentative subdivision of the undistinguished Group III pastes. The statistical treatment of the data, however, fails to distinguish clearly the remaining eight sherds into the two groups recognised by the other analytical techniques, although it is interesting that on reassessing the data for individual elements it would appear that the presence of detectable Be could prove to be a suitable criterion. This illustrates both the possible value of "tracer" elements as distinct from overall composition, and also the advantages of feedback that occur in a multi-analytical approach.

Referring again to figure 2a it is of interest to note that the two separate sets of samples [Anglesey and Bryn y Gefeiliau] overlap [particularly (e)]. This illustrates the danger inherent in such classifications where reference to petrographic or mineralogical information on the geological background is lacking. Finally, it should be noted that amongst the elements with more variable concentrations were Pb, Cu, Zn, which raises the possibility of contemporaneous contamination (the site is within an area of Zn/Pb mining). However, these elements do not figure prominently in the grouping obtained by Principal Component Analysis, nor did their exclusion change it significantly.

5.4 Summary

The study of pottery from Bryn y Gefeiliau illustrates clearly the advantages of a multi-analytical approach in dealing with a sample of such petrological heterogeneity. Each technique produced its own independent classification, and these are in good general agreement: no one technique provided the complete subdivision, however, and this was only achieved by the combination of their results.

As might be expected, the geochemical study was of distinct value in classifying the sherds but shed little light on the provenance of the materials involved. Information on the potential sources of the latter emerged from the complementary petrographic and mineralogical analysis, especially when related to the more comprehensive parallel investigation of Roman coarse wares that has recently been made (Peacock, 1973).

However, the most interesting problem relating to provenance was that posed by the presence of a grog fragment in one of the thin sections of the black burnished paste [sherd (d)]. This grog fragment contained a variety of rock and mineral grains (slate, rhyolite and clinozoisite) which reflect closely the geology of the Snowdonian region. Its presence in a pottery paste whose heavy mineralogy matches that of Tertiary beds in Dorset is intriguing: of the various possible explanations one might be that the tourmaline-rich clay of the Poole region of Dorset was transported along the Roman supply lines into Snowdonia where it was subsequently transformed into a black burnished I pot, but picking up in the process a fragment of the local pot filler. The implication that there was transport of unworked clay would gain further credibility if the rounded Triassic sand grains observed in association with Snowdonian clastic material in sherds (b) and (c) were to establish an extraneous source for these clays also.

6.   Hellenistic glazed wares from Apulia, southern Italy

The archaeological and geological contexts of this study are very different from those of the two preceding investigations. The study involved the analysis of Hellenistic glazed wares from six sites mainly centred on the Bradano valley in Apulia (Prag *et al.*, 1974). The pottery is the product of a wheel-thrown, mass-production economy using very fine textured clays to produce thin-walled, well-fired wares. Geologically, the region is a depression uniformly infilled with calcareous marine and fluviatile clays, sands and conglomerates, mostly of Pliocence age, and resting upon limestone. Thus both the solid and superficial geology are particularly restricted and repetitive, conditions that are not conducive to the successful application of petrographic techniques.

The aim of the investigation was to establish composition patterns for the Apulian glazed wares, using as a control group a sample of Attic glazed wares from the Athenian Agora. By this means it was hoped to recognise trading patterns within Apulia, and, in particular, between the colonial cities on the coast and the native settlements in the hinterland. The study was initially conceived as one for spectrographic analysis, in view of the fine texture of the pottery and the geological homogeneity of the region: it was undertaken in the Oxford Research Laboratory by Dr. F. Schweitzer. However, it was considered that petrographic and heavy mineral techniques might also be used to supplement the trace element data in an attempt to define more precisely the provenance of the pottery.

6.1 Geochemical studies

The concentrations of nine elements (Al, Mg, Fe, Ti, Mn, Cr, Ca, Na and Ni) were estimated by arc spectrographic analysis and the results treated first by simple statistical tests to produce composition patterns and secondly by computer classification using an Average-Link-Cluster-Analysis program. The composition patterns established that the Attic control pottery could be distinguished from the Southern Italian wares on their Ni, Cr and Ca content. It also showed how the composition patterns of the south Italian wares overlapped for all the elements, although certain variations of composition suggested that they were not all homogeneous.

For the A.L.C.A. program only four elements were used (Na, Ca, Mn, and Mg), for which high analytical precision had been obtained. The results showed that the composition of the Athenian control pottery was not homogeneous but that the three resultant clusters did not overlap with any of those established for the south Italian wares. The analysis was not able to discriminate between the glazed wares from the different Apulian sites, although it did establish the separate existence of a particular type of ware known as "Gnathian". This ware formed a separate cluster from those established for the remaining Apulian glazed wares; for the latter no correlation could be obtained between pottery from individual sites and specific clusters. The results of this analysis produced a classification system based on geochemical data which, although characteristic of sherds from a particular geographical locality, does not indicate or suggest the likely provenance of the ceramic raw materials.

6.2 Petrographic studies

The petrographic study suffered from the limitations imposed by the geological environment. In thin section all the pastes were characterised by very uniform mineral suites composed of detrital quartz, feldspar, mica, calcite and containing a few foraminiferal tests. Indicator rock fragments were invariably absent, but the assemblages did represent the products of a highly calcareous marine environment, similar to that which existed in the Bradano valley, but equally reproduced in many other marine sediments along the shores of the Mediterranean. The limitations of petrographic analysis under these conditions are best exemplified by the fact that it was not able to differentiate the Athenian from the Apulian pottery. On the positive side, however, the analysis did reveal how calcite of different origins (shell, limestone, foraminiferal, calcareous mudstone) and in varying quantities was present in the pottery and, most important of all, that some calcite could have been deposited from solution within the pottery fabric after burial in the soil. Such information is directly relevant to the spectrographic study and emphasises the care that should be taken in evaluating calcium results, particularly since heavy reliance has been placed on this element in some investigations (Hennessey and Millett, 1963).

6.3 Heavy mineral studies

A purely exploratory heavy mineral study was also undertaken in order to discover whether this technique could establish further subdivisions within the Apulian pottery. The results showed that a mineral assemblage, characterised by zircon, rutile, amphibole, pyroxene, apatite, garnet, etc., was present in all the pottery; such a suite might be consistent with that of an area like the Bradano. Unfortunately, comparisons with assemblages from the Attic control pottery could not be undertaken within the framework of the project, neither was it sufficiently comprehensive to allow interpretation of observed variations in the detailed composition of assemblages from individual sites (e.g. garnet, sphene, epidote, hypersthene, etc.) In general, however, the study did indicate that heavy mineralogy could be employed more extensively to provide an appropriate kernel of geological data in spectographic projects where the aim is to obtain information on the provenance of the ceramic raw materials (Millett and Catling 1967). In the Mediterranean area, however, such projects might suffer from the lack of pre-existing detailed heavy mineral studies on archaeologically important sedimentary deposits.

Two other technical difficulties might also become manifest. Firstly, hard, well-fired pottery, like the Apulian glazed wares, are very difficult to disaggregate; secondly, the texture of the paste body governs the availability of the heavy mineral fraction. Thus, in very fine textured pastes, such as in Mycenaean or the Hellenistic wares described here, only minerals in the 60μ-20μ size range will be present in amounts sufficiently large to allow quantitative studies; moreover problems of identification increase in this small size range, as was illustrated in Table 4.

6.4 Summary

This study illustrates the contrasting effectiveness of the three analytical techniques in a region of extreme geological uniformity. Trace element analysis emerges as the most relevant technique, the statistical clustering of its data producing a classification of the pottery into both chemically and archaeologically viable groups. The petrological observations are relevant, but totally inconclusive concerning the provenance of the ceramic raw materials since the technique is operating at the extreme limits of its applicability: neither do the results allow for even a rudimentary classification of the pottery into groups. Of the three techniques applied, the heavy mineral analysis produces the most relevant geological information, although it is difficult to establish, in view of the limited scope of the study, how it should be best interpreted. Nevertheless, its potential contribution to ceramic projects involving the study of fine textured pottery from undiagnostic geological terrains should be seriously considered.

Conclusion

In this paper at attempt has been made to evaluate the various advantages and limitations inherent in the application of selected analytical techniques to ceramic problems. The dependency of each technique on the geological context has also been illustrated by reference to three separate studies. As in the analysis of sediments, the results of petrographic, mineralogical and trace element analysis are interrelated geochemically, and this provides a theoretical basis for understanding and interpreting the data, with the important proviso that in ceramic studies the sediments have become artificial mixtures through human intervention. It is, however, this underlying geochemical unity within the ceramic raw materials that makes a multi-analytical approach a feasible and worthwhile proposition. In theory each ceramic problem should be tackled on a broad analytical front though in practice the choice of technique reflects more often the availability of equipment and expertise, and the amount of sample that can be spared as balanced against the destructiveness of the technique and the sampling error than can be tolerated. In many instances the choice also reflects the lack of familiarity on the part of the archaeologist with the potential or limitations of the various techniques in relation to his own particular problem.

Of the three techniques considered here, petrography is the most valuable in the initial formulation of an analytical problem, though not of necessity the best adapted for the development of that project in depth. Petrographic analysis of thin sections provides a visual characterisation of the paste body in terms of its overall composition as well as its individual constituents, whether they be clay, rock fragments, mineral grains, grog, or organic debris, etc. This information is fundamental to the subsequent interpretation of heavy mineral and trace element data in terms of the geographical provenance of the raw materials as opposed to a provenance determined by the geographical distribution of the finished pots. It is this latter provenance which has been stressed in most spectographic projects to date, and it is only recently that some aspects of the former have attracted attention (Millet and Catling, 1967).

Petrographic data thus contribute directly to the environmental aspects of provenance, although their sensitivity is very dependent on the nature of the geological setting: this has been illustrated by the contrasting studies in Anglesey and Apulia. More dramatically, the delicacy of the technique is illustrated in the identification of the single grog fragment in the sherd of black burnished I ware from Snowdonia. Upon this chance observation hinges the possibility that there was transport of clay, as well as of pots, with all the archaeological implications that this would have; it re-emphasises the significance both of the sampling error and of the human factor when interpreting petrographic analysis of sherds.

The contribution of heavy mineral and trace elements analysis to the solution of ceramic problems also varies, again depending largely on the geological background. However, both techniques have distinct advantages over petrography in certain situations − the former in extracting relevant information from petrographically undiagnostic pastes, and the latter in classifying and providing a direct chemical comparison of pottery pastes over wide areas, particularly for fine textured paste compositions.

In general, there is no one technique suited to each individual ceramic project. Where a multi-analytical approach can be employed, even if only on a small scale or for the elucidation of a specific aspect, the results are generally complementary and a valuable feedback of information may result. Exploratory studies of the type described in this paper, whether developed empirically or conceived strategically, can initiate a productive line of research, this being neatly expressed in the Welsh aphorism 'gorau cam, cam cyntaf'.

Acknowledgments

We wish to thank Professor W. Charles Evans and his staff, particularly Mrs. E. Cardwell, for facilities and help in carrying out analytical studies in his Department. Dr. F. Schweitzer has generously allowed us to quote the findings of his spectrographic study prior to publication. Professor R. Hodson kindly gave advice and permission to quote his A.L.C.A. analysis, and we are grateful to Miss F. Lynch and Mr. R.G. Livens for their general advice and encouragement. Financial assistance from the Board of Celtic Studies for the Anglesey and Snowdonian projects is gratefully acknowledged, and we thank Dr. H.N. Savory and Professor A.H. Dodd for permission to analyse pottery from their collections in the National Museum of Wales and the Bangor Museum respectively.

REFERENCES

Asaro, F., M. Dothan and I. Perlman (1971), 'An introductory study of Mycenaean IIICI ware from Tell Ashdod', *Archaeometry* 3, 169-76.

Brewer, R. (1964) *Fabric and Mineral Analysis of Soils*. New York.

Catling, H.W. and A. Millett (1965) 'A study of the inscribed Stirrup Jars from Thebes', *Archaeometry* 8, 3-85.

Cornwall, I.C. (1958) *Soils for the Archaeologist*. London.

Farrar, R.A.H. (1973) 'The techniques and sources of Romano-British black-burnished ware,' *in* A. Detsicas, ed. *Current Research in Romano-British coarse wares*. C.B.A. Research Report 10, 67-103.

Gillam, J.P. (1970) *Types of Roman Coarse Pottery Vessels in Northern Britain*. Newcastle-upon-Tyne.

Greenly, E. (1919) *The Geology of Anglesey*. Mem. Geol. Survey. H.M.S.O. 2 vols. London.

Griffiths, J.C. (1967) *Scientific Method in Analysis of Sediments*. New York.

Hall, J.P. (1923) *Caer Llugwy; excavations at the Roman fort between Capel Curig and Betws y Coed*. Manchester.

Hennessy, J.B. and A. Millett (1963) 'Spectographic analysis of the foreign pottery from the Royal Tombs of Abydos and Early Bronze Age pottery of Palestine', *Archaeometry* 6, 10-17.

Hodson, F.R. (1970) 'Cluster analysis and archaeology: some new developments and applications', *World Archaeol.* 1, 299-320.

Howells, M.F., B.E. Leveridge and C.D.R. Evans (1973) *Ordovician Ash Flow Tuffs in Eastern Snowdonia*. Inst. Geol. Sci. Rept. 73/3.

Jenkins, D.A. (1964) 'Trace element studies on some Snowdonian rocks, their minerals and related soils', unpublished Ph.D. thesis, University of Wales.

Jenkins, D.A. (1968) 'The trace element content of soils, with special reference to Snowdonia', *Welsh Soils Disc. Grp. Rept.* 9, 6-16.

Lynch, F. (1970) *Prehistoric Anglesey*. Studies in Anglesey History, 3, The Anglesey Antiquarian Society, Langefni.

Millett, A. and H.W. Catling (1967) 'Composition patterns of Minoan and Mycenaean pottery: survey and prospects', *Archaeometry* 10, 70-7.

Mitchell, R.L. (1964) *The Spectrographic Analysis of Soils, Plants and Related Materials*. Comm. Bur. Soils, Tech. Commun. 44A Harpenden.

Peacock, D.P.S. (1967) 'The heavy mineral analysis of pottery: a preliminary report', *Archaeometry* 10, 97-100.

Peacock, D.P.S. (1970) 'The scientific analysis of ancient ceramics: a review', *World Archaeol.* 1, 37589.

Peacock, D.P.S. (1973) 'The black-burnished pottery industry in Dorset'. *in* A. Detsicas, ed. *Current Research in Romano-British coarse pottery*. C.B.A. Research Report, 10, 63-5.

Peacock, D.P.S. and C. Thomas (1967) 'Class E imported Post Roman pottery: a suggested origin', *Cornish Archaeology* 6.

Prag, A:J.N.W., F. Schweitzer, L.R. Llewellyn and J. Ll.W. Williams (1974) 'Hellenistic glazed ware from Athens and Southern Italy: analytical techniques and implications', *Archaeometry* 16, 2, 153-87.

Shepard, A.O. (1965) *Ceramics for the Archaeologist*. Carnegie Inst. Washington. Pub. 609.

Smithson, F. (1953) 'The micro-mineralogy of North Wales soils', *J. Soil Sci.* 4, 194-210.

Williams, H. (1922) 'The igneous rocks of the Capel Curig district (North Wales)', *Proc. Liverpool Geol. Soc.* 13, 166-202.

Williams, J.Ll.W. (1975) 'A petrological examination of the prehistoric pottery from the excavations in the Castello and Diana Plain of Lipari – an interim report', *in* L. Bernabo Brea and M. Cavalier, eds., *Meligunis Lipara 4*. Palermo, in press.

Williams, J.Ll.W., D.A. Jenkins and R.G. Livens (1974) 'An analytical study of the composition of Roman coarse wares from the fort of Bryn y Gefeiliau (Caer Llugwy) in Snowdonia', *J. Archaeol. Sci.* 1, 47-67.

DISCUSSION

Dr. D. Peacock expressed himself as worried by the possibility that ion exchange could take place between pottery and soil during burial. Dr. D. Jenkins and Dr. J. Williams considered that this might well be a problem in specific contexts with unstable mineral components. However, in general the proportion of trace elements added by ion exchange was small in comparison with the elements already there, except in calcareous materials with strontium, barium and manganese. However, these elements could be recognised during analysis, and the need for a corroboration of geochemical results by petrography and heavy mineral analysis was emphasised.

The origins of some particular 'grog' fragments in black burnished ware found in Snowdonia was then discussed, and Dr. D. Peacock suggested that since the sediments composing this type of fabric had been transported from Cornwall, the grog fragments might originate in Cornish rather than in Snowdonian rocks. The discussion was concluded by some general remarks made by the speakers and Mr. G. de G. Sieveking concerning the origin of the potting clays on Anglesey and the reasons for distinguishing them.

H. Laville

Deposits in calcareous rock shelters: analytical methods and climatic interpretation

A rock shelter is generally characterised by a shallow cavity developed to a greater or lesser depth, surmounted by a convex rock face and a vertical overhang forming its roof. The deposits accumulated in the interior of the rock shelter constitute its fill, which is formed by the superimposition of varied strata whose nature, texture and structure are linked to the numerous climatic variations which succeeded each other during the corresponding periods. This means that the sedimentological study of these deposits particularly favours the reconstruction of the palaeoclimate of prehistoric man at various stages of his evolution.

Numerous analytical methods have been proposed. This paper will explain and justify those used in the study of the fill of calcareous rock shelters in Périgord.

A.   The fill of rock shelters: formation and evolution

The sedimentological characteristics of each deposit are the result of two fundamental stages.

(a)   The sedimentation, which (besides frost shattering whose role remains predominant) is achieved by other depositional agents such as run-off, solifluxion and man.

(b)   The physical and chemical alterations which have affected the deposits during and subsequent to their deposition, including secondary thermoclastism, cryoturbation, erosion, chemical alteration and the presence of man.

Before starting an exposition of the analytical methods which are used systematically in the sedimentological study of the fill of rock shelters, the different factors which may have affected the formation of the deposits and their evolution will be briefly dealt with. Critical review of these different factors has conditioned the choice of analytical methods, as well as the interpretation of the results.

1.   The deposition of the sediments

1.1 Thermoclastism

A volume of water tends to expand during the process of freezing, the act of freezing of the water completely filling up the cavities contained in the interior of the rock, causing its fragmentation. According to Malaurie (1968) the expansion of the occluded water can produce, when frozen, forces of pressure which may attain several hundred kilograms per square cm.

It is quite clear, therefore, that in a periglacial environment thermoclastism is considered as one of the most important natural phenomena, and that it is reponsible, more than any other, for the formation of rock shelters, and consequently for their fill.

Influence of moisture. All authorities are agreed in allowing that the role of water action is predominant in the fragmentation of calcareous rocks. Laboratory experiments have shown that fragments of the same rock, subjected to the same temperature variations and to freeze-thaws of the same intensity, during the same period and to the same degree have reacted differently according to the mode of water action on the samples (Tricart, 1953; Guillien and Lautridou, 1970). Experiments on freezing cubes of calcareous rocks carried out in the Centre de Géomorphologie of the C.N.R.S. at Caen have shown in particular that for a daily rhythm of freezing, with the same temperature variations (-5°C – +15°C), the total or partial immersion of the samples conditions the speed of fragmentation of the rock (Guillien and Lautridou, 1970).

The role of climate. Beside moisture, the rhythm, intensity and the duration of freezing are also the factors whose variations define the climates of different types, conditioning the degree of thermoclastism and therefore the development of the deposits resulting from it. Generally speaking the experiments of Tricart (1965) and Malaurie (1968) have established that under different climatic conditions the same rock may give different kinds of products of varied fragmentation. Thus in experiments involving mild freezing there is above all an abundance of fine debris, whereas under more severe freezing coarse debris is produced.

The respective roles of macro- and microthermoclastism are a function of the climate. Macrothermoclastism can only occur when the ice can penetrate deep enough to freeze the water contained in the crevices and diaclases. On the other hand, microthermoclastism can happen when frequent mild but regular and repeated freeze-thaw action occurs. Different observations have led Tricart to distinguish two major types of climatic variations:

(i)   The daily cycle: very frequent alternation of freeze and thaw, when the ice does not penetrate deeply. The rock, constantly absorbing water, is frequently and regularly subjected to a mild freezing, ending in a complete thaw. These conditions are very favourable for the production of fine debris, fragments, granules, gravels, sands and silts. This can only result from microthermoclastism (Tricart 1967, p. 257).

(ii) The annual cycle: the number of freeze-thaw oscillations being inversely proportional to the annual variation, the sediments are subjects to a more or less intense freezing, always prolonged and able to exploit the diaclases of the rock by macrothermoclastism. The reduction of the number of annual cycles results in a reduction of the relative abundance of the fine products.

Influence of lithology. The experiments carried out at the Centre de Géomorphologie of the C.N.R.S. at Caen have shown that, in identical experimental conditions (the same moisture, freezing rhythm and temperature amplitude) rocks of different texture do not react in the same way, each category of rock providing, after the same number of cycles, debris of different grain sizes. Each rock type corresponds, in effect, to a certain number of characteristics, including porosity, which conditions the absorbence capacity of the material, therefore its degree of moisture and finally its sensitivity to freezing.

From these different considerations it appears, therefore, that thermoclastism is a complex phenomenon controlled by several essential factors:

1. the moisture of the rock
2. the frequency of freeze-thaw cycles
3. the intensity of freezing
4. the duration of freezing
5. the texture and structure of the rock.

The different textures which characterise the deposits of a stratigraphic sequence are directly linked to the intensity and to the patterns of occurrence of the gelivation phenomena which took part in the formation of the detritus material, with two fundamental stages, weathering and erosion of the walls of the rock shelter followed by secondary fragmentation of the congelifracts.

The variables which constitute the periglacial climate condition first the thermoclastism of a rocky mass of which the volume that is available for freezing can be considered as a constant (both on the walls and roof of the rock shelter); it is there a process of ablation which, after Hamelin and Cook (1967), we shall call *gélivation initiale* (primary thermoclastism). In addition we propose the term *gélifraction de reprise* (secondary frost shattering) which tends progressively to modify the grain size and morphological characteristics of the primary thermoclastic product, during and after its formation.

1.2 Solifluxion

The process of solifluxion consists of a slow translocation, on gentle slopes, of fine grained saturated material. It presupposes a surface thaw in the soil, and a certain rise in temperature, seasonal or episodic. Considered generally as a manifestation of very cold climate, solifluxion appears to have been particularly active during periods which it is convenient to call 'episodes of climatic instability'. In Périgord these effects are never observed in deposits built up during a period of intense cold, and they seem to have occurred either at the beginning or at the end of glacial phases. Solifluxion is often manifested in the fill of rock shelters by the sedimentation of allochthonous materials. The accumulation of these deposits is accompanied by characteristic grain size and morphological features of the sediment.

1.3 Run-off

In the general way run-off is not considered as an agent of deposition characteristic of cold climates. Inactive on the plateaux and the slopes during cold periods, it also remains generally insignificant during the thaw, the gaps created by the increase in the volume of mobile rock during the period of freezing mobilising the greater part of meltwater and seasonal rain. This is to say, confirmed by these researches, that the presence of material deposited by run-off in the fill of rock shelters may be attributable to an episode of climatic amelioration. This form of accumulation could have occurred in three different ways:

1. By the washing of sediments and the leaching through of the finer elements.
2. By the sedimentation of originally allochthonous material, as is the case with material redeposited from ancient fluviatile terraces and the residue of weathered outcrops.
3. By the sedimentation of material redeposited from the fill of ancient caves related to the rock shelters.

Evidence of run-off is detectable by granulometric, morphological and mineralogical analyses.

1.4 Fluviatile sedimentation

This is a form of sedimentation which may be encountered in rock shelters situated at low levels. The products deposited may either be derived from true alluvial terraces or be the deposits of seasonal inundations.

1.5 Man

The occupation of rock shelters by man is an agent of accumulation which must not be ignored. Flint knapping debris and tools sometimes accumulate in considerable thicknesses. One may add to this the products of occupation or of the reorganisation of the habitation soil by Palaeolithic man (river pebbles used to construct fireplaces, pebbles and limestone slabs arranged as floor tiles) as well as accidental imports. Obviously all sedimentological analysis of the deposits belonging to occupation levels presupposes the preliminary removal of products accumulated by man.

2. *Modification of the deposits*

The second stage in the process which will end in the formation of the fill consists of all the physical and chemical modifications just enumerated which have affected the deposits during and subsequent to their deposition.

2.1 Secondary frost shattering of the congelifracts

The fragmentation of the calcareous congelifracts after their deposition theoretically corresponds, better than the primary frost shattering of the rock outcrop, to the fundamental laws of thermoclastism as evidenced by cryoclastic experiments. It would be tempting to relate the experimental results directly to the identification of secondary frost shattering of calcareous screes. However, in nature, many factors intervene which combine with exterior climatic conditions to regulate the speed of accumulation, redistribution, and morphology of the debris. The secondary fragmentation of congelifracts will be conditioned by the intensity of the frost shattering phenomena which will have produced them, as well as by their own volume and the sedimentary context in which they are found.

Two borderline cases may be imagined.

1. A cold period, characterised by an intense and prolonged annual cycle. The result will be a frost shattering of the roof of the rock shelter into large fragments; there will be concurrent sedimentation of these congelifracts, which will result in a protection more rapid and efficient with the intensity of the freezing. Secondary thermoclastism of the congelifracts will be otherwise limited by a reduction of their time in water, after the immediate evacuation of the meltwater from the interstices of the rock, owing to the fact that the rising by capillary action of the water which impregnates the underlying deposits will be hindered by the presence of these interstitial voids.

2. The case of minor ablation may now be considered relating to the same rock outcrop, and therefore on a material of the same texture and structure as the preceeding one. The ablation is limited by moderate daily freezing. There will be fragmentation of small pieces and a crumbling of the rocky surface. The sedimentation is not very active and the congelifracts will be more directly exposed to exterior forces, in particular frost. There will be found a high proportion of sands, silts and clays, the presence of which will help in maintaining the dampness of the calcareous remains. These are favourable conditions for a new grain size distribution of the debris. The experiments on frost shattering of samples buried in humid loess,

described by Tricart (1967, p. 124) are particularly illuminating on this point. Morphological examination of thermoclastic scree also confirms this interpretation.

It is possible to distinguish among calcareous pebbles two kinds of fragmentation which correspond to the two extremes described above:

1. *frost shattered pebbles*, characterised by a total fragmentation into two or three large bits. This effect is common among large congelifracts. This sort of fragmentation presupposes an intense cold, but is rapidly brought under control by the very characteristics of the frost which have favoured it.

2. *fissured pebbles*, characterised by a superficial network of fine cracks. This form has been identified in fine grained sediments, with a matrix chiefly composed of sandy clay. The collection of sedimentological characteristics of the deposits with which these are associated leads us to an interpretation of the product of the occurrence of moderate freeze-thaw, favoured by the presence of an abundant fine-grained matrix.

The evaluation of the state of calcareous scree in a fill in one of these categories, associated with granulometric data, permits, in certain cases, an appreciation of the relative intensity of gelivation phenomena.

2.2 Solifluxion and cryoturbation
Solifluxion has often led, in rock shelters, to the accumulation of sediment of allochthonous origin. It is also manifested by the resorting of previous deposits, in which case it is often accompanied by cryoturbation, and it is usually impossible to differentiate between the two phenomena. However, solifluxion and cryoturbation are often shown by the weathering and, in certain cases, by the total removal of the deposits.

Besides a particular arrangement of the deposits, detectable by stratigraphic observation, the two phenomena are identifiable in the laboratory by a characteristic distribution of the fine sediments, as well by the modelled and polished nature of the congelifracts. In archaeological levels the two phenomena have led to a fragmentation of lithic material: worked flint, totally defaced, often presenting a blunted appearance.

2.3 Erosion
All the phenomena which contribute to the removal of all or part of the fill are classified under this heading. The erosion of a deposit may have three origins.

1. Erosion by solifluxion. This is an aspect which has already been dealt with, and to which is attributable the absence of the superficial horizons of interglacial or interstadial palaeosols.

2. Erosion by the action of circulating water down a fissure. These effects are limited.

3. Erosion by the arrival of a mass of water at the mouth of a channel developed in the limestone, in relation to the rock shelter. This action has occasionally been considerable and explains, in certain cases, some vast stratigraphic lacunae.

2.4 Chemical alteration

This is essentially associated with moisture and temperature. In warm or temperate periods the combined effects of warmth and humidity leads to an intense biological activity, vegetable as much as microbial, adding to the power of the water and thus favouring increased chemical reactions leading to true diagenesis in the sediments. The chemical alteration of sediments, identified by the analytical methods described here, seems to characterise periods of increasing temperature, which are always accompanied by an increase in humidity. Evidence of solution under rigorous climatic conditions is seldom found. In any case of superficial modifications or elaborations of complex pedological profiles, the greater part of the weathering phenomena identified coincide with periods of less frost action, and even with pauses in sedimetation.

In calcareous material the most immediately perceptible effect of chemical alteration is the solution of the carbonate components of the rock. These solution processes, which are favoured by an increase in the acidity of the water in warm or temperate periods due to the presence of humic or organic acids, are demonstrated by:

(i) the solution of carbonates
(ii) the liberation of insoluble elements in the rock
(iii) the migration and, above all, the illuviation of the soluble products and colloidal material from the rock
(iv) the precipitation of carbonates.

Many granulometric, morphological and mineralogical criteria are demonstrated which permit the identification and evaluation of these different phenomena.

2.5 Human presence

Apart from taking a role in sedimentation, the occupation of rock shelters by man has had profound repercussions on the evolution of the sediments. One must take into account all the changes attributable to the building of dwellings, and those which disturb the original composition of the deposits (for example hollowing out of ditches or hearths or reorganisation of the soil). There is also the question of the secondary fragmentation of material by the action of fire. The accumulation of organic material of animal or vegetable origin contributes, in certain cases, to the precipitation of carbonates into solution in particular horizons.

B.   Methods of study and analytical techniques

The points just described with relation to the fill of rock shelters illustrate the diversity and complexity of the phenomena set in motion in the process of accumulation of the deposits, in addition to the sedimentary characteristics linked to each phenomenon. The climatic intepretation of rock shelter sediments makes it necessary to determine, first, what has been the respective role of sedimentation and diagenesis in the sedimentary facies of each deposit. Next one must establish the different climatic factors which have occurred during both these stages, and finally one should consider the relative intensity of the different climatic components responsible for the build-up of the deposit and its evolution.

The precise details which are required during research into minor climatic fluctuations, and in the understanding of the chain of climatic events, have led to the refinement and completion of the methods suggested by many authors, particularly by Bonifay (1955, 1956). They are not fundamentally different from those used by Miskovsky (1970) in the study of caves in the south of France.

1.   Taking samples

After the stratigraphic and descriptive study of the deposits is completed, samples for sedimentological analysis are taken from several sections of the layer at every possible opportunity. Since the sample should be representative of the level it characterises, its volume should be determined by the texture of the sediment. Therefore, for layers composed of scree and whose elements are to be the object of morphological analysis, samples of 20-30 kg are taken.

2.   Global granulometry

For methodological reasons only particles of a diameter <10 mm are studied in the laboratory, the proportion and characteristics of larger fragments and collapsed blocks being evaluated qualitatively in the field. Global granulometry consists of a division of the sample into three grain size fractions:

1. The coarsest fraction, that which contains the pebbles or congelifracts: 10-100 mm.
2. The intermediate fraction corresponding to gravels and granules: 2-10 mm.
3. The fine fraction of sands, silts and colloids, less than 2 mm.

A graph is drawn with the different layers and horizons on the ordinate and the cumulative weight percentages (for the three fractions considered) on the abscissa. This permits a study of the granulometric evolution of the sediments in the sequence being studied. This analysis gives a first approximation of the global composition of the sediment and its evolution in the stratigraphic

A

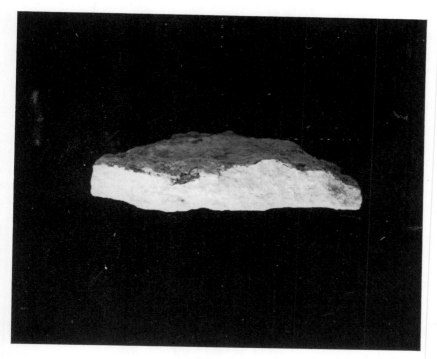

B

Figure 1. Frost slab (photo Ian Johnson).

series. It permits, in addition, a preliminary identification of the phenomena active during the course of the filling.

The respective roles of the different agents of accumulation and weathering should next be precisely defined, as well as the relative intensity of the different climatic phenomena which occurred in the process of filling. The presence in the sediment of a large quantity of congelifracts can be attributed to a phase of frost shattering (thermoclastism), but it is necessary to determine its degree and intensity. A large proportion of fine sediment may be associated with a lesser contemporary gelivation phase, and with sedimentation of a different type whose origin and mode of deposition must be precisely determined (solifluxion or run-off). This may equally well be the result of chemical alteration of the products of thermoclastism. Accurate details may only be obtained by appropriate analyses carried out on all three fractions isolated by global granulometry.

3.   Study of the pebbles (10-100 mm)

The principal object of the study of the coarsest fraction of the sediments is to indicate the relative intensity and the degrees of thermoclastism and weathering which have succeeded each other during the course of filling. Since both are strictly dependent on the texture and structure of the rock it is convenient to classify congelifracts of different lithological characteristics. Each analysis is carried out separately on each category of pebbles thus isolated. This procedure must be adopted wherever the textural differences are sufficiently marked to be appreciated with the naked eye.

3.1 Research on evidence of thermoclastism

(a) *Grain size analysis.* This is carried out on the whole of the fraction between 10-100 mm. The pebbles are divided into nine grain classes according to whether their greatest size is between 10-20 mm, 20-30 mm, . . . 90-100 mm etc. A diagram indicates the variations in weight percentage of the different categories, placed on the abscissa, against the layers of fill, placed on the ordinate. This analysis contributes towards an appreciation of the modalities of thermoclastism.

(b) *Research on frost slabs.* Under certain conditions of temperature, moisture, exposure and lithology, thermoclastism of the rock is conducive to the formation of frost slabs. These frost slabs, as they appear before any disturbance, present a characteristic appearance with angular outlines, one face (which was inside the rock before detatchment) being fresher than the other. The pebbles corresponding to this definition are extremely rare in the fill of rock shelters, as a result of the physical and chemical changes that they have undergone after their formation. It is for this reason that the less angular examples, occasionally blunted but identifiable by flattening, are included in this category (Fig. 1). Each pebble whose thickness is less than a quarter of its length, without taking account of its

A

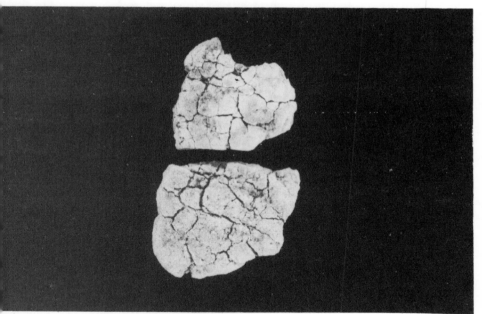

B

Figure 2. A Thermoclast (photo Ian Johnson).
B Fissured pebble (photo J.Ph. Rigaud).

appearance, is therefore classed under the heading of frost slabs. The calculation of the percentage of frost slabs is carried out on the best represented size class in the assemblage of fill being studied, generally those elements between 30-60 mm in diameter.

(c) *Research on frost shattered pebbles.* The action of intense cold on calcareous thermoclasts can lead, under certain conditions described above, to a new fragmentation. The products of this fragmentation are frost shattered pebbles. It is a result of the total shattering, generally of the largest elements, into 2 or 3 large fragments. These fragments only remain conjoined if the deposit to which they belong has not been subsequently disturbed. They generally appear slightly disturbed in relation to each other and are sometimes made to adhere by secondary cementation of diagenetic calcite (Fig. 2A). Polygonal rock fragments with one face with sharper *arêtes* still having the appearance of a fracture are also classed in this category, that is to say fragments of frost shattered stones dispersed by later disturbance or separated at the time of sampling. The percentage of those elements is calculated with reference to the total weight in the classes chosen. Frost shattered pebbles must not be confused with pebbles that have been fragmented by the action of fire. The latter, frequent in archaeological levels, are characterised by a less defined fracture whose edges are blunt. The material is, however, extremely friable and often reddish in colour.

(d) *Research on fissured pebbles.* Secondary gelivation of calcareous rock fragments has led, in minor periods of alternately repeated mild freeze and thaw, to the formation of fissured pebbles. These are generally of small size, often blunted, with the whole surface permeated by a network of fine cracks of different sizes. They present the appearance of a mosaic of angular fragments, separated by fissures 1-3 mm in depth (Fig. 2B). The weight percentage is calculated with reference to the total weight of the chosen grain size class.

3.2 Research on weathering phenomena

The most obvious weathering phenomena, as has already been illustrated, are those which are linked to solution processes. The combined action of moisture and temperature leads to the solution of the carbonate component of calcareous materials, which is shown in a grain size analysis of the sediment. On thermoclastic fragments the phenomenon is at first shown by a morphological modification.

The solution of carbonates causes the liberation of detrital constituents of the rock which is followed by a smoothing of the cryoclastic fragments and the progressive degrading and wearing down of surface irregularities. This superficial smoothing will be more accentuated in proportion with increased intensity of the solution phenomena, or as their action becomes more prolonged. As well as surface smoothing, the solution of carbonates happens equally in the interior of the rock, by the progressive enlargement of pores of the material and by the development of *méats* whose development is

A

B

Figure 3. A Pebble blunted by chemical alteration.
B Pebble faceted by cryoturbation.
(photos Ian Johnson)

proportional to the intensity of the solution phenomenon. The final stage of this process is the formation of 'phantom' pebbles, totally decalcified and consisting only of insoluble residues; these forms are particularly common in weathered soils.

The evaluation of the degree of blunting of congelifracts, the estimation of their corrosion and their friability, and of their degree of porosity, as well as the evidence of concretion processes which affect the sediment, permit conclusions to be drawn from a study of the weathering processes which occurred during the build-up of the deposit. It also permits measurement of their relative intensity.

(*a*) *Study of 'blunting'*. Using the 'bluntness index' created by Bonifay, the pebble is placed in a grain size class, the assemblage studied being represented in four groups:

1. Non-blunted pebbles, very angular and with sharp *arêtes*.
2. Slightly blunted pebbles, still angular but with smoothed *arêtes*.
3. Blunted pebbles, with no angularities but with curved sides.
4. Very blunted pebbles in which the original shape is difficult to distinguish and where the surface is entirely curved.

Each category is weighed, its weight percentage calculated and multiplied by an index corresponding to the degree of blunting:

0 for unblunted pebbles
$\frac{1}{3}$ for slightly blunted pebbles
$\frac{1}{2}$ for blunt pebbles
1 for very blunted pebbles.

The sum of these three values represents the bluntness index.

(*b*) *Study of corrosion*. Following the same principle which was used to calculate the bluntness index, it is possible to define the degree of corrosion of the congelifracts, by classifying the different particles of given grain size class according to the state of their surfaces.

(*c*) *Study of friability*. The pebbles of any representative grain size class are classified into four groups distinguished by increasing friability. The weight percentage of each category is then multiplied by a coefficient increasing according to the degree of friability, taken between 0 and 1. The sum of these values represents the weathering index as defined by Bonifay. Since the estimation of the degree of friability of the material is particularly subjective, and the results obtained are less conclusive than those of the measure of porosity, it is no longer used.

(*d*) *Measuring porosity*. The quantity of water a calcareous pebble can absorb is proportional to the development of the *méats* in the interior, and thus of its porosity. For each sample analysed, a representative amount of pebbles is chosen, which is immersed in water for 24 hours. The weight difference between the dry and saturated samples is calculated in percentages with respect to the original weight of the dry sample.

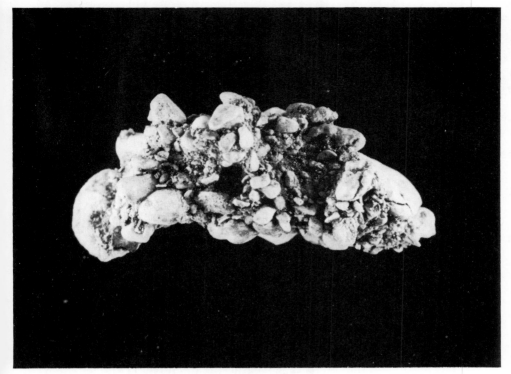

A

B

Figure 4. A Illuvial calcareous concretion (photo J. Ph. Rigaud).
B Slab with a coating of stalactite (photo Ian Johnson).

The results obtained by this analytical method not only supply useful information on the intensity of the solution phenomenon, but their comparison with the values of the blunting index also permits the estimation of the respective role of mechanical abrasion, and in particular of solifluxion or cryoturbation and chemical solution, in the formation of calcareous pebbles (Figs. 3A and B).

(*e*) *Study of the carbonate concretions.* Leaching of carbonates in solution leads to the formation of varied concretions within the sediment. These concretions are identifiable in all grain size classes, including the fraction between 10-100 mm. Many varieties of concretions are distinguished:

1. Aggregates of sand, gravels, granules and pebbles, even flint and faunal debris joined by a more or less indurated calcareous cement (Fig. 4A).
2. Encrustations of calcite on the outside of the particles or a coating of finer sediment with calcitic cement.
3. Developing concretions present on the lower face of calcareous particles and principally identifiable in deposits of large pebbles.

The comparison of results obtained by the calculation of percentages of the different types of concretions, which those of the blunting index and the measurement of porosity, provides useful information on the intensity of weathering phenomena and leaching.

3.3 Stalactites

This heading includes all forms of concretions which develop on the roof and walls of caves and rock shelters, and which begin to grow following the precipitation of calcite from water exuded from the rock. In this category are placed all the typical forms which are habitually encountered in the deepest galleries of caves, including coatings and concretions which develop particularly on the roofs of rock shelters.

The genesis of these concretions presupposes a stabilisation of a walls and an increase in moisture. Their presence in a deposit indicates a phase of increased moisture availability immediately preceding it.

4.   Study of the medium sized fraction (2-10 mm)

The examination of particles between 2-10 mm in size completes the picture obtained from a study of the coarser fraction, and may, in certain cases, make apparent a certain number of characteristics which are not identifiable solely by the use of the above-described procedures (notably evidence of solifluxion, run-off and illuviation). With this aim the elements of grain sizes 2-10 mm are divided into two fractions; granules 5-10 mm in diameter, and gravels 2-5 mm in diameter. Each fraction thus isolated is the object of an investigation for four categories of characteristics:

1. Particles brought by man, flint flakes, bone splinters and fragments of charcoal, whose weight is subtracted from the weight of the sample analysed.
2. Fragments of calcareous rock. Only granules are studied, and these are subjected to a morphological examination, in the course of which the proportion of frost slabs and blunted and fissured pebbles is evaluated.
3. Illuvial calcareous concretions (aggregated in various ways).
4. Elements of allochthonous origin such as quartz gravels, sandy ironstone nodules and so on, forming part of the material brought by run-off or solifluxion.

5.   Study of the fraction finer than 2 mm.

A small number of simple methods is used here, including grain size analysis, calcimetry and pH. These may profitably be completed by X-ray diffraction of the clay minerals and by chemical analysis of the sediments, methods which required more sophisticated equipment.

(a) *Grain size analysis.* After pre-treatment with oxygenated water (H_2O_2) each sample is subjected to two grain size analyses, on the raw and on the decalcified sediments.

Each analysis is divided into two operations: densimetry of the particles finer than 50 μm, by the method of Meriaux (Meriaux, 1957), and sieving of particles between 50 μm-2 mm. The results are expressed as cumulative curves on 4-cycle semi-logarithmic graph paper and comprise 23 points cumulated between 0.002 and 2 mm. This analysis permits the grain size evolution of the sediment finer than 2 mm to be followed, taken globally in each stratigraphic sequence, and conclusions may be drawn concerning the mode of deposition (mechanical disaggregation, run-off, solifluxion) or contingent modifications. For this purpose we rely on the comparison of cumulative curves, the calculation of medians and the sorting index. The comparison of cumulative curves established from grain size analysis of the sediment before and after decalcification permits a consideration of certain minor processes of leaching and illuviation. By this procedure, associated with the examination of different fractions under the binocular microscope, it is possible to distinguish secondary carbonate enrichment, not identifiable by the sole analysis of the coarse and medium grade fractions of the sediment.

Finally, by completing global granulometric diagrams, obtaining and plotting the stratigraphic levels on the ordinate and the cumulative percentage of sands, silts and clays on the abscissa, and after grain size analysis of both the fresh and decalcified sediment, one may follow and summarise the granulometric evolution of each sequence. Gradual variations in the rate of clay weathering (degree of pedogenesis) in the illuvial horizons of palaeosols are more accurately evaluated by these means.

(b) Calcimetry. This is done by decalcifying 100 g of the initial sediment by adding hydrochloric acid, and weighing the insoluble residue. The comparison of the variations in the amount of fine carbonaceous sediments with values of the blunting index, the porosity of the pebbles and the illuvial concretions constitutes the principal means for the interpretation of leaching and weathering phenomena.

(c) Morphoscopy. The quartz grains around 0.7 mm in size are systematically examined under a binocular microscope, and the different categories of wear are counted. All the elements constituting the sandy matrix between 0.5-2 mm in size are also examined. This analysis supplies useful information concerning the origin of the material or its contingent modifications, although the survey is particularly concerned with illuvial calcareous concretions and with particles of allochthonous origin.

(d) Measurement of pH and Δ pH. The pH of each sample is measured in the laboratory using an electronic pH meter. The comparison of the different pH values between two suspensions of the same sample, one in distilled water and the other in a normal solution of potassium chloride (Δ pH, Thibault 1968), permits, in certain cases, the detection of weak weathering phenomena and defines those which may have been suggested by other sedimentary methods.

Unspectacular but effective, these methods offer the advantage of great simplicity combined with ease of interpretation. They had been successfully used for research into the climatic chronology of the Palaeolithic in the classical region of Périgord. The systematic confirmation of the results of the analyses, and of the climatic interpretations deduced from them, by palynological data, supports their validity.

REFERENCES

Bonifay, E. (1955) 'Méthode d'étude du remplissage des grottes', *Bull. Soc. prehist. Franç.* 52, 144-45.

Bonifay, E. (1956) 'Les Sédiments détritiques grossiers dans le remplissage des grottes – méthode d'étude morphologique et statistique', *L'Anthropologie* 60, 447-61.

Guillien, Y. et J.P. Lautridou (1970) '1 – Calcaires des Charentes. Essai au gel de quelques calcaires charentais'; *in* Recherches de gélifraction expérimentale du Centre de Géomorphologie. *Bull. Centre Géomorph. C.N.R.S. Caen,* 5, 7-45.

Laville, H. (1964) 'Recherches sédimentologiques sur la paléo-climatologie du Würmien récent en Périgord', *L'Anthropologie* 68, 1-48 et 219-52.

Laville, H. (1969a) 'Le remplissage des grottes et abris du Sud-Ouest de la France', *in* Etudes françaises sur le Quarternaire, présentées à l'occasion du VIIIe Congrès INQUA, Paris, 1969. *Suppl. Bull. Assoc. franç. Et. Quaternaire* 77-80.

Laville, H. (1969b) 'Paléoclimatologie du Würm ancien en Périgord: données sédimentologiques', *in* Etudes sur le Quaternaire dans le monde. *Union intern. Et. Quaternaire,* VIIe Congrès INQUA, Paris, 1969, 513-16.

Laville, H. (1973a) 'The relative position of Mousterien industries in the climatic chronology of the early Würm in the Périgord', *World Archaeol.* 4, 321-29.

Laville, H. (1973b) 'Climatologie et chronologie du Paléolithique en Périgord: étude sédimentologique de dépôts en grottes et sous abris'. *Thèse de Doctorat ès Sciences*, Université de Bordeaux I, No. 400.

Malaurie, J. (1968) 'Thèmes de recherches géomorphologiques dans le Nord-Quest du Gorenland, *Centre de Tech. Docum. cartogr. Géogr.*, Ed. C.N.R.S.

Meriaux, S. (1957) 'L'analyse granulométrique par densimétrie', *Assoc. franç. Et. sol* 90.

Miskovsky, J.Cl. (1970) 'Stratigraphie et paléoclimatologie du Quaternaire du Midi méditerranéen, d'après l'étude sédimentologique du remplissage des grottes et abris sous roche (Ligurie, Provence, Bas-Languedoc, Roussillon, Catalogne)'. *Thèse de Doctorat ès Sciences*, Université de Paris.

Thibault, Cl. (1968). '*Δ* pH et Paléosols dans le Quaternaire du sud-ouest de la France', *C.R. Acad. Sc. Paris* 266, 2215-18.

Tricart, J. (1953) 'Résultats d'expériences sur la désagrégation de roches sédimentaires par le gel', *C.R. Acad. Sc. Paris*, 1296-98.

Tricart, J. (1967) 'Le modelé des régions périglaciaires', *in* J. Tricart et A. Cailleux 'Traité de Géomorphologie'. *Ed. S.E.D.E.S.*

Section 2

Coastal and Lacustrine Environments

I.A. Morrison

Comparative stratigraphy and radiocarbon chronology of Holocene marine changes on the western seaboard of Europe

For those interested in ancient coastal sites or settlement patterns, the land and sea-level changes of the Holocene are important both in environmental and in chronological terms.

Firstly, a reconstruction of the original relationship between settlements and coastal geography is often a necessary starting point for any adequate interpretation of their economy and form (e.g. Evans and Renfrew, 1968, 5; Morrison, 1968, 1973a, 1973b).

Secondly, changes in ocean level ('eustatic' changes) are essentially synchronous on a worldwide scale, and although changes in the level of land masses vary in nature and amount from place to place, they frequently follow orderly patterns over considerable areas. Events characterised by as widespread geographical distributions as these offer considerable potential for the development of chronological frameworks of value to prehistoric archaeologists and geographers. It is with this chronological aspect that the present paper is concerned.

From time to time, relatively simple chronological models of sea level change have gained currency (the concept of 'the 25-foot raised beach' of northern Britain is an example of this, discussed by Sissons, 1967a and Morrison, 1969, among others). As the complexity of Holocene shoreline changes has become apparent, the passing of these simpler models has been viewed by some with regret. The complexity and, all too frequently, the obvious incompleteness of the shoreline change sequences now being established in many areas is certainly somewhat daunting. Nevertheless, the simpler models offered only a very limited number of dating horizons. What the complex models lose in convenience, they stand to gain in their potential for developing chronological frameworks of considerably higher resolution, due to the multiple horizons that they offer.

That potential is, however, as yet far from being fully realised. At present, it is often by no means clear to the fieldworker whether the submergence or emergence affecting his particular site is likely to form part of a sequence similar to or quite different from events elsewhere. It therefore seems worth attempting firstly to define how far this may be so, and secondly to develop some provisional time-table for any marine transgressions and regressions that appear to be widely recognisable.

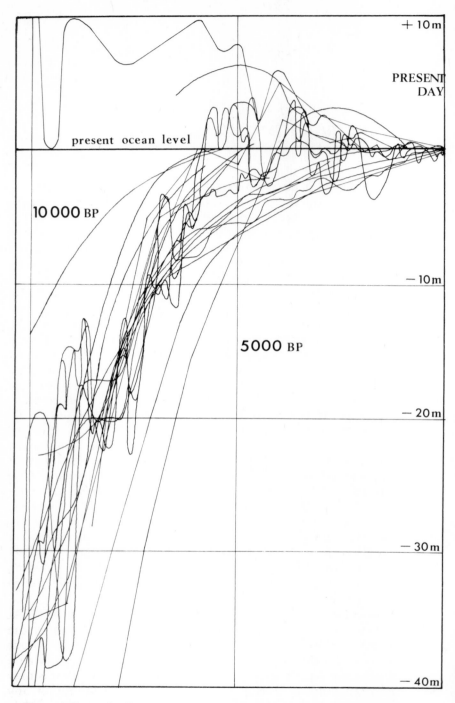

Figure 1. Examples illustrating the range of opinion on Holocene eustatic change.

1. The nature of the problem

If local factors tended to predominate, the value of shoreline changes for developing chronological frameworks for archaeological material would be minimal, whereas the opposite would be the case if ocean level variations ('eustatic' events) tended to synchronise the changes in different areas.

The field evidence surviving in any one area tends to record the changes of relative sea level resulting from the interaction of movements of the ocean's surface with changes in the configuration of the land, but without giving any immediate indication of their individual rôles in the interplay there. The considerable amount of controversy still surrounding the analysis of this interplay is illustrated in Fig. 1, which shows curves of Holocene sea level change published in recent years, superimposed on standardised coordinates. The problem is complicated by the number of factors involved, both in eustatic ocean variations and in local coastal influences.

It is well established that glacio-eustasy (i.e. the abstraction of water from the oceans to form ice sheets and glaciers, and its return on each deglaciation) has been the dominant process in ocean level variation in the Quaternary. There is general agreement (reflected in the graphs of Fig. 1) that ocean level was still low at the start of the Holocene about 10,000 years ago, but rose rapidly as world ice cover decreased in volume. Calculations based on changes in glacier ice volume (cf. Farrand, 1966; Moran and Bryson, 1969; Andrews, 1970) appear to offer a widely acceptable general picture of the overall eustatic trend in the Holocene. However, although dominant, glacio-eustasy is only one of a whole complex of eustatic mechanisms, and as yet insufficient quantitative information is available for the detailed nature of the interplay between oceanic and local factors to be evaluated by this means. Mechanisms producing eustatic changes, by changing the volume of marine water or the volumes of the ocean or sea basins, have been reviewed comprehensively by writers including Charlesworth (1957), Flint (1957) and Fairbridge (1961). More recently, Bloom (1967), Schofield (1967), Matthews (1969) and Walcott (1972) among others, have drawn attention to further complexities on both theoretical and observational grounds.

Faced with these difficulties, many workers have sought to identify areas of crustal stability, which would serve as direct 'measuring marks' against which eustatic changes might be traced. In recent years they have included Auer (1958), Fairbridge (1961), Shepard (1963), Scholl (1964), Milliman and Emery (1968). Opinion, however, appears to be turning against this approach. Newman and Munsart, for example, have gone so far as to state '. . . we suspect that coastal stability is a myth. We doubt that any coast has remained stable during the Pleistocene epoch' (1968, p. 95). Recent geophysical work on hydro-isostasy (response of the earth's crust to water loading) would appear to endorse such doubts by indicating that the continental margins respond to loads of the order involved in Holocene sea level change (e.g. Bloom, 1967; Broeker and Kaufman, 1965; Higgens, 1968; Mörner, 1969). At the present state of knowledge it would therefore seem advisable not to rely on the concept of 'stable areas', but to consider it probable that all available data may involve movements of both land and sea surfaces.

The mechanisms that may give rise to changes in the configuration of the land (through movements of the earth's crust itself, and changes in superficial deposits) are again both numerous and difficult to evaluate in quantitative terms. They were discussed comprehensively in the Symposium on Quaternary Changes in Level (*Geol. en Mijnb.* 1954, N.S. 16, No. 6) and the Symposium on Recent Crustal Movements (*Ann. Acad. Sci. Fenn.* 1966, A. III. 90). In addition, such factors as the creation or erosion of barrier beaches (e.g. Fisk, 1956; Leontyev and Nikiforov, 1965) may give rise to stratigraphy erroneously suggestive of a change in level.

In view of the considerable uncertainties that may thus exist in any one locality, it would seem desirable to approach the problem of assessing the rôle of the eustatic component by comparing the patterns of change recorded at a large number of sites. In such an approach, rather than attempt to select sites of supposedly uniform characteristics, it seems positively advantageous to embrace the complexity of reality, and frame the study in terms of an area characterised by diverse land-movement regimes and varied coastal environments. If the histories of change prove dissimilar across the area, the influence of eustatic variations on the details of coastal change may be assumed to be over-ridden by the local variables. If, however, the tendency is towards marked similarity, then there would appear to be a strong case for believing that the ubiquitous influence of eustatic change in ocean level is responsible.

The use of an approach based on geomorphological measurements would raise considerable problems in a geographically extensive comparative study. It has become clear that intensive, and hence expensive, programmes of measurement tend to be necessary to ensure that shoreline fragments are not miscorrelated (see, for example, Sissons 1967a and b), even when these fragments are made up of similar features. Often this is not the case, and various attempts have been made to codify the characteristic height relationships of different types of shore features to mean sea level (e.g. Hideo Mii, 1962; Fairbridge, 1961). However, because of the complicating influence of variations in exposure, tidal range, terrace composition and gradient, and such factors as the availability of constructional materials, the possibilities for confusion arising while attempting to allow for such factors in the course of a geographically extensive comparison would seem to be great, even if it was economically feasible to obtain adequate measurements of the present heights of the ancient shoreline features.

An approach based primarily on direct investigation of the timing of changes in marine influence would seem to offer a way of circumventing many of these problems. Since eustatic ocean level variations characteristically occur simultaneously everywhere, if a high degree of synchronism in coastal changes was present overall, across a diverse range of local conditions, this would suggest strong eustatic control. On the other hand, if this was not so and local factors predominated in determining the details of changes in marine influence, any geographical patterns in the timing of changes might help to identify which of the non-eustatic factors tended to be most important in different types of area.

2. Procedure

For such an approach to be viable, an adequate method of dating is required. Pollen or diatom analyses, and typological studies in archaeology, share the disadvantage of dealing with material often characterised by broad time bands in any one area, and by geographical distributions with boundaries that vary in date from place to place. Radiocarbon dating at present appears to offer the most effective system of relative dating generally available for Holocene material that can not be dated historically.

Since, for present purposes, radiocarbon is required only as a system of relative dating, it was decided to work throughout in 'radiocarbon years', and to await further progress with dendrochronology before essaying calendar calibration. The overall distortion of the radiocarbon scale in terms of calendar years does not affect the validity of C14 determinations for investigating whether geographically widespread events were synchronous. On the other hand, the possibility of confusion arising from short-term perturbations in the C14 input curve ('Suess wiggles') requires assessment. Indicators of the reliability of the dates for the intended study were therefore sought. One empirical check that suggested itself was to look at the incidence of inversions of C14 date order relative to stratigraphic order. The result was reassuring: in a sample of over 600 determinations from closely spaced superimposed stratigraphic contexts, over 95 per cent of the dates conform to their stratigraphic order (Morrison, 1973c). In several of the anomalous cases, the submitters had noted local contamination that could well have explained the result.

In passing, it may be noted that the remaining samples involved in inversions tend to fall in groups in C14 time, notably in the 8th and 9th millennia. The number of anomalous sites is too small for any definite conclusion to be drawn from this in itself, so the extension of dendrochronological calibration to that period is awaited with interest.

The reliability of the available dates may be considered not only at the level of internal consistency at individual sites, but also in terms of the overall pattern of the evidence. As indicated when considering Fig. 2 there are grounds for concluding that the reliability of the evidence is in fact adequate for the study in hand.

No single stretch of coast is likely to have registered and preserved a complete record of Holocene sea level changes. When Rhodes Fairbridge wrote his much-quoted 1961 paper on eustatic sea level, he sought to collate radiocarbon data on sea level changes on a world-wide basis. There are both theoretical and practical problems in this (e.g. Schofield, 1967; Munk and Revelle, 1952; Young, 1953; Jones, 1961). Fairbridge's graphs are based principally on circa seventy C14 dates, and he commented on the shortage of relevant determinations, even when material was drawn from both hemispheres. Since then, the situation has changed radically and made possible the type of study pursued here.

It was decided to work on the western seaboard of Europe. This area at present contains the largest concentration of radiocarbon determinations relevant to Holocene marine change now available in the world (over twelve hundred dates), and the long established interest in coastal research in Scandinavia, the North Sea countries and Britain meant that an extensive literature was available to aid the analysis of the data. In addition to its direct archaeological interest, the particular attraction of the area lies in the way that it embraces the desired wide range of land movement regimes and coastal environments. The maximum distortion of a synchronous ancient waterplane indicated directly by the data shows differential land movement in excess of 250 m during the the the Holocene. The conditions range from exposed Arctic beaches on islands off Finnmark to coastal swamps in Biscay; from clear fjords to the muddy estuaries of Elbe and Rhine; from the tideless Kattegat to the storm surge tract of the southern North Sea; and include a considerable variety of Baltic as well as British shores. The seaboard thus offers a very considerable spectrum of conditions, and any substantial synchronism of coastal transgression observable across this range would seem difficult to explain in terms other than control by ocean level change.

Clearly, with so many C14 determinations available, the danger exists that any approach giving scope for eclectic handling of the data might result merely in the selection of dates that accorded with preconceptions of the nature of postglacial marine change. A holistic approach, with a strictly stratigraphic classification of the evidence and the use of statistical techniques seemed indicated.

The radiocarbon determinations used were obtained primarily from the Radiocarbon Supplement of the *American Journal of Science* (vols. 1 and 2, 1959 and 1960) and its successor the journal *Radiocarbon* (vol. 3, 1961 to vol. 15, pt. 3, 1973), and were then followed up in the literature. Libby's original half-life value (5570 ± 30) has been retained throughout. Following an international programme of laboratory cross-calibration, a retrospective revision of all dates published internationally between 1950 and 1965 was issued as a check list with the title 'Radiocarbon Measurements: Comprehensive Index 1950-1965' (Deevey, Flint and Rouse, 1967); dates published in *Radiocarbon* from 1966 onwards have conformed to that standard. The deletions and modifications set out in the Index have been fully incorporated in the data for the present study.

It is important to emphasise that the writer has refrained from making any other modifications to dates. It is clear that the particular geological or botanical nature of some of the evidence and the problems of sampling are likely to have made some dates earlier or later than the events to which they were intended to relate. In some cases the sense and magnitude of such discrepancies may be suspected. At present however, so much concerning Holocene sea level change remains in doubt that the validity of allowances proposed to compensate for differences of this kind remains difficult to assess. At this stage it seems preferable to avoid special arguments concerning individual sites, and to concentrate rather on establishing whether, despite such variations, the evidence contains any overall pattern at all.

It was decided that it would not be advisable to make the basic classification of the dates in terms of the formations to which they had been attributed in the literature. The usage of terms such as 'Litorina', 'Tapes' or 'Duinkerke' proved variable, and some of the preconceptions inherent in the traditional names are of doubtful validity. Furthermore, errors of attribution were often admitted in the date lists. It therefore seemed preferable to confine the classification to basic stratigraphic description, with division of the data in terms of the geological reliability of the samples as diagnostics of marine transgression and regression.

Thus, for example, shell dates got a low rating because of the problems in equating shell determinations with dates of other types of organic material, together with the common difficulty of assessing whether individual shellbeds relate to transgression or regression phases. Some peat-bed dates, too, were of limited value. For instance, some workers failed to specify whence they drew their samples from thick beds, and it was not clear whether the date represented the start of a regression, some indeterminate point within it, or the start of the succeeding transgression. Dates that appeared to offer clear evidence individually but which referred to unusual circumstances were also regarded as falling at a secondary level of geological reliability, since it was considered that apparent agreements between rare and diverse cases would be less reliable than correspondences between systematically defined homogeneous classes, each made up of large numbers of dates.

It was considered that the remaining dates (approximately five hundred in number) offered the clearest evidence available. Firstly, they fell into unambiguous stratigraphic categories, appearing to relate directly to local transgression or regression, and secondly they represented classes which each contained enough dates to yield an independent pattern of evidence covering the greater part of the Holocene.

The form of classification used thus did not involve preconceptions regarding sea level change sequences. The aim was merely to make a clear distinction between information which required further discussion, and more clear-cut data that might offer a starting point for that discussion. The initial hypothesis, once formulated from material of the primary level of geological reliability could then be examined in terms of the full range of the evidence.

At this primary level, four classes could be distinguished: T, W, R and I. The T and R classes contain respectively dates that refer directly to what appear to be direct transgression and regression contacts in stratigraphy. Class T thus refers to dates from the top of peat beds where they are immediately overlain by a marine deposit, and Class R correspondingly to the bottom of such a bed where it overlies a marine deposit. In all cases where pollen or diatom evidence suggested an unconformity, the date was relegated from these classes. Such checks were not always available, and though Classes T and R thus objectively reflect the stratigraphic location of the sample dated in descriptive terms, they do not guarantee an absence of unconformities. Dates were not eliminated except where pollen or diatom evidence was specific. As indicated above, if an objective inventory and perspective of the data currently available was to be achieved, it seemed very necessary to eschew any judgements which might merely reflect preconceptions regarding the pattern of Holocene change.

Class W (standing for 'watertable') refers to a basal peat, overlain by marine clay but itself most characteristically lying unconformably on Pleistocene sands. In a detailed study, Jelgersma (1961) showed that its growth and subsequent preservation was due to the rise of water-table caused by sea level reaching its level. Class I ('isolation') refers to the isolation of shallow arms from the sea and their conversion into fresh water lakes, as established by diatom and vegetation studies.

Once the stratigraphic classification was complete, a form of statistically reliable visual description of the patterns of the various classes through time was sought. The aim was to produce a display that might be considered analogous in general characteristics to a pollen or diatom diagram, i.e. variations through time within each class should be portrayed objectively and with numerical accuracy, and major relationships between the patterns of variations in different classes should be accessible on inspection.

A simple histogram based on counts of dates falling within given

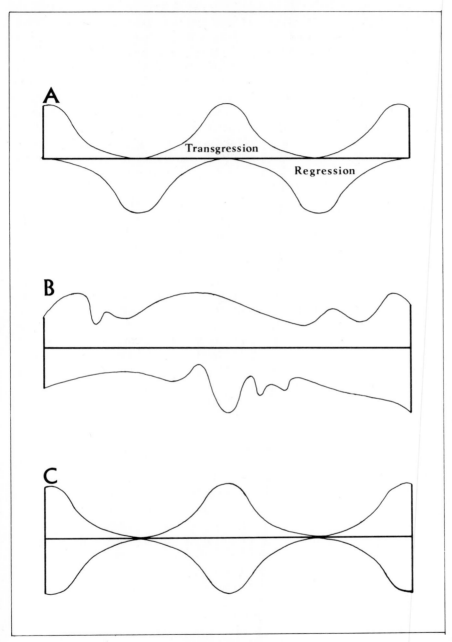

Figure 2. Possible patterns for the graphs of radiocarbon dates.

divisions of time has dangers in this type of enquiry, in that on present knowledge it is difficult to evaluate whether the chosen width of division might not interact in complex ways with any cyclical components in the timing of the sea level changes, or indeed in the postulated C14 perturbations. Rather than group the dates in arbitrary divisions, they were therefore plotted along the time axis of the graphs in terms of the actual mean values of their determinations. It seemed reasonable to give all dates classed at the primary level of stratigraphic reliability the same weight, in geological terms, but to take due account of the varying statistical uncertainties attributed by the laboratories to their measurements of the individual dates. These were therefore plotted as separate Gaussian curves with their forms defined by the counting errors of each date, so calculated as to give an equal area under the Gaussian of every determination. It was clear that because of the larger number of dates involved, diagrams of this form would be complex. To derive an objective expression of the pattern of the data in visually simpler terms, the ordinates were summed to produce a single summary curve for each class. The validity of this approach was accepted in discussion at the Royal Statistical Society (*J. Roy. Statist. Soc.* A118, 1955, p. 291).

Other things being equal, with a fluctuating sea level the curves representing the amounts of transgression and regression evidence could be expected to have an inverse relationship, in other words during a rise of sea level the curve of transgression evidence would tend towards a maximum, while the curve of regression evidence tended towards a minimum. In the form of display used here, the latter curve has accordingly been shown inverted.

Thus, in an ideally simple situation in which movement of the sea's surface was the only factor, the graphs could be expected to take the form shown in Fig. 2A.

If instead the curves showed no clear pattern but were characterised by unrelated fluctuations (Fig. 2B, say), one possible explanation would be that the influence of eustatic fluctuations had been outweighed by purely local factors such as local variations in land movement and local environmental changes such as the creation and breaching of protective sand barriers, etc. Alternatively, the various types of uncertainty and error inherent in radiocarbon dating could have combined with imperfections in field sampling to reduce the fidelity of the dating below the level of resolution required by the prevailing rate of the sea level changes. A graph of this nature might well reflect a complex interplay of both these types of factor.

If, rather, the curves tended towards Fig. 2C, it would seem conceivable that the imprint of eustatic changes had been so weakened by factors of the kinds indicated above that radiocarbon fluctuations of the types described by Suess had emerged as the dominant element in the pattern.

3. Results

When the actual curves for the data of the survey were derived, it was found that Classes T, W, R and I each exhibited marked saw-tooth forms. T and W, both considered to represent increase of marine influences, did in fact tend to show accordant patterns. Similarly, R and I, both representing marine regression, also tended to accord with each other, while being out of phase with TW. As Fig. 3 illustrates, if 'conflict' is defined in terms of the gradients of the curves, then there is no conflict between the TW and RI curves in more than three-quarters of the Holocene, and none lasting longer than a century and a half in nine-tenths of the period. It hence seems reasonable to claim that the evidence shows a strict alternation of episodes of marine transgression and regression.

The effects of local coastal factors and of dating errors will both undoubtedly be represented. Nevertheless, the unambiguous alternation through time of the peaks of transgression and regression evidence certainly appears more closely analogous to the situation illustrated in Fig. 2A than to those in B or C. The overall pattern as well as the internal stratigraphic consistency of the C14 evidence would thus seem to give grounds for concluding that the data are dominated neither by the shortcomings of field sampling nor by the limitations inherent in radio-carbon dating. Furthermore, the pattern bears a strong suggestion that eustatic control has tended to dominate and synchronise the detailed timing of coastal events.

The validity of each phase in the hypothetical sequence of marine events suggested by the TWRI graphs was assessed individually in a somewhat protracted analysis taking into account the stratigraphy and areal distribution of the full range of the evidence, not only the dates falling into those classes. The possibility existed that a rapid succession of events might be stratigraphically distinct at individual sites, yet perhaps be blurred into a single peak on the graph by 'noise' effects. It was clearly advisable to investigate this specifically for each phase suggested by the graphs. The same was true of the converse, the possibility that some of the groupings of dates appearing as distinct peaks on the graphs might not in fact be represented by distinct

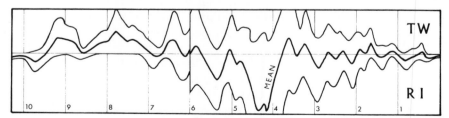

Figure 3. The actual pattern of the evidence falling in Classes TW and RI. Millennia in radiocarbon years B.P. are numbered. Baltic evidence prior to 6000 B.P. is omitted, lake stages there requiring separate discussion.

features or strata on the ground. Maps were then prepared for each phase to examine whether the evidence was sufficiently widespread, and distributed over sufficiently diverse conditions, to imply eustatic control.

The outcome of these investigations is outlined in the Appendix. It is emphasised that what has been attempted is an overall perspective of complex material, and the conclusions summarised there are accordingly provisional. Indeed the material of the survey was processed according to easily reproducible procedures and entered on edge-punched cards specifically with a view to facilitating periodic reassessment as fresh data become available.

Although subject to detail modification, the results nevertheless appear sufficiently clear-cut in their overall nature for certain general conclusions to be permissible.

For example, although local and regional considerations clearly condition the susceptibility of any particular stretch of coastline to change at any particular period (and condition, too, the likelihood of evidence of a change being preserved), it does seem that recorded shore-level changes tended to coincide in their detailed timing, conforming to a ubiquitous sequence involving a strict alternation of short duration transgression and regression phases. This would seem to offer strong support for the view that not only is eustatic control evident but that the Holocene eustatic curve has been characterised by short term oscillations rather than by smoothness. That these oscillations appear sufficiently well marked to synchronise events across the wide range of conditions represented from Scandinavia to the Low Countries would seem to suggest that, areas of catastrophic volcanic and tectonic movement apart, elements of the same time pattern are likely to be represented elsewhere in the world too. This is under investigation.

The cause of these oscillations, and the specific nature of their relationship to Holocene glacier variations and climatic events, would seem more accessible now that they can be defined according to a substantial number of C14 dates. It is intriguing in this context that according to Sherman's statistic Omega, the likelihood of a periodicity as regular as that observed in the dates of the transgressions and regressions arising by chance is less than one per cent (the figures used in this calculation were in postulated calendar years, for the latter part of the Holocene where most C14 calibration data are available).

Defining the number and timing of the oscillations may also help to dispose of some of the 'noise on the model' apparent in Fig. 1 with regard to the vertical amplitude of the eustatic changes. Whether this proves to be so or not, the maps constructed for the survey with reference in all to some twelve hundred C14 dates offer a direct basis for a reconstruction for archaeological purposes of the changing geography of the western seaboard of Europe, and this is now in hand.

Acknowledgments

This study was undertaken under the joint auspices of the Departments of Geography and Prehistoric Archaeology at the University of Edinburgh, with support from the Science Research Council and the Natural Environment Research Council.

APPENDIX

This list is provisional, and is offered for discussion. The dates proposed are quoted in radiocarbon years B.P., on the Libby half-life. They reflect the radiocarbon dated evidence considered as a whole, rather than the TWRI classes alone.

a : 10300-9750 : possible transgression reaching maximum c. 10000 and ended by regression c. 9750, but evidence indecisive.
   9750-9400 : evidence unclear.
b : 9400-8750 : transgression (rapid 9400-9100 then reduces) distinct from earlier and later events, and probably eustatic.
   8700-8300 : regression, probably eustatic.
c : 8300-7750 : transgression.
d : 7500-7000 : transgression.
   Neither *c* nor *d* can be further subdivided, but because of confused evidence (inversions) between 7750 and 7500, it is not clear whether *c* and *d* are distinct; *c/d* is however separated clearly from earlier and later periods of transgression, and considered either individually or as a single phase, the *c d* period transgressions are sufficiently widely distributed to seem unambiguously eustatic.
   7000-6650 : eustatic regression.
e : 6650-6400 : separate eustatic transgression.
   6400-6050 : eustatic regression, perhaps interrupted at 6250 by a brief transgression.
f : 6050-5950 : separate eustatic transgression.
   5950-5300 : eustatic regression.
g : 5300-4950 : separate eustatic transgression.
   4950-4800 : brief but apparently eustatic regression.
h : 4800-4600 : minor but distinct and apparently eustatic transgression.
   4600-4300 : eustatic regression.
i : 4300-4200 : minor but apparently eustatic transgression.
   4200-3850 : eustatic regression.
j : 3850-3600 : separate eustatic transgression.
   3600-3500 : brief but apparently eustatic regression.
k : 3500-3150 : separate eustatic transgression.
   3150-2900 : eustatic regression.

l :   2900-2700 : eustatic near stillstand (slight rise?).
        2700-2450 : eustatic regression.
m :   2450-2200 : slight but probably eustatic transgression.
        2200-2100 : brief eustatic near stillstand (slight fall?).
n :   2100-1900 : eustatic transgression.
        1900-1800 : brief eustatic near stillstand (slight fall?).
o :   1800-1650 : eustatic transgression.
        1650-1250 : eustatic regression.
p :   1250-1000 : evidence indecisive.

REFERENCES

Andrews, J. (1970) 'A geomorphological study of post-glacial uplift with particular reference to Arctic Canada', *Inst. Br. Geogr. Spac. Publ.* No. 2.

Auer, V. (1958) 'The Pleistocene of Fuego-Patagonia', *Ann. Acad. Sci. Fenn. Ser. A.* III, 50.

Bloom, A. (1967) 'Pleistocene shorelines: a new test of isostasy', *Bull. geol. Soc. Am.* 78, 1477.

Broeker, W. and A. Kaufman (1965) 'Radiocarbon chronology of Lake Lahontan and Lake Bonneville', *U.S. Geol. Surv. Prof. Paper* 454E.

Charlesworth, J.K. (1957) *The Quaternary Era.* (2 vols.) London.

Evans, J. and C. Renfrew (1968) *Saliagos.* Brit. School at Athens, Suppl. Vol. 5.

Fairbridge, R. (1961) 'Eustatic changes in sea level', *Physics Chem. Earth*, 99.

Farrand, W. (1966) 'The deglacial hemicycle', *Geol. Rdsch.* 54, 385.

Fisk, H. (1956) 'Near surface sediments of the continental shelf off Louisiana', *Proc. 8th Texas Conf. Soil Mechanics and Foundation Engineering*, Univ. of Texas.

Flint, R.F. (1957) *Glacial and Pleistocene Geology.* New York.

Higgins, C. (1968) 'Isostatic effects of sea level changes', *Proc. VII Congress Int. Assoc. Quat. Res.* 16, 141.

Jelgersma, S. (1961) 'Holocene sea level changes in the Netherlands', *Meded. geol. Sticht.* Ser. C, VI No. 7.

Jones, H. (1961) 'Variations of the earth's rotation', *Physics Chem. Earth*, 4, 186.

Leontyev, O. and L. Nikiforov (1965) 'On the causes of the global occurrence of barrier beaches', *Oceanology* 4, 653.

Matthews, R. (1969) 'Tectonic implications of glacio-eustatic sea level fluctuations', *Earth and Planet Sci. Lett.* 5, 459.

Mii, H. (1962) 'Some ancient shore features', *Sci. Rpts. Tokoku Univ. Sendai 2nd Ser. (Geol.)* Spec. Vol. 5, 361.

Milliman, J. and K. Emery (1968) 'Sea levels during the past 35,000 years', *Science N.Y.* 162, 1121.

Moran, J. and R. Bryson (1969) 'The contribution of the Laurentide ice wastage to the eustatic rise of sea level, 10000 to 6000 B.P.', *Arctic Alpine Res.* 1, 97.

Mörner, N.A. (1969) 'Late Quaternary history of the Kattegat Sea and the Swedish west coast', *Sver. Geol. Unders. Afh.* Ser. C, 640.

Morrison, I. (1968) 'Sea level change in the Saliagos area since the Neolithic', *in* J. Evans and A.C. Renfrew, 1968, 92.

Morrison, I. (1969) 'Some problems in correlating archaeological material and old shorelines', *Scottish archaeol. Forum* 1, 1.

Morrison, I. (1973a) 'Geomorphological investigation of marine and lacustrine environments of archaeological sites, using diving techniques', *Proc. 3rd Sc. Sympos. of C.M.A.S.*, 41.

Morrison, I. (1973b) *The North Sea Earls.* London.

Morrison, I. (1973c) 'Radiocarbon dating of sea-level changes during the Holocene', 1973 Archaeometry Sympos., Oxford, Abstracts.

Munk, W. and R. Revelle, (1952) 'Sea level and the rotation of the earth', *Am. J. Sci.* 250, 829.

Newman, W. and G. Munsart, (1968) 'Holocene geology of Vachaforeague lagoon, eastern shore peninsula, Virgina', *Mar. Geol.* 6, 149.

Schofield, J. (1967) 'Postglacial sea level maxima a function of salinity?', *J. of Geosciences, Osaka* 10, 115.

Scholl, D. (1964) 'Recent sedimentary record in mangrove swamps and rise of sea level over the south-west coast of Florida', *Mar. Geol.* 1, 344.

Shepard, F. (1963) 'Thirty-five thousand years of sea level', in *Essays in Marine Geology in honour of K.O. Emery*, Univ. Calif. Press.

Sissons, J.B. (1967a) *The Evolution of Scotland's Scenery*. Edinburgh.

Sissons, J.B. (1967b) 'Comments on Synge and Stephens 1966', *Trans. Inst. Br. Geogr.* 40, 163.

Walcott, R. (1972) 'Past sea levels, eustasy and deformation of the earth', *Quaternary Res.* 2, 1.

Young, A. (1953) 'Glacial eustasy and the rotation of the earth', *Monthly Notices, Roy. Astron. Soc. Geophys. Suppl.* 6, 453.

DISCUSSION

(Dr. I.A. Morrison was unable to attend the symposium, but his paper was read on his behalf by Miss C. Delano Smith. The questions were sent to Dr. Morrison and his written replies follow.)

Mr. G. de G. Sieveking: I would like to hear a geological comment from a specialist in sea level changes. It would seem to me that the method used in this paper tends to conceal the real field situations by statistical smoothing of many C14 dates from different areas and types of sea level evidence.

I.A.M.: (i) Much of the work behind this paper has in fact been examined in detail by Dr. J.B. Sissons and Professor W. Kirk.

(ii) The aim of the method was to produce a systematic inventory of the radiocarbon dated evidence in a substantial area, so that we might have some indication whether broad patterns existed. The graph is a convenient visual aid, but it represents only one stage in the analysis. The ensuing stratigraphic study of the full range of the data was aimed specifically at evaluating its apparent implications in terms of the individual field situations, as indicated in the paper.

Professor A. Straw: It would appear that all 'Transgression' dates have been consolidated (also 'Regression' etc.). This assumes that transgressions in the Western Seaboard occurred at the same time (regressions likewise) – however, there have been differential crustal movements, and transgressions in some places would have been contemporaneous with regressions in others. How does this affect the reliability of the curves demonstrated?

I.A.M.: Had the approach involved consolidating all dates that seemed readily attributable to, say, each of the traditional Litorina or Tapes stages and rejecting others, the results would inevitably have reflected assumptions inherent in preconceived models. As I think the full text makes clear, however, the curves merely provide a summary of the distribution through time of radiocarbon dates from objectively defined stratigraphic contexts. It was not until their pattern had been examined in terms of both individual site stratigraphy and geographical distribution that it was concluded that a widespread synchronism of events was apparent on the Western Seaboard of Europe. This was the finishing, not starting, point of the survey.

It seems important to distinguish between (a) widespread similarities in the timing of events and (b) contrasts in the dominant regimes of different areas. Thus, during the Holocene, emergence has been dominant in Scandinavia, submergence in the Low Countries. In neither case has the process been unrelieved, however, and what the present study indicates is that there is a strong tendency for synchronism in the detailed changes of marine influence throughout the seaboard. For example, phases involving the abandonment of strandlines or the isolation of arms from the sea to become freshwater lakes in gradually rising Scandinavia tend to coincide closely with periods of temporary drying out of Tidal Flat deposits and the development of peat upon them in stable or sinking Holland. Similarly, the widespread onset of submergence elsewhere consistently tends to coincide with temporary transgression of coastal peat beds in Scandinavia and with the reappearance of brackish flora and fauna in lakes there as these are temporarily reconnected to the sea – and so on.

Overall, the topographical importance of the events varies markedly from region to region according to land movement regime and coastal conditions, but their timing certainly seems to follow a common pattern.

W. G. Jardine and A. Morrison

The archaeological significance of Holocene coastal deposits in south-western Scotland

The marine coastal environment has always been an important habitat of man for reasons which are fairly obvious: the sea provides food both in mobile vertebrate and invertebrate forms (e.g. fish, clams, lobsters, crayfish) and in sessile invertebrate forms (e.g. limpets, mussels, oysters); transport by sea, or along river estuaries, or in large coastal embayments often is easier than across adjacent tracts of rough, densely-vegetated country; habitable natural caves and rock shelters are commoner near the coast, especially a raised former coast, than inland; flint and other raw materials suitable for the manufacture of tools and weapons may be concentrated in coastal deposits by marine processes although the original source is diffuse, is submarine, or is located on another shore many kilometres from the site concerned.

The seaboard of Dumfriesshire, Kirkcudbrightshire, Wigtownshire and southern Ayrshire – south-western Scotland of this discussion – has been the habitat of man during approximately the last 6000 or 7000 years (Fig. 1). With the emergence of new land from the sea after the maximum of the Flandrian marine transgression, not only were new habitable coastal sites available, but the navigable distances across the North Channel and Solway Firth were very much shorter than formerly. The recession of the sea and the slight rise of land which followed the marine transgression left a variety of raised coastal deposits which contain within them or bear upon them the evidence of human occupation. It is with the nature of these deposits, with the chronology of environmental and archaeological events to which they bear witness, and with their evidence of man's occupation that this paper is concerned.

1. Holocene coastal deposits

The deposits considered here comprise a variety of different kinds of sediments, each the product of a particular coastal environment. Although accumulated close to the contemporaneous coast, the sediments now may be located several tens or hundreds of metres from the present shore, and may occur up to approximately ten metres above Ordnance Datum (O.D.). The deposits were laid down during

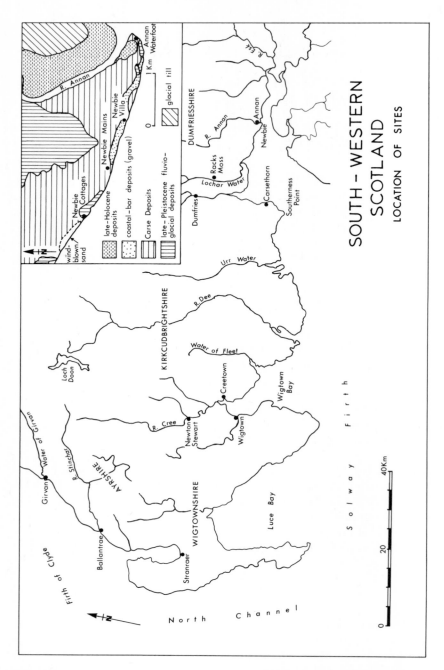

Figure 1. Map of south-western Scotland showing major towns of the area and main localities mentioned in the text. Inset: Map of the Newbie area, south of Annan, showing localities mentioned in the text, and surface deposits of the area.

the most recent geological time interval, the 'Recent' or 'Post-glacial' of some accounts, but now internationally designated the Holocene Epoch, and generally agreed to have begun approximately 10000 radiocarbon years B.P. (before present, i.e. before the internationally-agreed datum, A.D. 1950; Godwin, 1962).

The various sedimentary environments of the area concerned – beach, gulf or open-bay, estuarine (including tidal-flat), lagoonal, coastal-bar, coastal-dune and coastal-marsh – may be recognised on the basis of a number of diagnostic features (*cf.* Jardine, 1967). These characteristics, systematically considered below in the case of all but one of the main Holocene coastal environments of south-western Scotland, are: shape of sedimentary body and location of sedimentary body in relation to the contemporary shore-line; grain size (texture); degree of sorting; degree of stratification; biogenic content. Coastal-marsh deposits, being largely organic, have rather different characteristics from the other sediments.

1. *Beach sediments.* Elongate narrow tracts sub-parallel to present and to contemporary shore-line; sand, gravel; commonly well-sorted; stratification moderate to good, generally horizontal, individual units thick and persistent; occasional mollusc shells or fragments, more often in sands than in coarser grades; occasionally shells form entire beach structure (Figs. 2a, 2b, 2c).

2. *Gulf or open-bay sediments.* Semi-circular tracts adjacent to present-day open-bays or less regularly-shaped tracts occupying sites between widely-spaced contemporary headlands; micaceous fine sand, sand, occasional thin layers of gravel; sorting moderate to good; stratification moderate, and rather variable in vertical succession at any one site; mollusc shells when present may be concentrated in thin layers or lenses which occur sporadically both vertically and laterally; tests of Foraminifera common to frequent; valves of marine Ostracoda occasionally present (Figs. 2a, 2c, 2d).

3. *Estuarine sediments.* Elongate or triangular-shaped tracts at the head of, or irregular tracts along the flanks of present-day estuaries, or tracts occupying hollows between fluvio-glacial mounds; clay, silt, fine sand, sand, gravel, with silt and fine sand commonest; generally well-sorted; stratification good, in places with strong development of alternate silt and fine sand laminae; sporadic concentration of wood fragments, leaves and seeds; occasional lenses of mollusc shells; tests of Foraminifera common and occasional valves of Ostracoda in medium size grades (Figs. 2c, 2d).

Component parts of some estuaries are tidal-flat areas, which are uncovered by water at low tide in contrast with the channels of the estuaries which are occupied by water throughout the tidal cycle. The main characteristics of tidal-flat sediments are: mainly clay, silt, fine sand; generally well-sorted; stratification good where not disturbed by the activity of fauna; occasional wood fragments and other plant debris; mollusc shells (sometimes in living position) common,

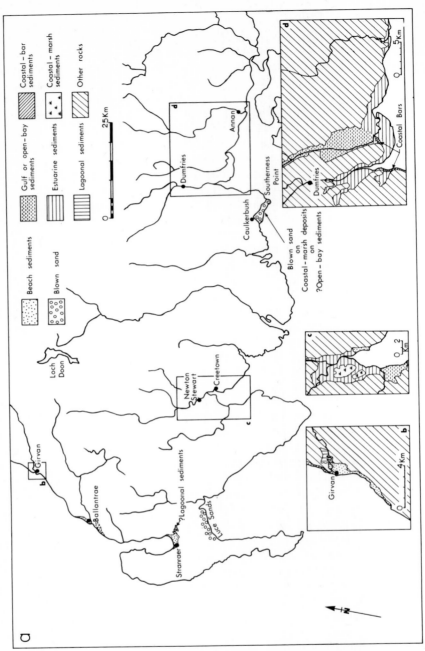

Figure 2. Representative examples of Holocene coastal sedimentary environments in south-western Scotland. Map 2a: Minor areas (slightly exaggerated) of surface sediments at selected localities, and locations of major areas of surface or near-surface sediments shown in detail in Maps 2b, 2c and 2d. Map 2b: Holocene coastal deposits of the Girvan area; beach and lagoonal sediments. Map 2c: Holocene coastal deposits at the head of Wigtown Bay; beach, open-bay, estuarine and coastal-marsh sediments. Map 2d: Holocene coastal deposits of eastern Kirkcudbrightshire and western Dumfriesshire; gulf, estuarine and coastal-bar sediments. For clarity, large areas of coastal-marsh sediments covering the inorganic sediments south-east of Dumfries are omitted.

occasionally concentrated in layers or lenses; bioturbation by Mollusca and Arthropoda (e.g. *Corophium volutator*) common, especially in upper tidal-flat areas (*cf.* van Straaten, 1961).

4. *Lagoonal sediments.* Thin lens-shaped sedimentary bodies occupying small embayments isolated from contemporary open sea by temporary barriers, generally of sand or gravel; mainly silt; stratification poor to moderate; abundant remains of plant debris; occasional mollusc shell fragments and tests of Foraminifera (Figs. 2a, 2b).

5. *Coastal-bar sediments.* Elongate single or multiple ridges approximately parallel to present shore, but commonly oblique to or transverse to contemporaneous coast; occasionally sand, but commonly gravel and cobbles or mixture of sand and gravel; sorting moderate; stratification moderate to good, generally horizontal; very occasionally whole shells or shell fragments of molluscs present (Fig. 1, inset, and Fig. 2d).

6. *Coastal-dune sediments.* Large irregular tracts overlying, or small pockets infilling hollows in, pre-existing Holocene or Pleistocene deposits, generally in vicinity of but not necessarily directly oriented in relation to the contemporaneous shore, or elongate ridges sub-parallel to present coast; commonly sand or fine sand; sorting moderate to good; stratification moderate to good; layers of peat of variable thickness up to 1 m, or thin humus-rich former land surfaces may be interstratified with the inorganic sediments (Fig. 1, inset, and Fig. 2a).

7. *Coastal-marsh sediments.* Accumulations of organic debris which originated mainly autochthonously as live marsh vegetation located close to contemporary High Water Mark, probably above High Water Mark Ordinary Neap Tides, but not necessarily above High Water Mark Ordinary Spring Tides. Located now on top of, or interstratified with, other coastal deposits, e.g. estuarine and/or coastal-dune sediments (Figs. 2a, 2c).

Each of the sedimentary facies described above is distinctive, but some may be grouped together as an association of contemporaneous deposits which for convenience may be termed *Carse Deposits* (*cf.* Carse Clays of Godwin, 1943; Lacaille, 1954; Nichols, 1967). These are mainly former tidal-flat or estuarine sediments, dominantly medium- or fine-grained in texture, which now flank the northern coast of the Solway Firth or form extensive relatively-flat tracts bordering the principal watercourses of the counties of Dumfries, Kirkcudbright and Wigtown. Some of the gulf or open-bay sediments of the classification given above constitute a minor facies of the Carse Deposits. The sediments which accumulated in south-western Scotland during the main Holocene marine transgression are largely Carse Deposits.

2.   *Chronology*

An important use of Holocene coastal deposits in relation to archaeology is in establishing a time scale for the last 10000 years. At present this may be achieved by two main methods: on the basis of pollen content of the sediments, and by radiocarbon dating of biogenic layers within the deposits.

2.1 Pollen analysis

In south-western Scotland relevant modern palynological studies of Holocene deposits are limited, as yet, to the work of Moar (1969) and Nichols (1967). Briefly, Moar established a series of pollen assemblage zones which also are recognisable in other parts of Scotland, but which differ slightly from the pollen assemblage zones of the 'standard system' for England, whilst Nichols related the pollen he found in the coastal deposits of the Racks Moss area (south-east of Dumfries) to the assemblage zones of the 'standard system'. Of importance to archaeological studies is the fact that the pollen assemblage zones are broadly equivalent to the climatic divisions suggested by Blytt and Sernander (Sernander, 1908) which, although still widely used, are now out-moded and should be abandoned. It should be noted, also, that the pollen assemblage zones are biostratigraphical zones (George *et al.*, 1967) and, as such, do not strictly define time limits. West (1970) discussed how these biostratigraphical pollen assemblage zones are related to chronozones (or time zones). The relationships between the climatic divisions of Blytt and Sernander, the pollen assemblage zones and the chronozones are shown in Table 1. Also incorporated in Table 1 is a scheme providing a link between the pollen assemblage zones and radiocarbon dating, a scheme suggested originally by Godwin and his co-workers in the late 1950s and modified as a result of later developments in Quaternary research (*cf.* Hibbert *et al.*, 1971; Smith and Pilcher, 1973).

2.2 Radiocarbon dating

Many of the disadvantages of a time-scale based on radiocarbon dating are well known, and have been discussed by a number of authors (e.g. McKerrell, 1971; MacKie *et al.*, 1971; Renfrew, 1970; Smith and Pilcher, 1973) but, provided it is understood that the divisions involved in such a time-scale are radiocarbon years, which individually neither may be exactly equal to calendar years nor may be precisely equal to each other throughout the Holocene epoch, the scale is useful for the time interval considered here.

At the time of writing (late 1973), radiocarbon dates are available for about thirty samples of biogenic material from the Holocene coastal deposits of south-western Scotland. In each case the stratigraphical position of the dated material is accurately known, so

Table 1. Comparison of Hologene pollen assemblage zones, chronozones, radiocarbon dates and climatic divisions.

Scotland (Moar, 1969)	'Standard System'	Chronozones* (West, 1970)	Radiocarbon dates (e.g. Hibbert et al., 1971)	Climatic divisions (e.g. Sernander, 1908)	Archaeological material+ (this paper)
V	VIII	Flandrian III		Sub-atlantic	Charcoal, Newbie area
					Canoe, Lochar Water
IV	VIIb		5010 ± 80 years B.P.	Sub-boreal	
III	VIIa	Flandrian II		Atlantic	Landnam, Racks Moss
	VI	c	7107 ± 120 years B.P.		Lochhill cairn
II	V	Flandrian I b	8880 ± 170 years B.P.	Boreal	Low Clone; Barsalloch
I	IV	a	9798 ± 200 years B.P.	Pre-boreal	Tentsmuir Sands, Fife

*The chronozones Flandrian I, II and III are time divisions of the Flandrian age which, in Britain, is the sole age constituting the Flandrian epoch.

+In this column, archaeological sites discussed in the text are shown in their probable chronological position.

valid conclusions regarding the sediments above and below the dated material may be reached. Most of the dates are already published in *Radiocarbon*, and the significance of some of the evidence has been discussed (Jardine, 1964, 1971). A detailed chronology of Holocene coastal events in south-western Scotland is in preparation for publication elsewhere; a summary of the chronology is given below.

In broad terms the dated biogenic deposits occur in one or other of two situations: (a) as organic detritus, wood, peat, charcoal, or mollusc shell debris which either accumulated during temporary interruption in the course of the main Holocene marine transgression, accumulated in the course of the recession of the sea from the maximum of the main Holocene transgression, or was deposited in a coastal but non-marine environment after the sea had receded from the locality concerned; (b) as organic detritus which accumulated on

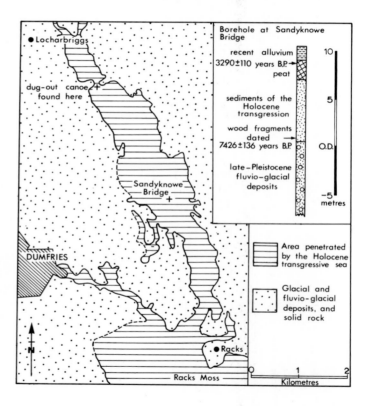

Figure 3. Map of the area to the east of Dumfries, between Locharbriggs and Racks, showing localities mentioned in the text and location of the dug-out canoe discovered in September 1973, in relation to the area penetrated by the Holocene marine transgression. For clarity, some areas of Pleistocene fluvio-glacial deposits between Locharbriggs and Sandyknowe Bridge are omitted. The section, of a borehole sunk at Sandyknowe Bridge in 1965, is typical of the sequence of Quaternary deposits present over most of the area penetrated by the Holocene sea.

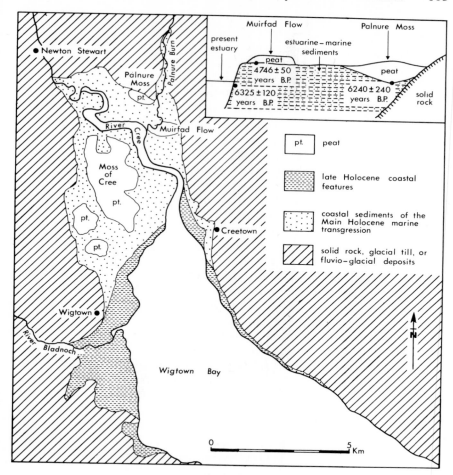

Figure 4. Area around the head of Wigtown Bay, south-western Scotland. Map: Distribution of Holocene coastal sediments. The late Holocene terrace, to the east and south of Wigtown and west of Creetown, partly due to erosive processes, partly to depositional processes (at approximately 2150 radiocarbon years B.P.), is also shown. Inset: Schematic section showing relationships between the peat deposits of Muirfad Flow and Palnure Moss, and the estuarine sediments deposited in the course of the main Holocene marine transgression. Vertical scale greatly exaggerated.

a pre-existing land surface which was inundated in the course of the main Holocene marine transgression.

The oldest dated samples demonstrate the presence of marine conditions in part of the Girvan area of southern Ayrshire and in the Carsethorn area of south-eastern Kirkcudbrightshire very early in Holocene times, followed by a short interruption around 9375 radiocarbon years B.P., and renewed marine conditions thereafter. Evidence exists also of the later penetration by the sea into parts of the

Girvan area and the Turnberry area (8 km north of Girvan), around 8400 radiocarbon years B.P.

In the eastern Solway Firth area, the Holocene sea first flooded the land surface at Redkirk Point, near Gretna, at approximately 8100 radiocarbon years B.P. At Newbie Mains, near Annan, marine transgression was interrupted temporarily at approximately 7800 radiocarbon years B.P., but the main inundation of the former land surface of this area by the Holocene sea, as demonstrated by the sedimentary evidence in the vicinity of Newbie Cottages, was between 7500 and 7200 radiocarbon years B.P. (Fig. 1, inset). The marine invasion of a northern extension of the Lochar Gulf, south-east of Dumfries, commenced around 7400 radiocarbon years B.P., approximately contemporaneously with the main transgression in the vicinity of Newbie Cottages, but it is not certain for how long before that time the Lochar Gulf itself had been occupied by the sea (Fig. 3).

Fifty kilometres west of Dumfries, at the head of Wigtown Bay, a situation comparable with that in the Lochar Gulf existed in early Holocene times. Wigtown Bay (formerly extending from the vicinity of Wigtown and Creetown northwards to near Newton Stewart, Fig. 1) probably was occupied by the sea continuously from before the beginning of the Holocene epoch. The valley of the present Palnure Burn, however, was first penetrated by the Holocene sea at approximately 7900 radiocarbon years B.P. (Fig. 4).

The above summary relates to the main Holocene marine transgression. In south-western Scotland by around 7200 radiocarbon years B.P. the sea essentially had attained its maximum lateral extent for the whole of the Holocene epoch. It had not, however, attained its maximum height by that time. Radiocarbon dating, palynology, and sedimentary evidence suggest that the Lochar Gulf was abandoned by the sea at approximately 6600 radiocarbon years B.P., possibly by growth of gravel bars across the mouth of the gulf (certainly the south-western part of the gulf is blocked by a gravel bar) while contemporaneously marine conditions persisted in the eastern Solway Firth area, for example near Newbie Cottages (Fig. 1, inset). There the main Holocene marine transgression − represented by Carse Deposits which in their lower part contain abundant remains of marine microfauna, in their upper part contain sparsely-distributed remains of brackish-water fauna − did not cease until around 5600 radiocarbon years B.P. At the head of Wigtown Bay, estuarine deposition occurred simultaneously with coastal-marsh formation nearby for sufficient time after 6400 radiocarbon years B.P. for three to four metres of estuarine sediments to accumulate (Fig. 4). At Muirfad Flow evidence from peat and inorganic sediments underlying it shows that estuarine sedimentation at the head of Wigtown Bay was succeeded by brackish-water conditions, and ultimately by a brief period of fresh-water conditions, before peat accumulation commenced around 4700 radiocarbon years B.P. (Fig. 4). The end of

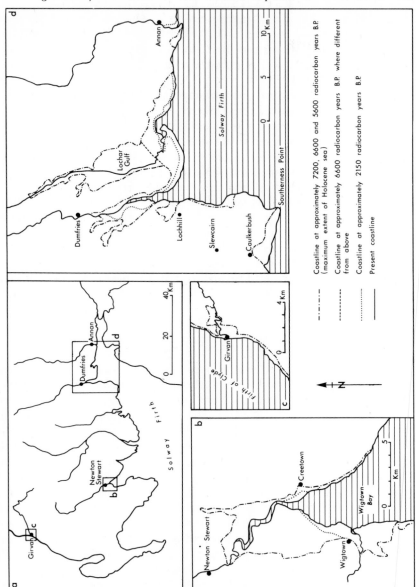

Figure 5. Major differences between present coastal configuration of south-western
Scotland and coastal configuration at approximately 7200, 6600, 5600 and 2150
radiocarbon years B.P. Map 5a: Present shore-line, and locations of areas shown
in detail in maps 5b, 5c, and 5d. Throughout the Holocene epoch the position of
the shore-line in the areas not included in maps 5b, 5c and 5d varied little from the
position of the present coast. Map 5b: Major changes in the coastal configuration
at the head of Wigtown Bay during the Holocene epoch. Map 5c: Major changes
in the coastal configuration in the vicinity of Girvan during the Holocene epoch.
Map 5d: Major changes in the coastal configuration of south-eastern
Kirkcudbrightshire and south-western Dumfriesshire during the Holocene epoch.

the main Holocene marine transgression in this area probably was around 5000 radiocarbon years B.P. Evidence of a temporary halt, around 2150 radiocarbon years B.P., in the recession of the Holocene sea from its maximum occurs on either side of Wigtown Bay (Fig. 4; see also Jardine, 1971).

The main contribution to archaeology of the chronological sequence of events summarised above is in the indication of the position of the coastline at various critical times during the last 10,000 years. Along much of the coast of south-western Scotland, the marine transgressions and regressions of the Holocene epoch caused changes of only a few metres or, at most, a few tens of metres in the position of the shore-line. Such was the case where true beach deposits accumulated or where the sea abutted against steep rocky cliffs throughout most of the Holocene epoch. Elsewhere, especially in such areas as that at the head of Wigtown Bay and in the Lochar Gulf, marked changes in the coastal configuration occurred in the course of the first part of the Holocene epoch, approximately between 10 and 5000 radiocarbon years B.P. The position of the coast of south-western Scotland at approximately 7200, 6600, 5600 and 2150 radiocarbon years B.P. is shown in Fig. 5.

3.   Early occupation of coastal sites

In south-western Scotland there is no evidence, as yet, of human occupation of coastal sites before the maximum of the Holocene marine transgression, although recent work by Coles (1971) suggested that Mesolithic groups of hunter-gatherers were already active in eastern Scotland at Tentsmuir Sands, Fife, between 8050 ± 255 (NZ-1191) and 6115 ± 110 (Q-928) radiocarbon years B.P., blown sand between archaeological levels indicating intermittent occupation during this period.

For more than a decade, mainly due to investigations by W.F. Cormack of Lockerbie, material evidence of Mesolithic man has been found in Wigtownshire along the coasts of Luce Bay and Wigtown Bay, and farther east in the counties of Kirkcudbright and Dumfries (Fig. 6). The finds were mainly from sites on top of an old marine cliff rising above the inner edge of the beach of the main Holocene transgression. The cliff varies in height from a few metres to over thirty metres, with a number of sites lying at heights of nine to fifteen metres above present sea level (Coles, 1964).

Typical of these locations, and perhaps therefore typical of the coastal Mesolithic cultures of the western Solway Firth, are sites recently excavated at Low Clone and Barsalloch, on the eastern shore of Luce Bay (Cormack and Coles, 1968; Cormack, 1970). The remains there provide evidence of the strandlooping and (probably) seasonal activities of groups which were also engaged in inland hunting and food-collecting. The sites, consisting of scooped areas or

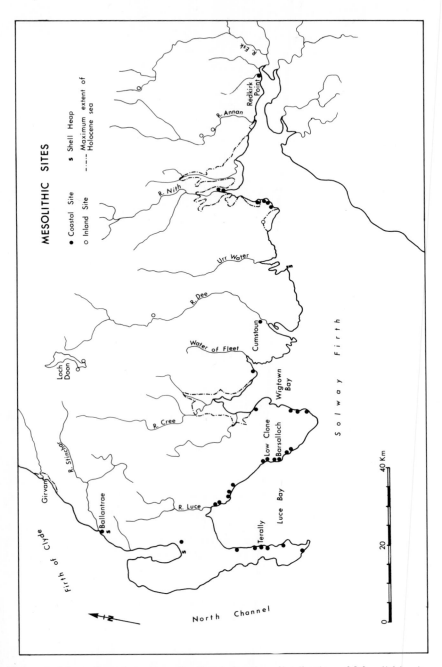

Figure 6. Map of south-western Scotland showing distribution of Mesolithic sites. The conjectural position of the coastline at the maximum extent of the Holocene sea is also shown.

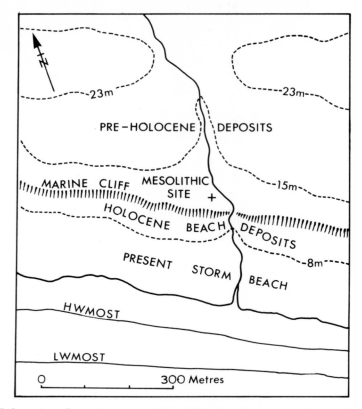

Figure 7. Location of a typical coastal Mesolithic site of the western Solway Firth at a
period of near-maximum Holocene sea level (based on evidence from Barsalloch
and Low Clone, Wigtownshire). HWMOST is High Water Mark Ordinary
Spring Tides, LWMOST is Low Water Mark Ordinary Spring Tides. Contours
are in metres above Ordnance Datum.

shallow pits which may have been covered by some form of light
shelter or windbreak, are located where fresh-water streams flow into
the sea (Fig. 7). The pH values of the soils of the area concerned
suggest that animal bones would not readily be preserved on
occupation sites, but a few burnt fragments of red deer bone were
discovered at Low Clone. This shows exploitation of the hinterland,
although a mainly fishing and food-collecting economy is indicated by
the flint industry at Barsalloch. The two sites were above the highest
level attained by the Holocene sea, and contemporaneity with a rising
or maximum sea level suggested by their location, was borne out at
Barsalloch by a date, from charcoal under a stone setting, of 6000 ±
110 radiocarbon years B.P. (GaK-1601).

At a site at Terally, on the western coast of Luce Bay, Mesolithic
material, mainly poor-quality artefacts manufactured from small
beach-pebbles, was discovered resting on former beach sediments
(Livens, 1957). There, between the present foreshore and the shore-

line of the main Holocene transgression, worked flints and waste flakes were found in the upper layer of raised beach deposits and within a thin layer of turf and humus immediately above those deposits. A fresh-water stream flows into Luce Bay about 125 m south of the site (*cf.* Barsalloch and Low Clone). The location of the flints directly upon and interstratified with the uppermost layers of the former marine deposits suggests coastal occupation after the sea had begun to withdraw. However, the water-rolled condition of some of the flint implements indicates that withdrawal of the sea may not have been complete by the time man had begun to frequent this area.

Connections have been suggested (Cormack, 1970; Mulholland, 1970) between the cultures represented at some of these coastal sites of the Solway Firth and those of inland Mesolithic sites at Loch Doon, Annandale, Eskdale and perhaps farther east in the Tweed valley (Fig. 6). These affinities seem strongest in coastal locations from the Nith estuary eastwards, where there is a large microlithic element in the range of artefacts, and more use had been made of materials other than beach-pebble flint.

4. *Sedimentary evidence of environmental changes*

In south-western Scotland, all of the varied coastal environments were not necessarily developed contemporaneously during the Holocene epoch. The foregoing statement is true in two senses: at any one time more than one environment existed along the long, diverse coast of the area concerned; at any one coastal locality the environment changed through time. Each of these variations was important in influencing man's choice of occupation site, his day-to-day movement and his more permanent migration. Obviously an aqueous environment newly created by marine transgression would be unsuitable for occupation and therefore would prompt migration, but a lagoon or recently-abandoned arm of the sea subsequently converted by natural processes to a fresh-water lake where dug-out canoes or other simple craft might be used, might ease certain transport problems of Mesolithic or Neolithic man.

The Holocene sedimentary sequences at a number of sites along the northern shore of the Solway Firth provide evidence of successive changes in environment which were exploited by man. For example, near Annan, Dumfriesshire, in the vicinity of Newbie Cottages (Fig. 1, inset), late-Pleistocene deposits are covered by a Holocene land surface represented by a thin layer of organic detritus and associated development of podzolic A_2 and B_2 horizons in the Pleistocene fluvio-glacial sediments. The former land surface in turn is overlain by Carse Deposits indicative of an estuarine-marine environment during the Holocene transgression. Blown-sand deposits, with several successive interruptions marked in places by layers of peat, in places by humus-rich temporary land surfaces, overlie the Carse Deposits. Associated with the former land surfaces are angular stones and fragments of charcoal (a sample from one surface near the middle of the sequence was dated 3480 ± 110 radiocarbon years B.P., Birm-218), suggestive of human activity nearby. A short distance to the east of Newbie Cottages, in the vicinity of Newbie Villa, the lower part of the sedimentary sequence is similar to that already described. At Newbie Villa, however, the blown sand is missing above the Carse Deposits, being replaced by gravels and cobbles of a coastal bar (Fig. 1, inset). Inferentially, during the early part of the Holocene epoch estuarine-marine conditions existed at both localities. Thereafter the whole area was potentially habitable but, while peat growth and human activity were interrupted from time to time in the vicinity of Newbie Cottages, contemporaneously occasional storms built up a gravel ridge in the neighbourhood of Newbie Villa.

Twenty-four kilometres to the north-west of Newbie Cottages, between Locharbriggs and Sandyknowe Bridge, to the east of Dumfries, the Holocene sedimentary succession records another interesting series of environmental changes (Fig. 3). In the lower part of the depositional sequence, late-Pleistocene fluvio-glacial sands and gravels are overlain by sands containing marine micro-faunal remains, evidence of the penetration of the Holocene sea into a narrow inlet north of the Lochar Gulf. Overlying the marine sands is a thick layer of peat which marks the end of marine conditions in this area at approximately 6600 radiocarbon years B.P. (as discussed above).

Pollen analysis of the upper part of the peat layer at Racks Moss (Nichols, 1967) indicated several periods of decline and regeneration of *Ulmus* (elm) immediately after the pollen zone VIIa/VIIb transition. This suggests that the Neolithic settlement or *landnam* in this area consisted of periods of selective clearance of elm, with probably grazing on the herbaceous plants which colonised the clearings, followed by abandonment and eventual recolonisation by elm. The suggestion of pastoral farming is supported further by the absence of cereal pollen.

North of the village of Racks, the peat is overlain by a variable thickness of up to approximately three metres of silts which are interpreted as representing fresh-water lacustrine conditions. A major piece of evidence in favour of this interpretation was the finding in these silts (at location NY 001801; see also Figs. 3 and 8) in September 1973 of part of a well-preserved dug-out canoe. The canoe was of shallow construction (the greatest depth was approximately 180 mm), such as would be most suitable for use in relatively still water.[1] The geomorphological evidence from the surrounding area, together with the distribution of the lake silts (conveniently exposed in 1973 by dredging operations in the Lochar Water which traverses

---

[1] A piece of wood from the port side of the dug-out canoe has subsequently been dated at the Scottish Research Reactor Centre, East Kilbride. The date is SRR-326, 3754±125 radiocarbon years B.P.

Figure 8. Dug-out canoe discovered on the banks of Lochar Water at location NY 001801, September 1973.

much of the area concerned) suggests that, following the withdrawal of the sea from the long narrow inlet north of Racks, fresh water was present in a number of large shallow hollows occupying the same area. In places these lakes were bordered by and underlain by minerogenic deposits (mainly late-Pleistocene fluvio-glacial sediments), elsewhere by peat accumulations resting on the earlier marine sands. It is unlikely that the whole area from Racks village northwards as far as Locharbriggs, approximately 6 km away, was at any time a single continuous body of fresh water, but at times some of the peat barriers between the lakes may have been breached to give temporary larger lakes. Early man occupying areas surrounding these lakes solved his problems of crossing the long valley between the present position of Sandyknowe Bridge and Locharbriggs by using dug-out canoes in the shallow lakes which in places were bordered by reedswamps and similar hazards.

A further example of changing coastal environments in the course of the Holocene epoch occurs in south-eastern Kirkcudbrightshire in the low ground between Caulkerbush and Southerness (Fig. 2a). There the early- to mid-Holocene deposits are mainly sands or fine sands, apparently the products of open-bay sedimentation. The marine sediments are overlain over much of the area by a variable thickness of peat, averaging around one to two metres, and the peat in turn is overlain by up to two or three metres of blown sand in which several former temporary land surfaces can be detected. The most significant change recorded by the sedimentary sequence and by geomorphological features is in the position of the coastline of this area. At the maximum of the Holocene marine transgression, the coast was located one to two kilometres north of its present position (Fig. 5d). The exact time of the maximum of the marine transgression in this area is not known on the basis of radiocarbon dating of local samples but, as discussed in an earlier section above, the culmination of the Holocene transgression in the Solway Firth area was between approximately 6400 and 5000 radiocarbon years B.P., probably nearer to the latter date than to the former. Recent investigations, by L. Masters, at Slewcairn (NX 924614), a long cairn near Boreland of Southwick, suggest that the cairn may be contemporaneous with a chambered cairn at Lochhill (NX 969651) which yielded timber, radiocarbon dated 5070 ± 105 years B.P. (I-6409). If the two cairns are of approximately the same age, the coastal configuration a short distance from Slewcairn at the time of its construction was very different from that of the present day. Also, the environmental changes which Neolithic, Bronze Age and Iron Age men living in that area must have experienced were marked. With the recession of the sea, the open bay between Caulkerbush and Southerness was replaced by a swampy coastal marsh which in turn gave way to a wind-swept area of low coastal dunes.

In contrast with the areas discussed above, where notable changes in coastal environment occurred in the course of the Holocene epoch, brief consideration is given now to an area where more stable conditions appear to have existed. Both north and south of the River Stinchar at Ballantrae, south Ayrshire, occur narrow tracts of beach deposits laid down in the course of the main Holocene marine transgression, and now raised to heights of about 4 to 10 m above O.D. Traces of the early Holocene coastal history are missing in the neighbourhood of Ballantrae where the river has deposited wide tracts of alluvium in recent times, and seawards from the village the river enters the sea through a gap in a large gravel bar, the configuration of which has changed markedly and dramatically in historical times (Ting, 1937). South of Ballantrae, more than one prehistoric period is represented; Mesolithic and later flint implements, mainly surface finds, have been collected from a stretch of beach lying between approximately 6 and 15 m above O.D. (Edgar, 1939; Lacaille, 1945, 1954). The raw materials used for artefacts, apart from beach flint, included quartz, chert, chalcedony and Arran pitchstone. The unrolled condition of most of the flints, and their location on the beach or in the overlying soil, point to occupation after the time of maximum sea level (Coles, 1964).

The significance of the Ballantrae area in a study of the coastal environment as a *locus* of early occupation of south-western Scotland is in its general stability once habitation had commenced. Unlike two of the areas already discussed, there was no growth of a coastal marsh in this area after the maximum of the Holocene marine transgression. On the other hand, the raised beach deposits north of Ballantrae have a thin cover 'of blown sand which may have threatened early habitation sites on the recently-abandoned beach, whilst in the more sheltered area of raised beach deposits to the south of the River Stinchar, where the main traces of occupation have been found, there is no such cover. The mouth of the river adjacent to the occupation area changed from time to time (if historical events may be accepted as a pattern), but the changes were localised and did not constitute a threat to human activity. It has been suggested that drift flint, carried perhaps by ice-action from Cretaceous rocks submerged beneath the Firth of Clyde, may have been part of the attraction of this area for early human occupation (Lacaille, 1945). Certainly the combined beach and river-mouth environment must have offered a wide range of food (*cf.* references to shell heaps or beds in the vicinity, Smith, 1895; Moore and Smith, 1885).

Included in the finds at Ballantrae were artefacts of Neolithic type, but no pottery or ground and polished stone implements. This suggests contact with later cultures but no great change in economy in the continuing occupation of this relatively stable coastal habitat by mainly hunter-fishers into Neolithic times and perhaps later, in what has been termed an 'epimesolithic' phase of culture.

Synthesis

The significance of Holocene coastal deposits in relation to archaeology is mainly in the provision of accurate indication of the position of the coastline at various critical times during the past 10,000 years, and in establishing a time-scale to which archaeological material may be referred. Such a scale may be established by accurate sampling and plotting of the thicknesses and extent of coastal sediments, by analysis of their pollen content, by radiocarbon dating of biogenic layers within the inorganic deposits and, occasionally, by the discovery of datable archaeological material stratified within or associated with the natural sediments. With notable exceptions at a few localities, the marine transgressions and regressions of the Holocene epoch caused only minor changes in the lateral and vertical position of the shore-line of south-western Scotland. Significant to archaeology, however, were the sequences of sedimentary environments produced by these changes, sequences which varied in both location and time, and which influenced the settlement, economy and movements of early man.

REFERENCES

Coles, J.M. (1964) 'New aspects of the Mesolithic settlement of south-west Scotland', *Trans. J. Proc. Dumfries. Galloway nat. Hist. Antiq. Soc.* 41, 67-98.

Coles, J.M. (1971) 'The early settlement of Scotland: excavations at Morton, Fife', *Proc. prehist. Soc.* 37, 284-366.

Cormack, W.F. (1970) 'A Mesolithic site at Barsalloch, Wigtownshire', *Trans. J. Proc. Dumfries. Galloway nat. Hist. Antiq. Soc.* 47, 63-80.

Cormack, W.F. and J.M. Coles (1968) 'A Mesolithic site at Low Clone, Wigtownshire', *Trans. J. Proc. Dumfries. Galloway nat. Hist. Antiq. Soc.* 45, 44-72.

Edgar, W. (1939) 'A Tardenoisian site at Ballantrae, Ayrshire', *Trans. Glasg. archaeol. Soc.* 9, 184-8.

George, T.N. *et al.* (1967) 'Report of the Stratigraphical Code Subcommittee', *Proc. geol. Soc.* No. 1638, 75-87.

Godwin, H. (1943) 'Coastal peat beds of the British Isles and North Sea', *J. Ecol.* 31, 199-247.

Godwin, H. (1962) 'Half-life of radiocarbon', *Nature, Lond.* 195, 984.

Hibbert, F.A. *et al.* (1971) 'Radiocarbon dating of Flandrian pollen zones at Red Moss, Lancashire', *Proc. R. Soc.* B, 177, 161-76.

Jardine, W.G. (1964) 'Post-glacial sea-levels in south-west Scotland', *Scott. geogr. Mag.* 80, 5-11.

Jardine, W.G. (1967) 'Sediments of the Flandrian transgression in south-west Scotland: terminology and criteria for facies distinction', *Scott. J. Geol.* 3, 221-6.

Jardine, W.G. (1971) 'Form and age of late Quaternary shore-lines and coastal deposits of south-west Scotland: critical data', *Quaternaria* 14, 103-14.

Lacaille, A.D. (1945) 'The stone industries associated with the raised beach at Ballantrae', *Proc. Soc. Antiq. Scotl.* 79, 81-106.

Lacaille, A.D. (1954) *The Stone Age in Scotland.* London.

Livens, R.G. (1957) 'Excavations at Terally (Wigtownshire), 1956', *Trans. J. Proc. Dumfries. Galloway nat. Hist. Antiq. Soc.* 35, 85-102.

McKerrell, H. (1971) 'Some aspects of the accuracy of Carbon-14 dating', *Scot. archaeol. Forum* 3, 73-84.

MacKie, E.W. *et al.* (1971) 'Thoughts on radiocarbon dating', *Antiquity* 45, 197-204.

Moar, N.T. (1969) 'Late Weichselian and Flandrian pollen diagrams from south-west Scotland', *New Phytol.* 68, 433-67.

Mulholland, H. (1970) 'The microlithic industries of the Tweed valley', *Trans. J. Proc. Dumfries. Galloway nat. Hist. Antiq. Soc.* 47, 81-110.

Moore, J.C. and J. Smith (1885) 'Notice of ancient graves at Dounan, near Ballantrae, Ayrshire', *Archaeological and Historical Collections relating to Ayrshire and Galloway* 5, 9-12.

Nichols, H. (1967) 'Vegetational change, shoreline displacement and the human factor in the late Quaternary history of south-west Scotland', *Trans. R. Soc. Edinb.* 67, 145-87.

Renfrew, C. (1970) 'The tree-ring calibration of radiocarbon: an archaeological evaluation', *Proc. prehist. Soc.* 36, 280-311.

Sernander, R. (1908) 'On the evidences of Postglacial changes of climate furnished by the peat-mosses of Northern Europe', *Geol. För. Stockh. Förh.* 30, 465-78.

Smith, A.G. and J.R. Pilcher, (1973) 'Radiocarbon dates and vegetational history of the British Isles', *New Phytol.* 72, 903-14.

Smith, J. (1895) *Prehistoric man in Ayrshire.* London.

Straaten, L.M.J.U. van (1961) 'Sedimentation in tidal flat areas', *J. Alberta Soc. Petrol. Geologists* 9, 203-26.

Ting, S. (1937) 'Storm waves and shore-forms of south-western Scotland', *Geol. Mag.* 74, 132-41.

West, R.G. (1970) 'Pollen zones in the Pleistocene of Great Britain and their correlation', *New Phytol.* 69, 1179-83.

DISCUSSION

Mr. K. Edwards remarked that the absence of cereal pollen at Racks Moss did not necessarily indicate a lack of cereal cultivation since cereals tend to produce little pollen. He considered that a stronger argument would be based on an absence of weed pollen since weeds are normally associated with arable activity. Mr. A. Morrison agreed with this view and referred to the work of Nichols (1967) who showed that there was an absence of weed pollen. The speaker felt that the early Landnam phase in the area was one of pastoral rather than arable activity. In answer to a general question posed by Mr. P. Wilks about the broader relationships of the work, Mr. A. Morrison said that the paper only represented a sample of what was planned. Dr. W.G. Jardine was at present working on the inter-relationships of the coastal deposits and the archaeological material. Dr. J. Evans asked whether the molluscan remains had been studied yet and the speaker replied that they had not, although such a study was planned. Blown sand deposits of the coast had still to be examined. Professor G.W. Dimbleby commented that the buried soils beneath the Luce sands had yielded pollen and that it was possible that pollen might be extracted from the sands.

C. Delano Smith

The Tavoliere of Foggia (Italy): an aggrading coastland and its early settlement patterns

1. The Tavoliere coastal area

That 'there are few wetlands left in southern Europe today' was an ornithologist's comment but it may be of interest to archaeologists too.[1] Planners' progress in most lowlying coastal areas is measured by the increase of cultivable land at the expense of marsh and lagoon. One such erstwhile coast fringes the largest plain of peninsular Italy, the plain of Foggia (just under 4,500 km²). From the impression gained while traversing it, of monotonous flatness, the Foggia plain is also known as the Tavoliere (the tableland).

Nearly 7 per cent of the Tavoliere lies at, or below, 5 m above sea level. Since early in the nineteenth century, at least 20,000 ha have been reclaimed or improved from the alluviums of the coastal hinterland and of the convergent river plains. Hence most of the alluvia marked on the new geological maps (2nd edition) are specified as *colmate* alluvia, that is, resulting from artificially induced siltation. Discharge from the Tavoliere streams has been directed into basins (*vasce*) where it is ponded until the entire sedimentary load ultimately settles out. The resultant loams have a higher clay fraction than do the natural flood-plain silts.

Even to the most casual observer, today's coastal landscape is a very new one. Dykes and drainage ditches bound huge rectangular open fields, ploughed by bright new caterpillar-tracked machinery. There always were big farmsteads (*massena*) on the occasional knolls and ridges, but today there are zones of neat white reform steadings (*podere*). There are still marshes, though, and even open water: Lago Salso at the mouth of the Candelaro and the saltpans of Salpi are both relics of ancient lagoons. Two extensive wild-fowl reserves have been created in the last few years as much, no doubt, to make good the failure to drain completely the coastal area as to preserve the region's cherished reputation as a wildfowler's paradise.

Inland, the limits of the Tavoliere are reached at 400 m above sea level where the highest Pliocene terraces meet the Appennine foothills. On the northern side of the plain the limestone promontory of the Gargano rises in a fault-line scarp to tower dramatically 1,000 m

1 Simms, E. 'Return to the Camargue', B.B.C. Radio, 30 September 1973.

above the plain and over the bay of Manfredonia. From the town of Manfredonia, at the foot of the Gargano, to the mouth of the Ofanto river 50 km further south, a low, narrow sandy beach lines the smooth curve of the bay. The dunes of this shoreline are not well-developed and are rarely higher than 3 m, but the sands are part of a barrier beach that is sometimes nearly a kilometre wide. The barrier is cut across by small river channels (the Candelaro and the Carapelle) and by *foce*, outlets from the saltpans. On its inner side are the alluviums of the former lagoonal hinterland. The coastal flats are lowlying but well inland higher ground is reached when Pleistocene terrain is exposed. Low promontories of this Pleistocene terrain project into the coastlands and advantage has been taken of the dryer land for settlement. In relation to the surrounding flats these *coppa* or *ischia* are seen as of exceptional altitude: *Mont*altino is 11 m above sea level, and the medieval city mound of Salpi, a mere 10 m in altitude, is known as *Mont*e di Salpi!

On the Tavoliere in general as well as in the coastlands, Pleistocene and Early Holocene terrain has a geomorphological peculiarity of immense archaeological significance — a well-developed sub-soil calcrete, sometimes six to eight metres thick, but usually less. Ditches, or any feature cut into this calcareous stratum, are revealed on aerial photographs in outstanding clarity and detail: prehistoric hut and settlement enclosures, Roman vine and olive trenches, roads, house outlines, medieval vineyards, suburban fields and tracks etc. (Bradford, 1957). Tragically, deep-ploughing has been introduced to break up the surface of this calcrete and thus the archaeological record is wiped clean the first time the tractor passes.

2.   *The archaeological problem*

Interest was focussed on the recent sedimentary history of the Tavoliere coastlands by a quite specific problem; that of accounting for the prehistoric and historic settlement pattern of the plain as a whole. Romano-medieval Siponto (in the north) and Salpi (30 km further south) have totally vanished and these once important cities (Siponto was described as '*satis opulenta*' in the seventh century A.D.) are today no more than dusty mounds in the ground. Moreover, while the Tavoliere has been hailed as having one of the densest and most remarkable prehistoric settlement patterns of all Europe (Bradford, 1957, 88; Trump, 1966, 41) exception had to be made for the coastal area, apparently as scantily settled in prehistoric as in recent centuries.

At first glance (Fig. 1) this lacuna in the settlement pattern is easily explained. One of the salient aspects of the distribution of just over 70 surveyed prehistoric sites (out of 1,000 or so cropmarks recorded to date) is that all are on the interfluves, virtually never in the valleys. The valley bottomlands are level and broad, over 3 km in some middle

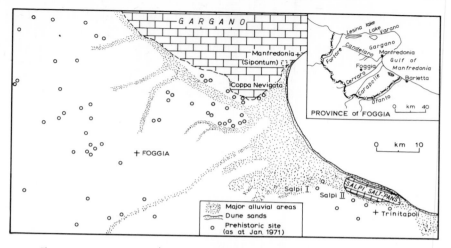

Figure 1. Distribution of Neolithic and Bronze Age sites on the Tavoliere.

reaches, but they have heavy alluvial soils and a cool-temperate type of climax vegetation (deciduous oaks, poplars, elms, etc.). They have always been considered unattractive or unsuitable for habitation. Likewise, towards the coast the alluvia and the marshy environs seem to have been avoided; malaria, as well as flood, is cited as a contributory factor to the negative settlement pattern.

A closer look, however, reveals discrepancies in this assumption. There are prehistoric sites in the marshes and they are, apparently, on *colmate* alluvia. The Neolithic site of Marandrea, for instance, discernible on an aerial photograph as two overlapping enclosure ditches, was the first to be described (Gambassini and Cesnola, 1967). These authors stated (p. 5) that 'the superficial geological formation is constituted by an alluvial deposit almost exclusively silty'. A field visit in 1971 showed this geological assessment to be incorrect (and that, incidentally, there is Bronze Age pottery in the vicinity too: see Delano Smith and Smith, 1973). The site lies at the extreme edge of a low calcreted terrace, separated from higher ground at Montaltino by a broad, shallow, formerly lagoonal channel. Certainly there are silts but only in this channel and the saline soils here remain waterlogged even during the dry summer months. In short, the Neolithic site of Marandrea, far from being found on *còlmate* alluvia, marks the inner edge of an outlying portion of an early Holocene terrace now 2-3 m above sea level.

Evidently, no assumption can be made that the *colmate* alluvia identify the precise outlines of the former lagoons. Aggradation, both natural and artificial, has blurred geological boundaries. More useful an indicator of early Neolithic *terra firma* in the coastlands is the presence of calcrete: where this is found it can be assumed that the locality has been dry land at least since the middle of the Holocene.

To explain the vacuum in the distribution of prehistoric and historic sites in the Tavoliere coastlands, two lines of enquiry are needed – a search for further settlement evidence and an investigation into the post-prehistoric infill that has created a misleadingly level landscape and a sedimentary homogeneity that is more apparent than real. The concept of the Tavoliere coastlands as an unattractive environment for permanent settlement and for arable farming needs reassessment.

3.   Evidence for post-prehistoric changes in the coastlands

The biggest single problem in historical studies is, of course, the inadequacies of sources. It was recently suggested that 'it is apparent that no single discipline will enable a precise statement of palaeo-environment [and its] relationship to the history of man' (Kraft, 1972, p. 15). In the present study of environmental change in the Tavoliere coastlands, as many sources of information as possible have been drawn on: cartographic, documentary, archaeological and sedimentary. In this paper, only the salient points from each source are introduced.

The use of ancient maps is always problematic. This is especially so when local detail is required but the maps are available only on provincial or national scale, as is the case for the Tavoliere. An excellent corpus of charts and maps covering the fifteenth to nineteenth centuries has been published for Italy: R. Almagia's *Monumenta Italiae Cartographia* (1929) contains over 150 maps but only 9 are of the Neapolitan kingdom, as opposed to the whole of Italy, and only 1 (manuscript) map is on a provincial scale.

The maps, taken individually, are disappointing. Most must be considered unreliable when it comes to coastal outlines (with which only the compilers of the portolans had any express concern, and are sometimes inaccurate as regards the precise location of settlements. Nevertheless, from the aggregate of maps, a consensus opinion emerges, about the general layout of the ancient lagoons since the late fifteenth century (Fig. 2). It would seem that there have been two quite distinct areas of water behind the barrier beach, separated by a broad deltaic zone:

in the south: Lago Salpi, with one or, more probably, two associated ponds at its northwestern end;

in the north: Lago Salso, again with two associated lakes, those of Versentino and Contessa.

Further north still, hard against the foot of the Gargano and bordering the deserted city of Siponto, was a third small lagoon, sometimes shown as open to the sea.

After examining the cartographic evidence for the early outlines of Lago Salso and descriptions of the coastlands from the classical period

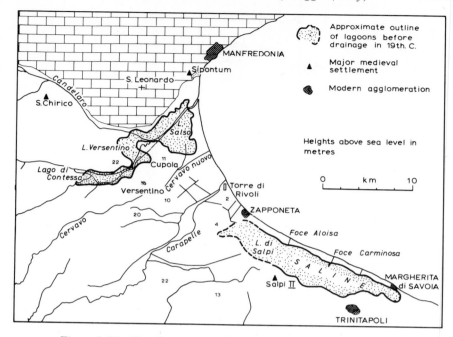

Figure 2. The Tavoliere coastlands early in the nineteenth century.

(notably the Peutinger and Antonine *Itineraria*, and Strabo) there is little doubt that formerly Lago Salso was not so much a lagoon as a *gulf*, open to the sea. As the biggest of the Tavoliere rivers, the Candelaro, would have flowed into this gulf (following a course that has changed little subsequently), Strabo's use of the word 'estuary' in relation to Lago Salso would seem justified. Until after the fifteenth or the sixteenth centuries it would have been impossible to travel south from Siponto (or Manfredonia) along the coastal dunes, as one can today.

From written evidence the names of a number of now-vanished settlements in the coastland can be listed. The sparseness of coastal settlement of recent times just was not typical in the Middle Ages or earlier. There may have been as many as five nucleated settlements where now (excluding Zapponeta, created in the Bourbon reforms of the 1770s) there are only two – Manfredonia and Margherita di Savoia. Again, whatever problems are encountered over the interpretation of each source individually, there is an overall consistency. The lost settlements are listed below (north to south):

Salpi and Siponto were ports in Roman times and during the Middle Ages: Lucan refers to 'the Salapinian pool and Sipus under the hills' (Pharsalia V v. 377). Salpi declined slowly throughout the fifteenth century and was probably deserted soon after 1547 when the bishopric was finally merged with that of Trani. It had been a New Town in the Roman period. Vitruvius relates that a new, healthier,

site was selected four (Roman) miles distant from the old Salpi (de
Architectura I.c.iv). This old city (Daunian or Iron Age Salpi)
survives as a spacious, ramparted site (Tinè, 1969). Siponto was
terminated by the creation of Manfredonia in 1250 (Palumbo, 1953).
Cupola and Rivoli were highly assessed for a special tax in 1300 A.D.
(Edigi, 1917).

Name	Found in Roman texts	Found in medieval texts
SIPONTO	x	x
CUPOLA		x
RIVOLI		x
ANXANUM	x	
SALAPIA	x	x
SALINE	x	x

Archaeological evidence adds topographical and chronological
detail for some of the localities but for others field evidence is elusive
still. Limited excavation has been carried out at Monte di Salpi,
Daunian Salpi, Siponto and Cupola. At Daunian Salpi, for instance,
three huts located on a knoll north of the ramparted area were found
to have been occupied from the ninth or eighth century B.C. (Tinè,
1969). Likewise at Cupola, occupation spanned most of the first
millennium B.C. but of the medieval village of that name, or of Rivoli,
there is no clue either on or offshore. (In 1973 fieldwork included a
submarine reconnaissance north of Zapponeta for the supposed
village of 'Santa Pelagina'; see Delano Smith and Morrison, 1974.)

But with these deserted Daunian, Roman and medieval cities in
mind, it comes as no surprise to find signs of a number of prehistoric
sites, many apparently on *colmate* alluvia. Within 4.5 km from
Marandrea, skirting the former western edge of Lago Salpi, recent
fieldwork (1971, 1973) resulted in the discovery of four quite new sites
(one a promising village of Bronze Age date – a rarity on the
Tavoliere), as well as what was probably a small Republican
farmstead. Also in that area, extensive vineyards, probably medieval,
are seen from the aerial photographs to lie on what the geological map
again classes as *colmate* alluvium, now under marsh scrub. From
further north comes much the same story. The site of a medieval
hamlet at Versentino, for instance, yielded an impressive variety, as
well as range, of Neolithic pottery. Fieldwork in and around the
colmate alluvia may be by no means yet complete but nor is the
distribution map of prehistoric sites in the Tavoliere coastlands.

There is, then, convincing evidence of some quite drastic changes in
the settlement pattern. There is no doubt too that during the long
period of time a variety of factors have been involved. All that need be
said here is that it would be misleading to look for mono-causal

factors. For instance, 'unhealthiness' (usually, but perhaps mistakenly, taken as referring to malaria) is a chronic complaint. Vitruvius is fairly explicit in allying the move from Old Salpi after 100 B.C., during the Roman Republic with the marshiness of the environs (the topic of his chapter), but in medieval literature the adjective becomes so commonplace that its environmental significance may be sometimes doubted. More pertinent to the decision to create a new town and port two kilometres north of Siponto would have been earthquake damage, a factor heavily stressed by Palumbo (1953, 372ff). However, he quotes the earthquake of 1223 as responsible, but from the cartulary of S. Leonardo it is seen that references to 'the ruined city of Siponto' start in 1155 (Camobreco, 1913). A third component of settlement desertion must be the decision itself. The Hohenstaufen were noted New Town developers and Frederick II had already created nearly a dozen new ports in Apulia before Manfred followed suit with the New Siponto (later to be called Manfredonia) in 1250 or 1253.

Nevertheless, the fact has to be faced that aggradation was present and presumably a pernicious influence. By siltation hollows were levelled and streams and channels became shallower, adding yet another contributory factor to changes in the settlement pattern. Certainly it was an effective process; the flatness of today's landscape is an unarguable testimony to this. The largest mound at the Bronze Age village of Carapellotto-Regina rises above the general level by no more than 70 cm; the site of the Iron Age huts at Daunian Salpi is but 150 cm higher than the surrounding flats. It may be that we are dealing now with an exceptionally flat landscape but this was not always so, and our interest in the environmental change affecting settlement location and siting must take us below the ground surface.

To demonstrate the relevance of the buried relief pattern to settlement siting in the coastlands the sediments of the post-prehistoric infill have been studied. For instance, a series of hand-augered profiles was investigated across part of a shallow depression within Daunian Salpi. This hollow, known as Marana di Lupara from its marshiness, is nearly three kilometres broad but has an altimetric range of no more than 150 cm. It is covered by *colmate* alluvia and remains inadequately drained. Towards the centre is the modest knoll on which were sited the three Iron Age huts excavated by Signora F. Tinè. Below the huts was about 30 cm of palaeosol, overlying calcrete. The sedimentary cross-section runs from this knoll south across to the ramparted city site (Fig. 3).

From the cross-section we see that the Marana contains an infill of about 2.50 m of loamy flood-plain silts resting on white lagoonal deposits. Over half of this infill is *colmate* alluvium but the silts are very similar throughout the profile and facies indicating depositional changes which might be revealed by a very detailed granulometric analysis. The stream responsible for the infill today enters the

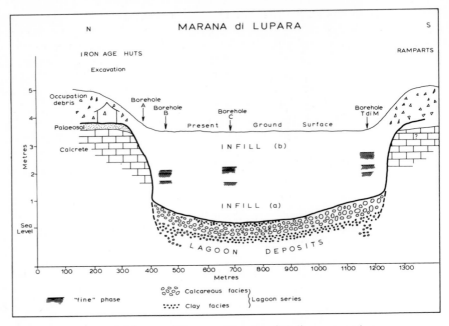

Figure 3. Marana di Lupara (Daunian Salpi), cross-section.

depression from the south-west and is very small, with a total catchment of less than half a dozen square kilometres. It is not surprising that the infill, too, is rather shallow. However, it is clear enough that formerly the hollow at Marana di Lupara was not only much deeper, but also water-filled, and that this was one of the ancillary ponds of Lago Salpi depicted in many of the ancient maps. That the water in the lake was well agitated and had good circulation is indicated by the whiteness of the bottom deposits, and signs of aerobic conditions and a good rate of plant decomposition. That it received some stream discharge may be indicated by its relatively sandy texture. Exact dating of the facies is not yet available but it is suggested that the occupants of the huts and of Daunian Salpi would have looked out across the flowing waters of à broad lagoonal complex of Lago Salpi.

4. *Sediments and archaeology at Siponto*

By definition, a lagoonal coast is transitional and a mixing of sediments is to be expected. At Siponto, in the northern part of the Tavoliere coastlands, some very different deposits were encountered. Two series of bore-holes were augered, about one kilometre apart, across the northern and the southern ends of the former lagoon. On the basis of detailed analyses of the samples a life-history of the lagoon – from its first appearance as a submerged bar to its recent transformation into dry land, land gained from the sea – can be outlined. Some of the salient points from this study of the recent sedimentary history of the Siponto lagoon are presented here. It is to be stressed, though, that neither fieldwork nor the study is yet near completion, and that although the outlines are presented with confidence, further work might well demand modification in respect of local details.

It has been pointed out (Morrison, 1973) that to recreate accurately a chronological sequence for buried sedimentary environments and to identify the major morphological changes in an area, is in itself a useful achievement. The finding of evidence giving a precise date for any buried sedimentary feature (other than by dating the deposits themselves, often a costly and then not necessarily convincing operation) is largely a matter of luck. As fortune had it, for the Tavoliere coastlands, the discovery of a first century A.D. *villa urbana* at Siponto provided the key to the date of the Siponto lagoon.

It may be useful to say something further concerning the analysis of the Siponto deposits before illustrating their interpretation, partly as background to the reconstructed sections and plans and partly from methodological interest.

It was found at Siponto that recognition of the original sedimentary environments depended on a threefold analysis, of texture, colour and mollusca content. Other characteristics analysed were merely corroborative. With regard to texture, it was noted that the early sedimentary environments in the northern part of the lagoon (S. Maria profiles) were substantially different from those in the southern part (Villa profiles). All the sediments in the north were wholly or very largely sandy. Those in the south, however, were more commonly clayey, the major contrast being between gravelly clays and clays without any gravel. There are sands at the Villa end of the ex-lagoon but only in the upper part of one borehole.

The colour of the deposits revealed further distinctions. These last-mentioned sands, in the Villa profile, were yellow-buff ·in hue and quite different from the sands further north. Here, in the S. Maria sections, they were either so dark as to appear black, or they were a lighter grey, or, finally, they were a rather dull buff. These colour differences relate to the origins and history of the sands in a most illuminating manner.

One of the peculiarities of the present-day shore of the Manfredonia Bay is that while dune sands are a dull buff colour, the beach below is grey or even black. This dark hue is due to an exceptionally high proportion of heavy minerals (pyroxene and magnatite mainly) derived from the volcanic rocks of Monte Vulture and brought to the sea by the Ofanto river (Sartori and Quaratesi d'Archiardi, 1966). Most of the Ofanto material is moved southwards (Admiralty, 1954), interfering with the port of Barletta, but some is evidently propelled north along the Tavoliere coast by currents, by winds and by tides.

While heavy minerals remain to colour the sub-tidal sands as far north as Siponto, they are quickly destroyed on exposure and so are rare in dune sands. The yellow Villa sands have a quite different origin and seem to come from the other side of Manfredonia, from the Gargano coast.

The Siponto molluscs fall into two groups, brackish water species, all gasteropods, and marine species, mainly bivalves. The dominant brackish species were undoubtedly *Phytia myosotis* and *Pseudamnicola confusa* (Ellis, 1926), and the dominant bivalve was a Donax, *Donax vittatus* (Tebble, 1966). In some of the Villa sands, *Mactra corallina* occurred in considerable numbers, together with *cardium*. The salinity tolerance of the brackish gasteropods is varied. *P. confusa* is normally associated with rather fresh water, together with the third brackish gasteropod encountered, *Hydrobia ventrosa* (Ellis, 1926) but for all three a lagoon environment is a normal habitat. Almost all the marine bivalves are normally associated with clean sand habitats in offshore and low-tide zones and are commonly found as empty shells on beaches. Those from the S. Maria profiles are indeed very fragmentary and were mixed with sub-tidal vegetation (*Posidonia sp.*) but the fact that the very thin and brittle *mactra* shells were, on the contrary, rarely damaged indicates that these had been deposited in water of some depth.

From the distribution of marine and non-marine species in the sedimentary profiles, a salinity curve may be inferred. This tells us something of qualitative changes in the lagoon. For instance, marine shells were predominant in all the basal deposits of the S. Maria profiles. But in profile III (the outermost; today, in the middle of the barrier beach) they cease as soon as the sands become grey instead of the dark subtidal variety. In none of the overlying dune sands were there any molluscs.

Quite a different story is deduced from the mollusca content of profile I (located in the former lagoon). After similar marine and highly saline beginnings, conditions became brackish. This phase was followed by a short-lived reversal to high salinity, with marine shells, followed by a return to brackish conditions. This might indicate merely local changes such as the shifting of a sandbank, but since it is paralleled elsewhere in the lagoon this may be evidence of a change in relative land/sea levels. Next, suddenly, the quantity of brackish shells in the profile increases quite dramatically and there is a good deal of plant debris (*Posidonia* again). Indubitably, this was an aquatic environment *par excellence*, where molluscal and plant life flourished in a good depth of brackish water. In the final phases of profile I sediments become progressively drier, to judge from the marked decrease in shell quantity, although the water remains brackish, as it is today. This was the period of infill, *sensu stricto*, and of *colmate*.

We can turn now from the individual sedimentary environments to attempt to recreate in plan the former lagoon. The entire sequence in

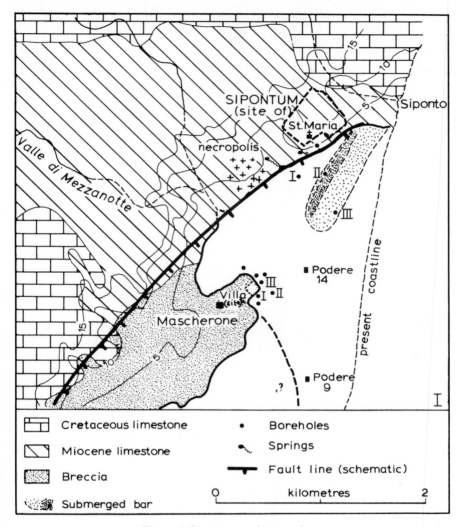

Figure 4. Siponto: pre-lagoon phase.

all its details cannot be presented here and so only the early stages are discussed. First, the pre-lagoon phase (Fig. 4). At this time there was no lagoon, only a submerged bar lying about 400 m off-shore. The coast had then an L-shaped outline demarcated by the limestone and breccia outcrops, only the Mascherone breccia shelf may have extended further east. On land, two streams united before debouching into the sea just opposite the northern promontory of the Mascherone terrace. By the following sedimentary period, however, the bar had emerged and, capped by dunes, is better termed a barrier beach. It enclosed brackish water in the north but it did not reach the opposite shore, the Mascherone terrace. Instead, water was kept circulating through this southern channel by outflow from the streams and, possibly more importantly, from the several springs that issued from the limestones. This current would have passed close to the Villa promonotory, hence the light-coloured river or channel clays, more or less gravelly, found at the base of the Villa profiles.

The rest of the lagoon's history, as recorded in the sediments, concerned at first only the land/sea-level changes that affected the quality of the environment inside the lagoon rather than its outlines. But the time came when a major influx of marine sands initiated the blocking of the channel. This was the process that led ultimately to the total closure of the lagoon, the shrinkage of its water-body, and an infilling process that recent reclamation merely accelerated.

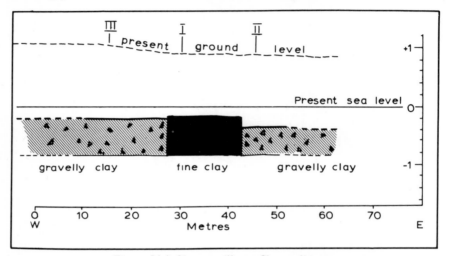

Figure 5(a). Siponto villa profiles: sediments.

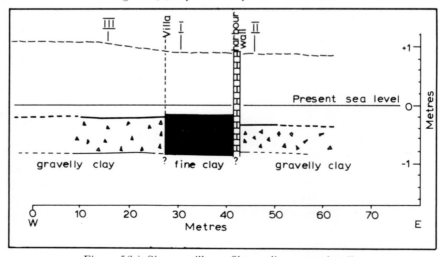

Figure 5(b). Siponto villa profiles: sediments and walls.

The appearance and then the closing of the Siponto lagoon can be dated, though in approximate terms only. Fig. 5 shows how the archaeological evidence fits, and explains the sedimentary sub-environments. This is one of the series of cross-sections from the Villa profiles. Two salient facts emerge: that while two of the profiles show gravelly clays, the third (central) one has a similar but gravel-less clay; and that there is a marked discrepancy in the depth of material accumulated. There are 80 cm of gravelly clay in profile III, a few centimetres more in profile I, 20 cm less in profile II. The gravel-less clay reflects deposition in quiet water, say a backwater, away from the channel current. As the borehole for profile I was in fact not in exact alignment with the other two but set back by about 30 m, this could be an acceptable explanation of the textural differences.

But the sub-environmental pattern is better explained if it is assumed that the villa itself had by now been built. In Fig. 5b the outer wall of the villa and the main wall of its little harbour complex have been added as appropriate. It is seen that the gravel-less clay was not deposited in a small creek, but in a moat-like channel in front of the villa itself. It is also apparent that while the accumulation of gravelly clays was greater on the landward, relatively sheltered, side of the harbour wall, rather less was deposited on the seaward side, perhaps due to scour.

The villa appears to have been built around the middle of the first century A.D. (Dr. E. de Juliis, personal communication). From then on, the sedimentary history on either side of the 'harbour' wall was different. The rest of profile II is composed wholly of marine sands, gravels and, finally, beach shingle; all this material blocked the lagoon channel. On the landward side the clays become silty, testimony to the infilling process that the blocking brought about. At this time, the sea must have been higher than it is today (perhaps by as much as 80 cm) but it is possible that upward land movement, perhaps associated with earthquakes, contributed to these changes.

The beach sediments of the marine infill (profile III) are predominantly sands alternating, however, with three thin layers of gravel in the middle section. This alternation of fine and coarse material reflects changes in depositional conditions: tidal, current, wind and wave patterns. Or once again the shift in beach-building equilibrium might be due to quite minor sea/land variations effected by local tectonics.

On the other hand it should be said that as fieldwork proceeds on the Tavoliere coastlands, so evidence is accumulating that strengthens the impression gained at the outset: a major factor in the aggradation of the Tavoliere coastlands may prove to be rises in the relative level of the sea as in the closing of the Siponto lagoon (datable, on cartographic evidence, to the early or middle sixteenth century A.D.).

5.   Conclusions and some economic implications

The transformation of the Tavoliere coastlands from lake and marsh to *terra firma* and arable land, so ardently striven for in recent years, is probably as complete as it ever will be. During the years of transition (essentially medieval but possibly also post-Roman) marshiness and flood hazard in the coastlands would have meant the area was viewed with dislike. This is the assessment that has passed into the literature and that influences our appreciation of the environment today. But in prehistoric times, when the lagoons would have been still well-defined water bodies with distinctive outlines (as at Marana di Lupara) much could be said for the economic advantages of such a coastland. Water from inflowing streams would have helped to maintain a measure of freshness in the lagoon water: the reputation of Lago Salpi as a fishing ground was lost only in the middle of the eighteenth century. On the surrounding interfluves and headlands, wherever there was subsoil calcrete, arable opportunities would have been every bit as good as further inland, where land is excellent for dry-farmed cereals. The lagoons themselves would have meant an opportunity for additional economic assets, in terms of diet (fish, for example), commercial exploitation (salt, shells) and navigation. To view the early coastal environment as merely an area 'of halophytic and hydrophilous vegetation [which] can have been of little importance for man, although locally reeds, rushes etc. were doubtless used for basketry and building' (Whitehouse, 1968, 339) is to come too near to false analogy. The only way to appreciate the human past in a 'dead flat' landscape is to look below its surface.

Acknowledgments

Grateful thanks are expressed to: the British School at Rome and the British Academy for financial aid; Mr. B. Ward-Perkins of the British School for many years of personal encouragement; Professor and Signora Tinè for generous collaboration and much help in Italy; the Soprintendente della Antichita della Puglia and Dr. de Juiliis at Foggia; the Department of Geography at Nottingham for assistance in many ways; Professor N.K. Jacobsen (the University of Copenhagen) for a most valuable lesson on the sedimentary history of the polders in East Jutland, and for an equally valuable tool, the soil auger; Mrs. E. Pyper for laboratory analysis and Mr. C.A. Smith for fieldwork in 1971; Nr. B.W. Sparks (Cambridge University) and Mrs. S. Whybrow (British Museum, Natural History, molluscs section) for the identification of molluscs; and to Dr. I.A. Morrison and to the rest of the 1973 fieldwork team – Don, Joan, Peter and Paul Smith, Martin and Judy Dean and Denis Mott.

REFERENCES

Admiralty (1957) *Mediterranean Pilot*, Vol. III.

Bradford, J. (1957) *Ancient Landscapes*, London.

Camobreco, F. (1913) *Regeste San Leonardo di Siponto*, Regesta Chartarum Italiae no. 10. Rome.

Delano Smith, C. (1973) 'Deserted settlements in Apulia (Italy): the Tavoliere', paper read at the *Conference Européenne Permanent pour l'Etude du Paysage Rural* at Perugia, Italy, May 1973. To be published in the conference proceedings as 'Villages desertés dans les Pouilles (Italie): la Tavolière.'

Delano Smith, C. and C.A. Smith (1973) 'A note on Bronze Age pottery from the Tavoliere, Apulia', *Proc. prehist. Soc.* 39, 454-6.

Delano Smith, C. and I.A. Morrison (1974) Report on 'Santa Pelagina', *Nautical Archaeology*.

Edigi, P. (1917) *Codici Diplomatici de Saraceni di Lucera 1285-1343*. Naples.

Ellis, A.E. (1926) *British Snails*. Oxford.

Gambassini, P. and A.P. di Cesnola (1967) 'Resti di villaggi neolithici a ceramiche impresse a Trinitapoli (Foggia)', *Rivista de Scienze Preistoriche* 12, fasc. 2, 1-18.

Kraft, J.C. (1972) *A Reconnaissance of the geology of the sandy coastal areas of Eastern Greece and the Peloponnese*. Technical Report No. 9. College of Marine Studies, University of Delaware.

Morrison, I.A. (1973) 'Geomorphological investigation of marine and lacustrine environments of archaeological sites, using diving techniques', *in* N.C. Flemming ed., *Diving Science* (Proceedings of the 3rd Symposium of the Scientific Committee of the Confederation Mondiale des Activitiés Subaquatiques), London, October 1973.

Palumbo, P.T. (1953) 'La fondazione di Manfredonia', *Archivio Storico Pugliese* Anno VI fasc. I-IV, 371-407.

Sartori, F. and L. Quaratesi d'Archiardi (1966 'Studio mineralogico di una sabbia della foce dell' Ofanto', *Atti. Soc. Tosc. Sc. Nat. Mem. A.* 73, 112-132.

Tebble, N. (1966) *British Bivalue Seashells*. British Museum (Natural History) Handbook. London.

Tinè, S. and F. Tinè (1969) '*La Magna Grecia e Roma nell'età Arcaica*', in *Atti dell' Ottavo Convegno di Studi sulla Magna Grecia, Taranto, 6-11 October 1968*, 233-241.

Trump, D. (1966) *Central and Souther Italy before Rome*. London.

Whitehouse, R.D. (1968) 'Settlement and economy in southern Italy in the Neothermal period', *Proc. prehist. Soc.* 34, 332-67.

DISCUSSION

The discussion on this paper was opened by remarks from several people concerning the speaker's use of an auger for sampling. Several speakers felt that the auger in question was inferior to standard models. Dr. M.L. Shackley· enquired about the degree of sample contamination when augering, but Miss Delano Smith felt that this was minimised since the obtained core was cut back to the auger for a clean break. The speaker further explained that although the auger was not generally a recommended way of obtaining an intact sample, for a preliminary study of this kind it was very useful, and that future work would include mechanical equipment.

Further discussion included comments from Mr. J.L. Bintliff concerning the causes of aggraduation in the area and to the effect of small streams. Mr. A. Bolt considered the effect of coastal dynamics and increasing shelter leading to infill and sedimentation. Dr. R.A. Webster enquired if it would be possible to draw maps showing the palaeophysiography and human settlement at various points in time. Professor A. Straw asked if the original harbour wall had been higher which seemed likely from the lack of pebbles on the other side. The speaker considered that the wall had originally been much higher and that it had been subjected to stone robbing. Professor A. Straw continued by asking whether the speaker had considered it possible that the materials outside the wall could represent one complex beach accumulation instead of four separate events. The speaker said that this was a hypothesis which had occurred to her, but that she felt that the recurrence of the pattern in areas 30 km apart argued against the deposits representing wave activity in a normal beach situation.

S. Limbrey

Tlapacoya: Problems of interpretation of lake margin sediments at an early occupation site in the basin of Mexico

The valley of Mexico was dammed by lava and tuff to form a close basin at some unknown time, perhaps in the Lower or Middle Pleistocene. Cumulo volcanoes, such as Popocatepetl, and numerous cinder cones, part of the volcanic belt which crosses Mexico at this latitude, have poured out ashes at intervals, blanketing the landscape, contributing to its soils and providing a great part of the lacustrine sediments which have accumulated in the basin, to a depth of at least 800 m, since it became closed. The finer of the ashes provide stratigraphic markers within and potentially far beyond the basin (Mooser, 1967; Cornwall, 1968), and a means of correlation between the lake sediments, with their high potential for preservation of evidence of environmental conditions in the lake and on the surrounding slopes, and the glacial features on the mountains of the enclosing wall (White, 1962). The region thus contains much information about the sequence of climatic changes at this latitude through at least the Upper Pleistocene.

In establishing a sequence of fluctuations in lake level which may be interpreted in terms of changing climate, the dumping of large quantities of volcanic ash into the lake from time to time introduces a complicating factor. Water level in the lake depends on the following variables which are climatically dependent and to varying degrees inter-dependent: rainfall; snowfall and melting regime on the mountains; evaporation from the lake and from the vegetation and soils of the slopes; absorption, runoff and erosion associated with the soils. It also depends on the amount of volcanic ash deposited directly into the lake, which is climatically independent, and the rate of redeposition in the lake of that which falls on land, which is dependent on the quantity and the characteristics of each ash fall and on the length of the interval between successive falls, and on the stabilisation of fresh ash falls by vegetation and the rate at which soil formation contributes structural stability to an incoherent ash, which are in part climatically controlled and in part dependent on the characteristics of the ash. Both the level of water in the lake and the salinity of that water, another potentially useful indicator of climate, and one which is being studied by means of analysis of diatom populations

(Bradbury, 1971), are also affected by the construction and erosion of volcanic barriers within the lake, forming sub-basins whose isolation or interconnection depend on the history of the barriers as well as on water level, so that continuity of lake level and mixing of water (whose salinity varies locally with position and volume of input streams and with depth and temperature of the water) are intermittent.

Thus, the sedimentary sequence at Tlapacoya can be interpreted only in local terms at present. Work done in future in other parts of the basin can be correlated with it by means of the ash layers and eventually an understanding of variability within the basin will allow interpretation of regional significance to be made.

1.   The site of Tlapacoya

The hill of Tlapacoya is the greatly eroded and fault-dissected remnant of an andesite volcano which existed in the valley before it was closed to form the basin. As sediments accumulated around its foot, the hill must at some times have been an island and at others a peninsular connected to the lake shore by a low marshy neck. It lies within the sub-basin of Lake Chalco, which is almost, and at times would have been completely, cut off from the main basin by a cindery ridge. Fig. 1 shows the location of Tlapacoya and the outline of the lake as it existed at the time of the Spanish invasion, when problems of salinity affecting food-growing areas around the shores, and of flooding, had already led to the initiation of the engineering works, designed to prevent mixing of fresh and saline waters and to drain parts of the lake, which continue to the present. Lake Chalco was drained probably in the last century.

Excavations at Tlapacoya began as a result of motorway construction. When material was scraped from around parts of the foot of the hill, lake margin sediments banked against the hill were exposed. Since the lake was drained, shrinking of the finer sediments out in the lake has been greater than that of the coarser, inshore deposits, and as these latter are supported by the solid rock, resulting tension has been relieved in a zone of complex faulting associated with increased tilting of sediments originally laid down at an angle established by shore processes. The effect of scraping around the skirts of the hill was therefore to cut across the tilted layers and sever the continuity between the more lacustrine sediments, now exposed in echelon in the flat surface, and the more terrestrial, exposed in a vertical section. In the cliff so created, among the sediments of predominantly terrestrial nature a layer of beach pebbles was noticed, and upon investigation there were found to be large numbers of animal bones, of an Upper Pleistocene fauna, together with artefacts and traces of hearths, on the beach and in the deposits immediately overlying it.

Work has been going on at Tlapacoya since 1965, carried out by

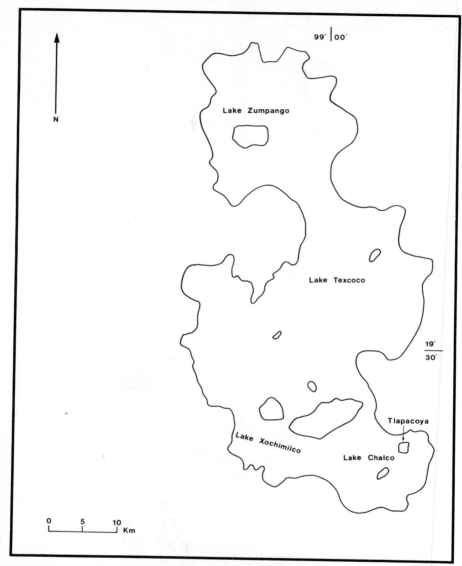

Figure 1. Outline of the lake in the Basin of Mexico, approximately at is was at the time of the Spanish invasion, showing the position of Tlapacoya in the sub-basin of Lake Chalco (after Bradbury, 1971).

archaeologists and environmental scientists of and associated with the Departmento de Prehistoria of the Instituto Nacional de Antropología y Historia, under the direction of Professor J.-L. Lorenzo. Excavations were directed in the first season by Michael and Elizabeth Goodlife, and in subsequent seasons by Lorena Mirambell. Interim reports have been published (Goodlife, 1966; Mirambell, 1967) and a monograph dealing with all archaeological and scientific aspects of the site is to be published (Lorenzo, in press).[1] The work on which this paper is based was carried out in 1969 and 1970, while the author was employed in the laboratories of the Departamento de Prehistoria, and involved detailed examination in the field and by microscope techniques of the lake margin and hillslope deposits.

Investigation of the more lacustrine deposits, in which stratification is generally sharp and detailed was carried out to provide stratigraphic control for the more terrestrial zone, in which the evidence of human occupation was found, in which stratification is blurred and lacks detail, as well as to provide evidence of environmental conditions throughout the period covered by the exposed deposits. Long trenches had been cut into the horizontal scraped surface, leading out into the lake from the hill slope against which the terrestrial deposits were banked. Because of the tilted arrangement, a sequence equivalent to about ten metres in depth was revealed in cuttings only two to three metres deep, but though each layer could be examined in a number of cuttings over lengths of several metres, in most cases direct comparison of deposits formed at the same point through a sequence of layers was limited to only part of the full sequence; a well shaft close to one of the cuttings was available for examination, and revealed the lower seven metres of the sequence, but the deposits at that point were different in some respects to those in the nearby cutting, and together with a number of subsidiary cuttings served to illustrate the degree of lateral variation. Cuttings were made on both sides of the hill, to the south-east where the original excavations were carried out, and to the north-west. The upper part of the deposits were also exposed in excavations of an occupation site dating from 7000 B.P. onwards to the east of the hill (Neiderberger, *in* Lorenzo, in press); work continues on the abundant evidence of occupation in these later periods in the area of the main cuttings (Mirambell, personal communication).

2.   *The deposits*

Fig. 2 contains a summary of the stratigraphy, as expressed in the main cuttings on the south-east side of the hill. The radiocarbon dates which are listed in Table 1 derive from samples taken from a number of cuttings; where layers have been dated more than once, results are consistent, and the sequence, with one exception, is stratigraphically

1. The monograph on Tlapacoya (Lorenzo, in press), contains reports of work on further aspects of the study of soils and sediments by A. Flores, and of tephrachronology by W. Lambert. Biological studies include pollen analysis by L. Gonzalez and faunal analysis by T. Alvarez, as well as the work on birds and on diatoms mentioned above. The results of these studies are essential for a greater degree of environmental interpretation than is attempted here, and will provide a basis for understanding the sequence of climatic variations in the Basin of Mexico through a period of over thirty thousand years.

    J. Liddicoat of the University of California, Santa Cruz, is using the sediments at Tlapacoya in a study of palaeomagnetic variation. (Liddicoat, pers. comm.)

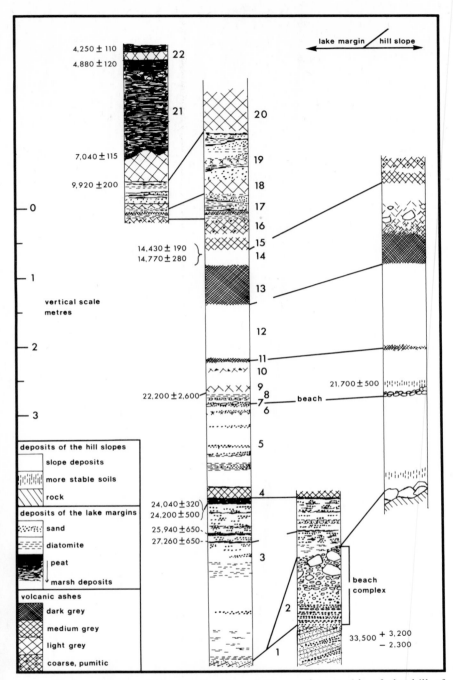

Figure 2. A summary of the stratigraphy at the south-east side of the hill of Tlapacoya. The diagram renders in vertical form a situation in which tilted layers are exposed over a horizontal distance of over 60 m, so that the foot of the long central column must be sought some 20 m to the right of its head, and the terrestrial situation shown in the column on the right is to be found another 20 m to the right; the upper part of the stratigraphy is shown on the left as it is to be found about 25 m to the left of the head of the central column. Dates are given in radiocarbon years B.P.

Table 1. Radiocarbon dates.

Stratigraphic position of sample	Date, in radiocarbon years before present	Laboratory number	Material, where known,
Base of unit 23	4250 ± 110	I 4404	charcoal
Top of unit 21	4880 ± 120	GX 1298	peat
Base of unit 21	7040 ± 115	I 4192	charcoal
Top of unit 19	9920 ± 200	I 6897	
Top of unit 14	14430 ± 190	I 5185	peaty marsh soil
Top of unit 14	14770 ± 280	GX 0878	peaty marsh soil
Hearth, in slope deposits above unit 7	21700 ± 500	I 4449	charcoal
Within unit 8	22200 ± 2600	A 790A	humic extract from soil and charcoal
Top of unit 3	24040 ± 320	GX 1362	wood
Top of unit 3	24200 ± 500	A 793	wood
Upper part of unit 3	25940 ± 650	GX 1364	wood (root)
Upper part of unit 3, below GX 1364	27260 ± 650	GX 1363	wood
Within unit 1, on the upper of two beaches	33500 $^{+\ 3200}_{-\ 2300}$	GX 1103	wood

Table 2. Volcanic ashes.

Unit	Hornblende	Hypersthene	Augite	Olivine	Biotite
Fine grained, light grey (almost white) ashes:					
6	x	x			
9	x	x			
18		x	x		
20	x	x			
22		x	x		
Medium grained, medium grey ashes:					
4	x	x			
15	x	x			
Coarse grained, dark grey ashes:					
1		x (little)		?trace	
11			x	x	
13		x			
Coarse, pumitic ash with fragments of andesite rock:					
16	x	x	x		x

consistent. The characteristics of the main marker layers of volcanic ash are given in Table 2. Ashes which appeared similar in the field could be distinguished by their heavy mineral associations. In the neighbourhood of Tlapacoya, where an exposure will usually contain a sequence of layers sufficient for the mineralogically similar ashes to be distinguished by their context, simple presence/absence criteria are adequate. More precise characterisation of the mineralogy, necessary if the finer ashes are to be used for long distance correlation, has been undertaken by Virginia Steen MacIntyre at the U.S. Geological Survey, Denver, Colorado.

The stratigraphic units numbered on the diagram were established as a convenient framework for reference and discussion. Well-defined layers, such as the main ashes and beaches, were given individual unit numbers; other units are complexes of interstratified materials which vary laterally around the hill and with the position of the lake margin, but each unit as a whole appears to represent an overall consistency of depositional conditions, in which fluctuations were of relatively minor importance. The deposits will be described from the lowest exposed upwards, mainly as seen in the main cuttings on the south-east side of the hill, and discussion will be concentrated on those features which present problems of interpretation in terms of lake margin processes and conditions rather than on interpretation in climatic terms. In the terrestrial zone, the equivalent of stratigraphic units established in the more lacustrine deposits could not always be distinguished; these deposits are discussed separately.

Unit 1

These coarse, dark grey ashes lie banked against the rock of the hill, at an angle of about 35° to the horizontal, in thin, rather uniform layers and elongated lenses. They show to a slight degree a characteristic of all beach materials at the site, and one by which a beach phase can be differentiated from a layer of coarse volcanic ash which has not been reworked at the lake shore: the cindery, vesicular shapes of the particles are filled out and coated with a colourless, highly birefringent material, which has not been identified. These ashes appear to have been washed off the hillside into the edge of a deep lake. There are among them at least two phases in which around parts of the hill beach activity was more strongly developed, and associated with these beaches are large numbers of bird bones: diving birds are abundant, confirming the impression of a deep lake (Phillips, *in* Lorenzo, in press).

On both sides of the hill the black ashes are faulted out to an unknown depth a little way from the shore. This is the only fault which took place during deposition of the sediments at the site: all the other faulting is a result of the drying out of the lake sediments in recent times. Here, beyond the fault there is a mass of unstratified ashes, suggesting that as the bank of ashes collapsed a cloud billowed out into the water and then settled.

Unit 2

The steeply banked ashes are truncated by the development of a major transgressive beach. If, as seems possible, the volcanic activity which produced them is the same phase as that which built the barrier almost cutting off Lake Chalco from the rest of the basin, it may be that this change in position and angle of at the shore is a result of the same episode.

The beach cuts right across the earlier deposits and into the rock of the hill. The sands are thickly coated with the birefringent material and there is a great abundance of a powdery deposit of amorphous silica, which lies among the sand and coats the pebbles; there is some encrustation and cementing of the beach materials with a combination of this and iron oxides. Traces of calcium carbonate are also present. During the period of activity of the beach, a rock fall occurred, perhaps because of the undercutting of the hill, or perhaps because of an earthquake: this latter suggestion is supported by the presence of a whole tree, a swamp cypress, *Taxodium mucronatum*, lying among the beach sands. Mixed with the sand at this level is an abundance of small wood fragments, not only around the tree, but in cuttings on the opposite side of the island. The tree would not have been preserved had it lain exposed on the beach; its rapid burial is consistent with a somewhat catastrophic process.

It is this tree which has yielded an anomalous radiocarbon date, 23150 ± 900 B.P. (GX 0959), which is younger than the sequence of dates obtained on samples from the upper part of unit 3.

Although the beach deposits of unit 2 occur on both sides of the hill, they are completely missing from the well shaft. Here, at this level, is a deposit unlike any other at the site. It is superficially like a rather muddy volcanic ash, and consists of predominantly silt sized particles of felspar and glass, and within it there are lenses or pellets of the powdery amorphous silica which is only otherwise seen among the beach materials. Pyrites has crystallised after deposition. A possible interpretation of the deposit is that it is the result of winnowing of fine particles from the beach and the settling out of the silt grade in an area which for some reason was protected from wave action and incursions of coarse materials. Throughout most of the sequence the deposits seen in the well shaft are reduced in thickness and low in mineral content, having in this respect more in common with those on the opposite side of the hill than with those in the neighbouring cutting.

Unit 3

Above the beach materials of unit 2 is a complex of layers of redeposited ash, with diatomite, showing evidence of channelling and infilling with current bedded sands and representing a transition between the terrestrial deposits and the marsh deposits which form the lake shore once regression from the beach had occured. For later periods, this transitional zone is missing, having been removed by the bulldozers, so it is not clear whether redistribution in this way by running water is always characteristic of the lake margin during the marsh phases.

Further out across the lake margin, these deposits lose their characteristics of sorting and differentiation and become the marsh deposits, or marsh soils, which form an important part of the subsequent stratification. They consist of partly weathered volcanic ash which has arrived at the lake shore, after some weathering and soil formation on the hill slope, as a result of continual or intermittent soil creep, hillwash and deflation. They contain a scatter of highly humidified plant fragments and abundant diatoms, and a content of amorphous humus which increases as the mineral content becomes finer and the layers thinner towards the lake. Within each layer there is little evidence of sorting or stratification, and the material is compact, showing no soil structure. There is much segregation of iron oxide around root holes, and some globular deposits of colloidal iron oxide, probably resulting from bacterial activity.

In the lower part of unit 3 in its more lakewards situation there is evidence of slumping of marsh deposits down the slope of the underlying beach, a movement perhaps facilitated by the slipperiness of a diatomite layer which seems to have overlain the beach in its lakewards part. Within the upper part of the marsh deposits of unit 3 are peat layers which persist, though becoming very thin towards the shore and are identifiable at different positions around the hill. These peats have been dated and yield the sequence shown in Fig. 2.

There appears to be a cyclic alternation here of sand and peat among the marsh deposits; this sand is not beach sand and may be the result of streams crossing the marsh zone. The association of peat with sand in this way does not occur at other levels, where sand, interpreted as beach sand, is almost invariably associated with diatomite.

Unit 4

This medium grained, medium grey ash is distinguished in the field by a veining with yellow deposits of iron oxides around root holes, giving it an overall colour which has lead to its being known as the 'grey-green ash'. It fell onto peat in the marsh zone, and the roots of marsh plants must have grown through it. The ash lies in two layers, a lower, unstratified one showing no sorting or stratification, and, after a very thin layer of very fine, white ash, an upper layer in which bedding is apparent. It is suggested that the lower layer represents the original fall, followed by an interval in which fallout of a fine component continued, and the upper layer is the result of washing in of ash from the hill slope. That which fell on the hillside certainly suffered sufficient disturbance either immediately after its fall or by slower slope movements later for it not to be identifiable as a distinct layer within the slope deposits. The bi-partite nature of the ash layer in the marsh zone is repeated in other ashes, notably the very similar one of unit 15; two of the pale grey ashes show a tripartite subdivision, imposed in part by the sequence of eruption, there being a initial deposit of coarse, glassy, pumitic ash followed by a fine layer, and in part by the same process of redeposition, the third subdivision being slightly weathered, and consisting of a mixture of the other two. The tripartiate character of unit 20 is so marked that it has been named 'the tripartite'; it shows, too, in its middle unit a regular cyclic pattern of minor variations in colour and texture of the fine ash. The ash of unit 22 is also tri-partite, but its basal coarse layer is much thinner; the similarity is sufficient to cause confusion in correlation between exposures in which only one is present, but luckily the two ashes are mineralogically distinct.

Unit 5
Peat accumulation was resumed after the deposition of the ash of unit 4, but soon gave way to deposition of marsh soils similar to those of unit 3, though at any point they are paler and coarser than those of the lower unit, indicating that the relative position of the marginal zone was further out. There are a number of interruptions of steady accumulation, marked by layers of sand and diatomite. In each case, whenever the association of sand and diatomite occurs, from a zone of interstratification of the two, the sand extends further inshore and the diatomite thickens and becomes an important layer lakewards. These materials appear to represent slight transgressions, the water deepening over the marsh and wave action sorting the upper part of the marsh soils, to leave clean sand, and forming a small beach. Diatomite accumulates in a zone further out, below the depth of wave action. After each of these episodes the re-establishment of the marsh occurs at a different position, indicated by a change in colour and texture of the marsh soils.

Unit 6
Within the top of unit 5 there is a thin layer of fine ash, which is indistinct, often invisible, being mixed with the marsh deposits. Since this ash and another one, unit 9, sandwich the important beach phase between them, they are potentially of importance in correlation between Tlapacoya and other localities. They are, however, both thin, difficult to identify in the field, and appear identical on the more simple mineralogical criteria. Closer mineralogical characterisation is particularly important here, since these ashes, being so thin, may be far travelled and valuable for long distance correlation.

Unit 7
A major transgression cuts across marsh deposits and truncates the

slope deposits of the terrestrial zone, so as to expose again the rock fall associated with the earlier transgression and bite into the rock of the hill. Some way from the shore, the beach is represented only by a thin layer of sand in a rather broad zone in which diatomite and some fine, humic, clayey materials occur. Similar fine material forms a greater component of the deposits on the north-west side of the hill, where they take the place of the more typical marsh deposits in units 3 and 5. There, they are interstratified with peats and with sand and diatomite layers which can be correlated in detail with those in the south-eastern cuttings, but the whole stratigraphy is compressed and lacks the content of partly weathered ash which contributes to the typical marsh deposits, though the position in which these deposits are exposed is no further from the hillslope.

As the beach becomes fully developed towards the shore, a close association of sand and diatomite becomes marked and there is a zone of complex interstratification of the two materials; abundant pebbles of the rock of the hill occur. In the former terrestrial zone the beach consists of pebbles and boulders, and it was here that a group of hunting people were present, some 22,000 years ago.

Units 8 to 10

Occupation continued as the lake receded from its position of maximum transgression and slope deposits crept over the beach pebbles. Charcoal derived from the hearths, whose involvement in slope movements leaves them as only traces in the slope deposits, is found in the deposits accumulating above the beach in the marsh zone; it is this charcoal which provided the radiocarbon date indicated in unit 8 (Haynes, 1967). In the marsh zone, the type of clayey and dark humic to fine peaty materials which occurred below the beach continue to form a major part of the stratigraphy in units 8 and 10, forming rather distinct, but laterally very variable layers. Diatomite occurs, and in places forms an important layer, a little way above the beach, but after this there are no diatoms in the materials of units 8 and 10. the deposits take on a distinctly terrestrial character upwards, their physical character being in accord with the lack of diatoms, but whereas terrestrial materials on the steep slope show little clear stratification, here on the flatter slope of the former marsh zone some distinction of layers is retained. Fluctuating water table within these materials is indicated by the development of bands of yellow staining by iron oxides, in addition to the deposition around and within root holes which is characteristic of the marsh deposits. The layer of fine light grey ash, mentioned below, forms unit 9; just below it there are traces of fine dark grey ash, and unit 9 merges upwards into a coarser, darker material at the bottom of unit 10. Within the upper part of unit 10 is a thin layer of a coarse, pumitic ash, which is much mixed with redeposited material and difficult to define in the field. It is unlike anything which occurs at lower levels, but similar to the basal materials of the ashes of units 20 and 22.

Unit 11

This is a basaltic ash, the only basalt occurring in an otherwise andesitic context, apart from a slight indication that the ashes of unit 1 contained traces of a olivine bearing material. Unit 11 provides a useful stratigraphic marker, and, unlike all earlier ashes, is distinctly visible as a blurred layer among the slope deposits.

Units 12 to 14

Diatomite reappears in small quantity among thin, peaty and yellow stained layers at the base of this unit, and soon marsh conditions very similar to those that gave rise to units 3 and 5 are re-established in the zone which had been rather more terrestrial for a time. Within a thick layer of marsh deposits are a number of slight occurences of sand and diatomite, with changes in colour and texture of the marsh deposits.

A very thick layer of coarse, dark grey ash, unit 13, which must have blanketted the landscape, and which shows much evidence of redeposition from the slopes into the lake, though much remains on the hillslope, makes, surprisingly, apparently little difference to the conditions at the lake shore once normal marsh conditions are re-established. Unit 14 is very similar to unit 12, and has similar occasional incursions of sand and diatomite.

Units 15 and 16

Two important ash layers, a medium grained, medium grey ash similar to that of unit 4, and similarly disposed in primary and redeposited layers, and a very coarse ash, consisting of large pumice particles and angular fragments of holocrystalline andesite. This ash has been named the 'pumice with andesites'. Between the two, a thin peaty layer becomes in places a distinct layer of marsh deposits.

The pumice with andesites provides a convenient point at which to terminate detailed discussion. It must have formed a very unstable surface, and in the former marsh zone there is evidence of its redeposition in a complex of torrent deposits, interstratified with peaty and diatomitic materials and with a number of sandy layers derived from earlier coarse ashes. There are traces of slight ash falls, and one important, fine, light grey ash, unit 18. The redeposited 'pumice with andesites' does not extend far towards the lake, and the other materials form a closely and sharply stratified complex extending out from the shore; in addition to the materials already discussed, the occurrence of thin layers consisting almost entirely of phytoliths and sponge spicules are worthy of note, and phytoliths become abundant at certain levels in the peats of the upper part of the section. These peats indicate the development of conditions from about 7000 B.P., which were very different from those of earlier periods, and will not be discussed here.

Slope deposits

The arrival at intervals of fresh, incoherent volcanic ashes on a steep, in places precipitous, hill side led to conditions of constant instability. The processes of soil creep, hillwash, colluviation and deflation moved the superficial materials, mixing fresh ash from each fall with the partially weathered material from all earlier falls. The deposits accumulated at the foot of the hill, tapering out upwards against the slope. Ash layers are diffuse, only the thicker ones being identifiable in the field. The partially weathered material has the character of an immature soil, with a low degree of alteration of primary minerals, a low content of secondary minerals, and poor development of soil structure. At certain periods, greater stability of the surface allowed more mature soils to develop in places. These soils have a higher humic content and show a much greater development of soil structure; in thin section the secondary material shows signs of preferred orientation and the primary minerals are more altered. Renewal of more active slope movements has removed all traces of the surfaces of these more mature soils, and they appear merely as darker bands within the slope deposits.

As the slope deposits reached the lake margin they were either attacked by wave action, sorted and redeposited as sand and as lake sediments, or gradually incorporated into the marsh deposits. The slope deposits form the source of materials of the marsh deposits and differ from them only in those characteristics imparted by situation. The marsh deposits have a higher content of colloidal humus and iron oxides, have diatoms and are strongly humified but still contain recognisable plant fragments. They lack soil structure and are more compact. Throughout the exposures the zone of transition between slope deposits and marsh deposits has been severed, and so the transition cannot be followed in detail.

REFERENCES

Bradbury, J.P. (1971) 'Paleolimnology of Lake Texcoco, Mexico. Evidence from diatoms', *Limnol. Oceanogr.* 16, 180-200.

Cornwall, I.W. (1968) 'Outline of a stratigraphical 'bridge' between the Mexico and Puebla basins', *Bulletin of the Institute of Archaeology, London*, 7, 89-140.

Goodlife, M. and E. (1966) 'Un sitio pleistocénico en Tlapacoya, Estado de México', *Boletin, Instituto Nacional de Antropología y Historia, México*, 23, 30-2.

Haynes, C.V. (1967) 'Muestras de C14, de Tlapacoya, Estado de México', *Boletin, Instituto Nacional de Antropología y Historia, México*, 29, 49-52.

Lorenzo, J.L. (ed.) (in press) *Tlapacoya: 35 mil años de historia de la cuenca de México*.

Mirambell, L. (1967) 'Excavaciones en un sitio pleistocénico de Tlapacoya, México', *Boletin, Instituto Nacional de Antropología y Historia, México*, 29, 37-41.

Mooser, F. (1967) 'Tefracronología de la Cuenca de México para los últimos treinta mil años', *Boletin, Instituto Nacional de Antropología y Historia, México*, 30, 12-15.

White, S.E. (1962) 'Late Pleistocene glacial sequence for the west side of Iztaccihuatl, Mexico', *Bull. geol. Soc. Am.* 73, 935-58.

DISCUSSION

Professor A. Straw commented on the early radiocarbon date (22000 B.P.) obtained for a hunting culture in the area and asked what relationship this date bore to others from similar contexts. The speaker replied that recent work showed that the date could not be regarded as controversial although it was the earliest in Mexico. Earlier dates have been obtained from parts of South America but few are published. Dr. D. Jenkins enquired about the sampling intensity employed in characterising such a rapidly alternating and extended succession of strata. The speaker said that sampling was carried out on a continuous basis, with constant reference back and forth between site and laboratory. Very large numbers of samples were examined, as problems as to the nature of materials and their variation from point to point arose, and sample interval and size depended entirely on the detail and complexity of stratification and the question to be answered at any particular point. The thin-sections prepared from these samples were particularly useful for the interpretation of marsh deposits and soils. Mr. P. Buckland commented on the compression of the sediments and the possibility of tectonic influence.

Section 3
Terrestrial Environments

A. Kirkby and M.J. Kirkby

Geomorphic processes and the surface survey of archaeological sites in semi-arid areas

The discovery of an archaeological site by surface survey is dependent upon at least four factors: (1) the original characteristics of the site; (2) the natural and cultural processes (including re-use of the site) that have occurred since its original occupation; (3) survey procedures and conditions; and (4) chance. This paper considers the surface recovery of sites in the light of the second of these factors.

The arguments and models developed here are for semi-arid areas where sparse vegetation enhances the likelihood of sites being recovered through surface survey and reconnaissance. It also, as will be shown, allows rapid surface erosion to reduce the topographic form of small mound sites to nothing more than a surface scatter of sherds within 500 to 2000 years, depending on the mean annual rainfall. In these areas the change in form of such sites is rapid and consistent enough to allow dating on the basis of slope profiles alone.

For earlier sites, however, recognition of the existence of a site as well as its dating is largely dependent upon the evidence of surface sherd scatters. Much of the evidence and analyses presented are therefore concerned with the alteration of sherd concentrations at the surface through time. These changes are particularly important in semi-arid areas where alternations of cut and fill in river systems are often large scale and within a short time period. Archaeological sites in these areas are therefore often subject to alluvial deposition as well as chance exposure through later incision.

The measurements and evidence used to illustrate and support the theoretical models presented, are based on work carried out over 1966-70 in the Valleys of Oaxaca and Nochixtlan in southern Mexico, and in Khuzistan (especially the Deh Luran Plain) in south-west Iran. Both areas are semi-arid but the Mexican valleys have higher rainfalls (550-800 mm per annum for Oaxaca compared to 200-300 mm for Deh Luran). The models are developed specifically for the evolution of form and sherd scatters from small, unconsolidated house mounds which are produced by the collapse of clay-brick structures. Some of the implications drawn from them about surface survey methods and the interpretation of population densities and changes over time are however, more generally significant.

The models and arguments developed are concerned with (1) the change in the *form* of the house mound through time; (2) the alteration of the sherd scatter on the surface over time through the three interacting processes of *accumulation, breakdown* and *transportation*; (3) the loss of sherds at the surface through natural alluviation and cultural accumulation over the original site; and (4) the combined effects of these and other factors on the probability of recovering early sites, or early periods, through surface survey, and thus the interpretations that may be made for apparent population change over time.

Clay-brick house mounds

Unbaked-brick houses formed from local clay containing pottery sherds are a widespread building form. They are common in both Mexico and Iran today, and in many periods in the past. In terms of their natural evolution since abandonment, two aspects of their composition are important; the amount of clay material and the rate of its erosion through time, and the concentration of sherds in the bricks.

The evolution of these house mounds is often cyclical because they are subject to re-use as building material for later structures. In terms of the topographic form, the process is one of gradual flattening over time after the walls have initially collapsed to form a circular or elliptical mound. Natural slope processes remove material from the mound until it disappears and the heavier sherds are left on the surface as a 'lag' deposit. If a new house is built in the same location before the mound is completely flattened, the building of a 'tell' may be initiated. The sherd cycle is a little more complex. Following the initial collapse of the clay-brick structure, the sherds are evenly dispersed throughout the mound material. Measurements of sherd frequently show that they range from the equivalent of 4 to 35 sherds of 2-4 cm size per cubic metre of clay bricks. After about 500 years the removal of a thickness of 1.5 to 2.0 m of fine material from the surface of the mound will produce maximum sherd concentrations at the surface of between 8 and 50 2-4 cm sherds per square metre. This means that a three-house group forming a mound of about 100 cubic metres in volume will contain between 400 and 3500 of this size sherds.

The immediate main source of these sherds is undoubtedly the clay-bricks of the walls. The question remains of how the sherds became incorporated into the bricks – whether they were deliberately added and introduced from outside the mound or were casually included as part of the local refuse and domestic debris. The concentrations of sherds found are consistent with the latter. Survey of a modern house in Nochixtlan showed a total of 700 sherds of 2-4 cm produced around the adobe house during a 14 year occupation. Observed sherd densities on archaeological mounds could therefore be achieved at similar rates over periods of 10 to 70 years occupation. Re-use of such mound material for new building would produce clay bricks with these sherd densities.

Over time the initial even distribution of the sherds within the mound will change. Erosion of fine material will tend to concentrate the sherds outwards from the mound centre, and later when the mound has disappeared as a topographic form, they will slowly be moved generally in a downslope direction. Alluviation will bury them and other surface processes (the action of plant roots, burrowing animals, ploughing) will rework them near the surface. The digging of pits or re-use of the material for building will mix the sherds up again to a greater depth. Thus through time the sherds become alternately concentrated and accumulated on the surface and redistributed through the mound material.

1. Changes in house mound forms

In semi-arid areas where vegetation (particularly grasses) is sparse, slopes in unconsolidated material become flattened within archaeological time periods rather than being 'fossilised' as imperceptibly changing vegetated forms. Within any one area therefore the angle of the declining mound slopes can be used to determine *relative ages* of archaeological sites. This provides an easy survey dating technique which is particularly useful in areas with poorly developed ceramic typologies or few diagnostic surface sherds available.

1.1 Theory

In a simple model the initial collapse of a house produces a mound with talus-like debris slopes at angles of rest at 30° (Fig. 1). When the walls are first breaking down, material from them will tend to accumulate higher on the inside than the outside of the building. As the walls are eroded lower, the debris from them on the inside will begin to be transported outwards across the original wall line. On recently formed mounds therefore the highest points will be just outside of the original wall positions. For typical wall dimensions of 60

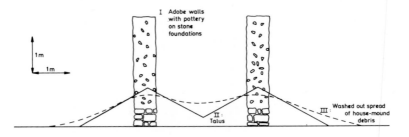

DIAGRAMMATIC HOUSE-MOUND COLLAPSE MODEL

Figure 1. Diagrammatic model of mud-brick house collapse.

cm width and 3 m height in a structure 7 m long by 3 m wide, the mound formed will have a cross-sectional area of about 4 m².

Geomorphic slope processes will reduce the mound from this initial form (Fig. 1) to a flat surface. The form of the mound at intervening times can be modelled using two equations; a statement of continuity of mass, and a statement of the rate of debris transport.

(1) *Continuity of mass*: that is, for any slope section, a deficit of debris inflow compared to debris outflow must produce a lowering of the slope section:

$$\frac{\delta S}{\delta x} + \frac{S}{x} + \frac{\delta y}{\delta t} = 0 \text{ for a circular mound (2 dimensional)} \qquad (1)$$

$$\frac{\delta S}{\delta x} + \frac{\delta y}{\delta t} = 0 \text{ for a ridge mound (1 dimensional)} \qquad (2)$$

where x is horizontal distance measured from the centre of the mound,
y is elevation above the basal surface,
t is time elapsed, and
S is the rate of debris transport (measured in cm³/cm/year or similar units).

(2) *Rate of debris transport*: that is, debris transport is directly proportional to the tangent of the slope angle, with a constant of proportionality D (the diffusivity measured in cm³/cm year).

$$S = -D \cdot \frac{\delta y}{\delta x}$$

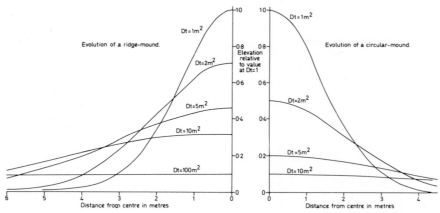

Figure 2. Mathematical model for the evolution of a house-mound by erosion over time for (a) a ridge-shaped mound (b) a circular mound.

The appropriate solutions to these equations for the given initial and boundary conditions are:

$$y = \frac{V}{4\pi Dt} e^{-x^2/4Dt} \quad \text{for a circular mound} \qquad (4)$$

$$y = \frac{A}{\sqrt{4\pi Dt}} e^{-x^2/4Dt} \quad \text{for a ridge mound} \qquad (5)$$

where V, A are respectively the mound volume and cross-sectional area for the two types of mound.

In both cases the solution shows a mound profile which lowers and spreads out through time as a normal (Gaussian) curve (Fig. 2). The absolute heights of the mound can be seen to depend on both its original size and shape but the spreading out of the mound (indicated by its variance $\sigma^2 = 2Dt$) is independent of these factors.

1.2 Evidence from Oaxaca, Mexico
The proposition that the mound form evolves as a normal curve

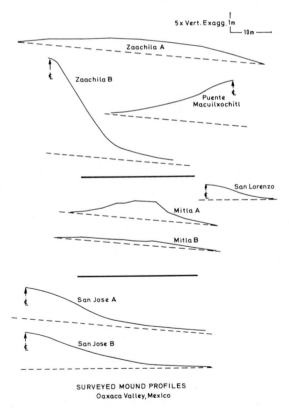

SURVEYED MOUND PROFILES
Oaxaca Valley, Mexico

Figure 3. Surveyed profiles of eight dated mounds from the Oaxaca Valley, Mexico.

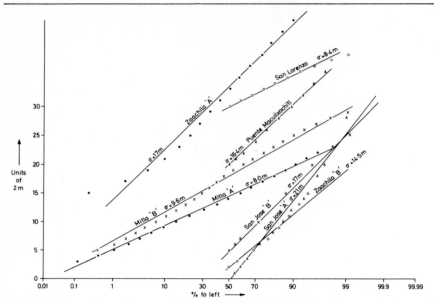

Figure 4. Comparison of surveyed mound forms with normal curves, on the basis of
    cumulative cross-sectional area.

through time can be tested by comparing the forms of mounds of
known different ages within one area. This was done for the Valley of
Oaxaca where earth mounds with established dates ranging from 400
to 1700 years B.P. were measured in the field. The resulting profiles
are given in Fig. 3. Compared to the model in Figs. 1 and 2, they are
mainly larger in size and both lower and more spread out than the
theoretical initial form.

How far the field evidence supports the normal-curve model of
evolution is shown in Fig. 4. Here cumulative areas beneath the
surveyed mounds are plotted on an arithmetic probability scale which
gives a straight line if the mounds approximate normal curves. The
slope of the line gives the 'variance' of the mound which is equal to
(2Dt) in equations (4) and (5). The quality of the 'fit' cannot be tested
statistically, but is considered to be good. Using the values obtained
from Fig. 4 for the mound variance ($\sigma^2 = 2Dt$) and plotting them
against age, a reasonable straight line relationship is found (Fig. 5),
corresponding to a roughly constant value of D over time. In Fig. 5 the
best-fit slope of a line through the origin is the mean value of 2D, and
gives a value of $D = 1070$ cm^3/cm year for the Oaxaca data.

1.3 Profile form as a method of relative dating
It is proposed that mound profiles could be used in this way to provide
comparative dating within a single climatic area, over the period
during which mound forms survive. The suggested procedure is as
follows:

(i) If there are some mounds dated by ceramic typology, radiocarbon or other means, then a calibration curve of σ^2 against age can be made as described in the previous section. If there have been appreciable changes in climate or land use, then the graph need not be the straight line suggested for Fig. 5, but may be curvilinear.

(ii) In the absence of a dated sequence, D may be estimated approximately as

$$D = 1070 \left(\frac{\text{Ann rf in mm}}{600}\right)^{2.2} \text{ x proportion unvegetated area}$$

$$= 8.3 \times 10^{-4} \text{ R}^{2.2} \text{ b cm}^3/\text{cm year} \tag{6}$$

This expression is based on the figure of 1070 cm³/cm year calculated for Oaxaca above; combined with correlations with rainfall and unvegetated area, as discussed by Kirkby and Kirkby (1974) on the basis of geomorphic work in semi-arid areas. The mound variance, σ^2, is then estimated as $\sigma^2 = 2Dt$ where t is the mound age in years. It should be stressed that in a fully vegetated area such as lowland Britain, the value of D drops as low as 10 cm³/cm year. With rates as low as this, mounds do not evolve appreciably *at all* over archaeological periods, so that this method of dating applies *only* to semi-arid areas.

(iii) Survey of the profiles of undated mounds in the same study area can then be made using hand-levels or clinometers and tapes. Calculation of their variance, using the cumulative area method of Fig. 4, can then be converted to an *estimate* of absolute age using the calibration curve described above. The range of validity of this method varies with both the climate and the size of mound used. For

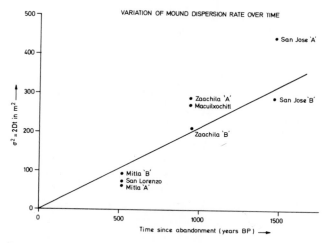

Figure 5. Mound variance as a function of time. Data for Oaxaca mounds shown in Figs. 3 and 4.

Table 1. Age estimates based on mound profiles, Oaxaca, Mexico.

Mound	Ceramic period		Assigned dates for period BP	Estimated date from mound profile BP
Mitla A	MA	V	400-650	300
Mitla B		V	400-650	430
Zaachila A		IIIb-IV	650-1250	1350
Zaachila B		IIIb-IV	650-1250	980
Puente Macuilxochitl		IIIb-IV	650-1250	1260
San Lorenzo		V	400-650	330
San Jose A		III a	1250-1700	2060
San Jose B		III a	1250-1700	1350

the Oaxaca valley, with 500-700 mm annual rainfall, individual house mounds survive with recognisable profiles for only 500-800 years; while larger earth structures survive up to twice as long. In south-west Khuzistan, with 200-300 mm of rain, individual house mounds survive for up to 2000 years, and larger mounds proportionately longer.

Within these time periods however, mound form is a fairly rapid and accurate means of comparative dating of sites which is especially valuable in reconnaissance survey of areas with little or no established ceramic typology.

2. Surface sherd scatters on eroding house-mounds

The evolving mound form affects the distribution of sherds at the surface. As it is lowered more sherds are produced which are then moved across the surface and reduced in size through natural weathering and erosion, and are broken by being trodden on. Models of sherd distributions through time must therefore take account of sherd accumulation, breakdown and movement. Here two approaches are considered. The first is relevant for larger sherds such as 4-8 cm sizes which are subject to accumulation and to breakdown but move very little across the surface. The second approach allows both breakdown and surface movement for several sizes of sherds. It enables the dispersal of different sized sherds out from the mound centre over time to be predicted. Basic to both models is some understanding of how sherds break down in size once they are exposed at the ground surface.

2.1 Sherd breakdown at the surface

Over long time periods natural processes of wetting and drying cycles or freeze-thaw alternations of the moisture within the sherd will lead to its disintegration. Over shorter, archaeological rather than geological, time periods, a more effective process is that of breaking by being trodden on by people and animals. Larger sherds are obviously more vulnerable to being both trodden on, and being broken in the process, so that the rate of breakdown will decline through time and with smaller sherds. A simple experiment was conducted to measure the rate of breakdown of sherds of given sizes when walked upon. The standard conditions were that of a person weighing 60 kg and wearing local Oaxacan flat sandals walking once across the sherds. This provided one 'treading event'.

The results of this experiment are summarised in Table 2 which shows the number of fragments of each size produced by one 'treading event' applied to 100 sherds of a given initial size. The cumulative effect of a sequence of five 'treading events' on the breakdown of 100 4-8 cm sherds to nearly 2000 sherds of different sizes is given in Table 3. These values which are derived from the data in Table 2 could be taken as the grain-size distribution of the sherd material at each stage of its evolution.

Table 2. Progress of experimental sherd breakdown. Each 'treading event' consists of being walked on once.

Size class before 'treading event'	Size classes after 'treading event': Number of fragments produced per 100 of original size					
	8-16 cm	4-8 cm	2-4 cm	1.4-2 cm	1-1.4 cm	<1 cm
16-32 cm	0	230	400	500		30
8-16 cm	11	252	450	165	100	45
4-8 cm	—	29	243	108	35	23
2-4 cm	—	—	38	143	50	19
1.4-2 cm	—	—	—	82	29	17
1-1.4 cm	—	—	—	—	95	13
½-1 cm	—	—	—	—	—	100

(1) In the centre of mounds, the field measurements correspond well with a model in which additional sherds are being supplied between treading events. This fits in with the description of surface lowering in the mound centre and continuous accumulated of sherds as a lag deposit.

(2) Near the periphery of the mound, observed sherd-size distributions in the field correspond to a model in which no new material is added.

(3) The field size distributions correspond to *a rate of one treading event per century* over the total area of the mound. This is estimated to be equivalent to ten people walking across the mound every year.

2.2 Model of surface accumulation and breakdown of large sherds

Rates of breakdown provided in the above experiment may be

Table 3. Breakdown of 100 4-8 cm sherds in a sequence of 'treading events'.

No. of treading events	Size class in cm: Values are numbers of fragments					Median diam (cm)
	4-8 cm	2-4 cm	1.4-2 cm	1-1.4 cm	<1 cm	
0	100	0	0	0	0	5.6
1	29	243	108	35	23	3.2
2	8	163	465	195	98	2.3
3	2	81	622	403	235	1.7
4	0.6	36	628	605	407	1.5
5	0.2	15	567	773	599	1.4

combined with the rates of mound erosion (which defines the rate of sherd accumulation) to model the frequency of sherds on the mound surface through time. At least in the early stages of mound lowering the exact positions of the house walls have considerable influence on the specific source areas for the sherds. Allowance is made for this by considering three cases: (a) a simple central mound formed from a single house; (b) the same size mound divided into two units space 3 m apart; and (c) a mound of indefinite extent but with one well-defined edge. Larger groups of houses may approximate cases (b) and (c).

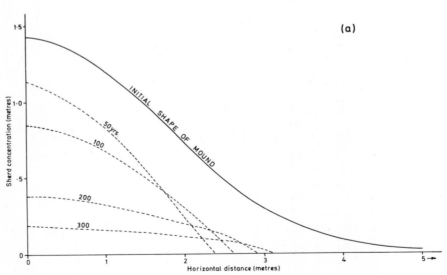

Figure 6. Rates of production and breakdown for 4-8 cm sherds on mounds of different shapes.
(a) single central mound.

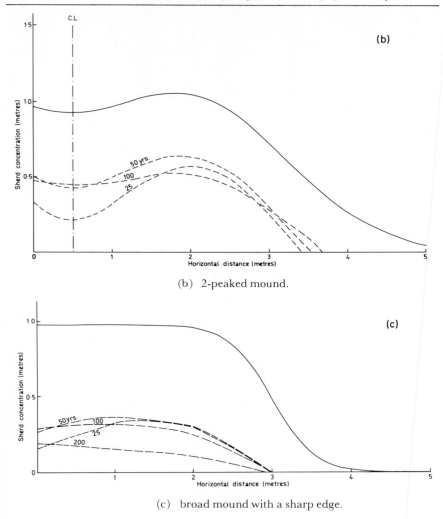

(b) 2-peaked mound.

(c) broad mound with a sharp edge.

In Fig. 6 are given the patterns of 4-8 cm sherds produced by the erosion of a house mound for the three conditions (a) to (c) described above, in which sherds are accumulated and broken down but not transported over the surface. In each case the model shows that sherd density increases to a maximum which is reached 50-100 years after initial collapse of the house. In all three situations the precise location of the highest sherd concentration is initially dependent on the positions of the original walls. Later however, sherd distributions are not only less localised within the mound but decline in overall density at the surface.

The loss of sherds at the surface over long time periods means that older sites will be represented by lower densities of sherds. When the model is applied to house mounds undergoing rates of erosion as

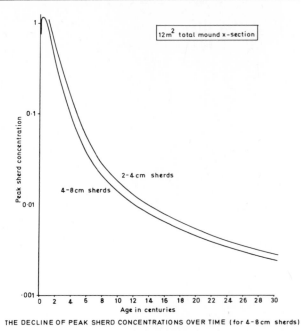

Figure 7. The decline of surface sherd concentration over time for 2-4 cm and 4-8 cm
sherds. The values refer to the depth (in metres) of mud-brick material required to
yield the same number of sherds.

measured in the Valley of Oaxaca (see Fig. 7) the same initial
concentrations of sherds in the wall material will produce surface
sherd densities of 0.056 after 500 years compared to 0.0074 after 1500
years (densities refer to the number of sherds originally in that depth
in metres of mound material). Thus compared to the latter sherds, the
early ones are under-represented by a factor of 7.6 times. A similar
pattern is found for sherds which start off at the surface at 2-4 cm
instead of 4-8 cm although the values of density are about 50 per cent
higher at all times for similar initial concentrations in the wall
material (Fig. 7). These declining surface densities of sherds through
time obviously affect the chances of early sites being recovered and
will be further discussed in this context later.

2.3 Sherd movement
Movement over the surface is not an important factor for the sherds
sizes usually involved in surface survey of archaeological sites. Surface
transport is related to the size of material with fine sizes moving much
more rapidly than coarse ones. Where the mound is on a level plain
sherd movement will be only dispersed outwards from the mound.
Where the mound is itself built upon a sloping surface, the whole
sherd distribution may be displaced downslope. Over a 500-year
period sherds less than 4 cm can be displaced by up to 5 m on slopes

as low as 1 to 3 degrees. Larger sherds of 4-8 cm are displaced on similar slopes by less than 1 m. These sherds therefore can be regarded as reliable indicators of the original location of the sherd source. The corollary is also true: that widespread sherd scatters indicate widespread sources rather than dispersal from a single source.

3.   Upward mixing of sherds

House sites may be buried either through geomorphic processes of deposition such as river alluviation, or through cultural accumulation of later material on top through re-use of the mound. The latter process, if repeated many times, can produce a sequence of occupation levels and the topographic development of a tell. Buried material can also become mixed with higher, younger levels through geomorphic and cultural processes. They thus provide the danger of assigning falsely early dates to surface material and the opportunity of locating early sites whose structures are entirely buried.

3.1 Mixing processes

Geomorphic processes of vertical mixing of material in semi-arid areas include wetting and drying cycles, small random disturbances, and the action of animals and plants. Wetting and drying cycles are similar to the freeze-thaw processes which produce sorting of coarse surface material into polygonal patterns and parallel stripes in periglacial areas. Wetting and drying cycles operate at a much slower rate and are usually confined to the top 50 cm of surface material. They are therefore likely to be of only slight importance to the vertical movement of archaeological material. Small, random disturbances tend to diffuse coarse material *in all directions* outwards from areas within the soil where they concentrated to areas where there is little coarse material. This leads to a more uniform mixing of sherds within the mound but the net movement is slow with estimated diffusivities of the order of 1 to 10 cubic centimetres per centimetre year. In total these geomorphic processes produce only very slow mixing of material which is probably always less significant than the rate at which the mound material is being eroded.

   Cultural mixing processes through the re-use of material for building and the digging of pits and wells into earlier material are much more effective in bringing early sherds up to the surface. Within the plough zone (20-25 cm below the surface) cultivation produces regular mixing of material. The construction of irrigation and drainage channels (and especially quanats in Iran) also bring up early material from greater depths.

3.2 A simple model of mixing

If the simplest model of mixing is taken, it can be stated as a depth y of new material added every n years to the top of a site to give an average accumulation of y/n per year. Subsequently thorough mixing is assumed to occur to a depth x. Clearly if x is less than y, no upward mixing of sherds will take place.

This model can be interpreted in at least two ways:

(1) annual flooding (n = 1 year) and deposition of natural alluvium to a depth y; followed by ploughing for cultivation to depth x.

(2) The rebuilding of old house ruins with an addition of new material over an n-year cycle. In this situation y represents the average depth of new material and x the total depth of new and old material needed to build a new house. The time n would be of the order of 20 to 100 years.

The model can be developed for either the alluviation or the rebuilding situations in the same way: after r cycles of burial and mixing, let the sherd concentration of (oldest) sherds at the surface be c_r. To keep the model simple, it is assumed that mixing involves no reduction in sherd size.

$$\text{Then:} \quad c_{r+1} = \frac{x-y}{x} \cdot c_r \tag{7}$$

Applying this relationship repeatedly to represent mixing cycles;

$$c_r = c_o \left(1 - \frac{y}{x}\right)^r \tag{8}$$

where c_0 is the initial sherd concentration.

This is the general form for the reduction of sherd concentration at the surface over a number of cycles of deposition and mixing. It thus represents an important relationship for the calibration of sites located by surface survey with the likely total number of sites buried for the same period.

The sherd concentration at the surface may also be expressed as a function of time. If there are r cycles in t = nr years then:

$$c(t) = c_o \left(1 - \frac{y}{x}\right)^{\frac{t}{n}}$$
$$= c_o \lambda^t \tag{9}$$

$$\text{where} \quad \lambda = \left(1 - \frac{y}{x}\right)^{\frac{1}{n}} \tag{10}$$

For small values of y/x (that is, depth of new material added over depth of mixing) which are less than about 0.3 in practice, then the rate of sherd concentrated at the surface declines through time as

$$\lambda = \left(1 - \frac{n\alpha}{x}\right)^{\frac{1}{n}} \simeq 1 - \frac{\alpha}{x} \tag{11}$$

This means that the *rate of attentuation of sherd concentration at the surface with time depends principally on the ratio of the annual average rate of accumulation,* α, *to the depth of mixing,* x. The importance of this relationship is that for calibrating surface concentrations of sherds with the number of buried concentrations of sherds with the number of buried concentrations, one needs to know only the *average accumulation rate* rather than, for example, the frequency of flooding. The average accumulation rate is much easier to obtain during the course of archaeological survey than flood frequency figures.

Lastly the mixing model can be considered from the point of view of how the sherd concentration at the surface is affected by the depth of material deposited above the original site.

For a total burial of $z = ry$, the sherd concentration at the surface is

$$c(z) = c_0 \ (1 - y/x)^{z/y}$$

$$= c_0 \beta^{z/x} \tag{12}$$

where $\beta = (1 - y/x)^{x/y}$

$$= p^{1/(1-p)} \tag{13}$$

where $p = 1 - y/x$

3.3 Application of the model to natural alluviation

Natural alluviation is a widespread and common way in which sites become buried over time. In the Valley of Oaxaca there is evidence that on the valley floor alluviation over a 1500 year period has produced maximum rates of deposition of 0.25 cm per year (Kirkby, 1973 and 1974). Cultivation by both *coa* (local form of hoe used in pre-Conquest times and rarely now) and oxplough (the main form of present tillage) mix up the surface material to a depth of 20-30 cm. Taking an average mixing depth of 25 cm, the decline of sherds at the surface from a site being progressively covered by 0.25 cm of alluvium per year will be:

$$c(t) = c_0 \ (1 - \frac{\alpha}{x})^t$$

$$= c_0 \ (1 - \frac{0.25}{25})^t \tag{14}$$

What this means is that after one year only 1 per cent of the original

surface sherds will be covered by alluvium. After a century the
number of original surface sherds still at the present surface will be 37
per cent, so the buried site still has a good chance of being detected. In
a thousand years, which is more like the time scale of archaeological
sites in the areas are being considered, *only 0.004 per cent of the sherds
from the buried site will be lying on the present surface* and available for
discovery by surface survey.

In a situation of continuous, small increments of alluvial deposition,
therefore, sites become buried within a short time and the likelihood
of their being detected at the surface is very small unless special
conditions bring the sherds up to the surface. Of these, modern
practices like deeper ploughing have not been continued for long
enough in the past to affect archaeological sites, and the most
important ones are re-use of the site in later archaeological periods
and past and present deep excavation for wells, quanats, building
foundations etc.

3.4 Application of the model to cultural mixing in tells
In a tell, the successive building of houses on the collapsed, and
presumably levelled, remains of earlier ones produces a sequence of
occupation levels which rises above the surrounding plain to form
man-made hills up to 30 m high and which can be more than a
kilometre across (Hole, Flannery and Neely, 1969). The addition of
new building material from outside the tell provides a deposit of
material on top of the existing house material in a way analogous to
alluviation by natural flooding, although the increments of material
are larger and less frequent in a tell situation. During continuous re-
use of a tell surface, rebuilding of individual structures occurs at about
15-100 years intervals. The depth of material, part new and part old,
which is mixed together to construct a new house, and which collapses
when it is abandoned, is about 2 m.

The mixing model, in the form given in equations (12) and (13),
can now be applied to a tell situation, given the simplifying
assumption that the area of the tell surface is *uniformly* covered with
individual house mounds. These are assumed to collapse (providing a
depth of 2 m of material each time) with little or no erosion because
they are in close juxtaposition on a *level* surface. This situation is
therefore very different from that discussed for isolated house mound
sites on an open plain whose form is subject to rapid erosion
approximating normal curves as discussed earlier. In the tell, the
proportion (p) of old material (collapsed) used in the new house
structure is:

$$p = (1 - \frac{y}{x})$$

If a value of x = 2 m for the depth of mixing is taken, then the rate of
attenuation of sherds from the original occupation level at successively
higher and later ground levels is that shown in Table 4. In this table

Table 4. Attenuation of sherds during mound growth by rebuilding, with a proportion p of material re-use ($x = 2m$).

Mound ht		2m	5m	10m	20m	50m	100m
p = 0.1 (y = 1.8m)	No of occupation levels	1.1	2.8	5.6	11	28	56
	Sherd attenuation	.077	.002	10^{-6}	—	—	—
p = 0.2 (y = 1.6m)	No of occ. levels	1.2	3.1	6.2	12	31	62
	Sherd attenuation	.134	.007	10^{-4}	—	—	—
p = 0.3 (y = 1.4m)	No of occ. levels	1.4	3.6	7.1	14.3	36	71
	Sherd attenuation	.179	.014	10^{-4}	—	—	—
p = .4 (y = 1.2m)	No of occ. levels	1.7	4.2	8.3	16.7	42	83
	Sherd attenuation	.217	.022	5×10^{-4}	—	—	—
p = .5 (y = 1.0m)	No of occ. levels	2.0	5.0	10.0	20.0	50	100
	Sherd attenuation	.250	.031	.001	10^{-6}	—	—
p = .6 (y = 0.8m)	No of occ. levels	2.5	6.2	12.5	25	62	125
	Sherd attenuation	.279	.041	.002	10^{-5}	—	—
p = .7 (y = 0.6m)	No of occ. levels	3.3	8.3	16.7	33.3	83	167
	Sherd attenuation	.305	.051	.003	10^{-5}	—	—
p = .8 (y = 0.4m)	No of occ. levels	5.0	12.5	25.0	50.0	125	250
	Sherd attenuation	.328	.061	.004	10^{-5}	—	—
p = .9 (y = 0.2m)	No of occ. levels	10	25	50	100	250	500
	Sherd attenuation	.349	.072	.005	3×10^{-5}	—	—

the survival of early sherds at later surfaces is shown in relation to different mound heights, the number of occupation levels involved, and for different proportions (p) of old brick material being re-used to build later structures (from p = 0.1 to 0.9). In fact evidence of the proportion of re-use is best obtained directly from the long-term rate of growth per occupation level, than from the extent of sherd attenuation. Evidence from tells in south-west Iran suggests that re-use is likely to be about 50 per cent (p = 0.5). This corresponds to a net accumulation of 1 m per level.

Table 4 shows that the decline in sherd concentration from earlier periods at the surface is *much more sensitive to mound height* than to the proportion of material re-used, or to the number of levels. Even more important, evidence of occupation from lower, earlier levels can only be expected within depths of about 5 m. At greater depths, *older periods will be represented at the present surface by less than 1 per cent of their original sherd concentrations*. Without extra-ordinary good luck or special conditions of erosion or excavation these earlier, buried occupation levels are unlikely to be discovered by surface survey.

4.　Implications for archaeology

4.1 Implications for survey methods

The loss of early sherds at the surface under conditions of natural alluviation and cultural accumulation to less than 1 per cent within 1000 years or 5 m depth of later material makes the recovery of such sites extremely hazardous. Survey techniques which will improve the chances of the 1 per cent of early sherds being picked up are therefore important. Similarly, at such low surface concentrations of early sherds, the condition of the ground surface at different localities and on separate days may become critical to the identification of early sites. It is here that improving the surface pick-up sample by scraping the surface to a depth of 1 or 5 cm to recover more sherds may become worthwhile.

Table 5. Expected sherd densities at the present surface for mounds of different ages. Values refer to the number of 2-4 cm sherds per square metre of ground surface (assuming 20 sherds per cubic metre of mound material (initially).

	Survey collecting strategies		
Age in years	(a) surface only	(b) surface + 1 cm	(c) surface + 5 cm
200	10.6	10.8	11.1
500	1.76	1.96	2.26
1,000	0.36	0.56	0.86
2,000	0.11	0.31	0.61

Fig. 7 shows the decline of surface sherd concentrations over time, taking into consideration sherd breakdown and surface accumulation through mound erosion. On the basis of this curve, Table 5 shows the densities of sherds available to surface pick-up from mounds of different ages (200 years to 2000 years B.P.). Sherd densities per square metre of mound surface are shown for (a) the ground surface only; (b) the surface plus 1 cm depth; and (c) the surface and scraping to a depth of 5 cm. All densities assume an initial concentration of 20 sherds per cubic metre of building material which is comparable for concentrations measured in Oaxaca. It may be seen that although surface scraping is more time consuming, it is rewarding in the much improved yield of sherds it produces. It also compensates a little for the rapid loss of early sherds with increasing depth so that the difference in sherd recovery between different aged mounds is lessened. Thus surface scraping produces more comparable results *between* sites of different ages as well as improving the chances of discovering an early buried site in any one locality.

It was suggested above that deeply buried material is more likely to be brought up to the present surface by the digging of pits or wells than through widespread and continuous but shallow processes of mixing. Table 4 shows that if the pits are more than 3 to 4 m deep they will produce more early material on the surface than mixing processes. Similarly a 10 m well shaft cutting through a 2 metre thick mound feature buried 8 to 10 m below ground level will bring sherds up to the surface at a concentration of 20 per cent of the original amount after being mixed with the rest of the shaft debris. Without being cut by a well, the surface concentration would be $\frac{1}{2}$ per cent at most. Pits have the further advantage for the archaeologist that the excavated material is usually deposited on the surface as a mound which then becomes subject to erosion and sherd concentration on its surface. Pits dug in later archaeological periods therefore show up in surface survey as local sources for early sherds.

Similar arguments apply to the use of specially dug test pits as a part of surface survey. Test pits may be considered as a series of random sampling points for buried sites. If it is assumed that when a

Table 6. Probability of detecting a scatter of house mounds, at average spacing D, from examination of sherds brought up in n pits.

D = spacing of mounds in metres	30	50	100	200
n = number of pits examined				
1	.792	.283	.071	.018
5	>.999	.810	.307	.085
10	>.999	.964	.520	.163
20	>.999	.999	.769	.300
50	>.999	>.999	.974	.590

test pit cuts through a buried house mound the sherds can always be identified, then the probability of detecting a scattered pattern of early buried sites can be calculated. Let us assume that (a) a house mound produces a circular scatter of sherds 30 m in diameter; (b) the pit is 1 m in diameter; and (c) that the buried mounds are randomly spaced at an average distance D apart so that there is an average of one mound in every D^2 of area. Given these assumptions the probability of n test pits dug within the survey area is shown in Table 6.

From the Table it may be inferred that any village with house mounds about 30 to 50 m apart is likely to be detected by the excavation of a few pits. However, a less dense scatter of mounds, (for example, of farmsteads) spaced more than 100 m apart is unlikely to be found even if 20 pits are dug. The alternative strategy of digging test trenches appears to show no advantages, from a statistical point of view, in locating a site initially, although trenches are clearly preferable once a site has been located.

4.2 Implications for interpretation of survey data

The previous discussion has shown that surface concentrations of sherds from early, buried sites may be so low (less than one per cent) that recovery is unlikely. It has also shown that the detection of early sites on the surface is dependent upon the location and density of later sites, wells or other cultural disturbance of the archaeological material. In this context the location of later sites in relation to ones from earlier periods is a similar statistical sampling problem to that described for test pits and shown in Table 6. The *interpretation* which archaeologists have commonly put upon the number and position of early sites in relation to later ones, has however less often stressed their *statistical interdependence* and more been in terms of *changes in population and settlement patterns*. Thus it can be argued that site 'evidence' for an exponential increase in settlement increase over time may be wholly or partly an artefact of progressive loss of information about older sites.

For example, in Fig. 8 are shown the number of sites revealed by surface survey against age for the Valley of Oaxaca, Mexico (surveyed by Bernal, Flannery and others) and for the Deh Luran Plain in Khuzistan, Iran (survey by Neely, Hole and Flannery). For each area, the number of sites from different periods declines with earlier sites, with the Valley of Oaxaca showing about twice the rate of change through time. This could be interpreted as an increase in both areas of population through time with the Valley of Oaxaca showing the greater population expansion. However, when the loss of early sites at the surface through geomorphic processes, as described here, is taken into account, the steeper loss of early sites in Oaxaca is consistent with the difference in rates of geomorphic processes and site loss since abandonment.

The variation in rainfall (200-300 mm per annum in Deh Luran

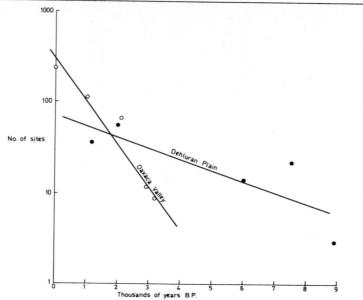

Figure 8. Apparent change in number of sites over time for surface surveys in Oaxaca, Mexico and Dehluran, Khuzistan, south-west Iran.

and 550-750 mm per annum in Oaxaca) leads to a difference in erosion rates which will account for about half of the differences in slope between the lines for the two areas. Thus the data do show some increase in population over time for Oaxaca even if the Deh Luran population was assumed to be relatively static, but the increase is likely to be less than half the 25 times indicated by the number of sites recovered. This consideration affects also our own earlier analyses of the changing efficiency of land use for food production through time (Kirkby, 1973, 145). It also suggests that perhaps residuals from the 'population growth' or 'site loss' line should be given as much significance as the overall trend. They may be interpreted either as changes in population for that period or changes in the type of settlement pattern (for example, from dispersed to nucleated) which effect site recovery in surface survey.

Fig. 9 illustrates this general argument with a hypothetical example in which the actual site density is the same for two, long separated periods. The early sites (broken line) are assumed to have surface sherd concentrations below one per cent (that is, they are about 1000 years older than the later sites in Oaxacan conditions and 3000-4000 years older in Deh Luran conditions). They will therefore *only* be recovered by surface survey if they are partly re-used and coincident with later sites. Two densities of settlement are considered in (a) and (b). In each situation, sites are randomly located. In condition (a) with ten early and ten late sites, more than half the early sites will be recovered (6). In condition (b) where settlement in both periods is

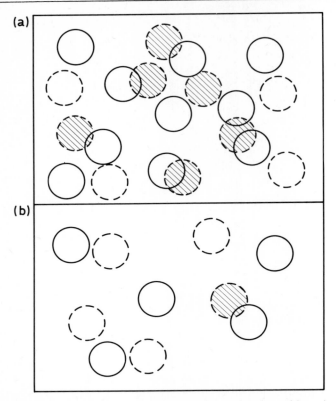

Figure 9. Hypothetical maps showing equal numbers of early and late sites, and the
rate of recovery of early sites over long periods.
(a) 10 early and 10 late sites.
(b) 5 early and 5 late sites.

sparser although equal (5 sites at each time) only *one* of the early sites
will be recovered. Thus although the population remains the same
throughout time for both the high and low settlement densities, the
apparent increase in population is much greater for the less densely settled
area (a 500 per cent increase compared to a 50 per cent one). Thus the
site recovery rate is also settlement-density dependent.

To generalise from this hypothetical example; if *each* site occupies a
proportion α (which will be very small) of the *total* survey area, and
there are n_1 early sites and n_2 late sites, the expected recovery of early
sites is approximately:

$$4n_1 n_2 \alpha \tag{15}$$

so that the best estimate of n_1 (the number of early sites) is:

$$\frac{1}{4\alpha} \times \frac{\text{Number of early sites recovered}}{\text{Number of late sites recovered}} \tag{16}$$

Thus the probability of missing the early site *completely* is
approximately:

$$e^{-4n_1 n_2 \alpha} \tag{17}$$

These approximations are valid provided that sites never cover more than about 10 per cent of the survey area. In practice, the overall density $(n_2\alpha)$ rarely exceeds 5 per cent even on a densely settled plain.

The important implications of this analysis for interpretation of surface site survey data are that within most archeological areas (other than urban situations) evidence of early periods will only be recovered *at all* with any certainty $(p > 0.95)$ if there are more than 15 sites in the area dating from that time. Likewise, if an early period is represented by only a single site even if the area is later densely settled, the chances of recovering that period is only 20 per cent *even with a complete surface survey*. Therefore the errors in estimates of settlement patterns due to site loss at the surface since abandonment are important to the interpretation of surface site data, since *the rate of site loss over time may be greater than the rate of population increase*.

Conclusions

This paper has been an attempt to quantify the changes that occur to house mound forms and their associated surface sherd scatters after their original abandonment. Analysis of these changes allows them to be used positively as a means of dating sites, and of calibrating the number of sites recovered against the probable numbers that are not. For the two areas studied in Mexico and in Iran, the combined effect of processes operating since the abandonment of earlier sites are great enough to significantly alter the chances of site recovery for different periods. Therefore in any interpretation of population increase or decrease, or related aspects of social organisation and cultural change over time, the processes and errors discussed here cannot be safely ignored. Measurement of their amount and direction is possible and will be valid within any local survey or climatic area. Understanding of them will improve the accuracy and interpretation of settlement patterns based on surface survey and quick reconnaissance data.

The main conclusions of the study may be summarised as:

(1) The declining angle of mound profiles approximates a normal curve and can be used to determine the *relative ages* of small mound sites within a given area. In Oaxaca with a rainfall of 600 mm rainfall per annum, mounds of 500 years are close to the limit of recognition for *single-house sites*. In Deh Luran with 200-300 mm annual rainfall, 2000 year-old mounds are at similar limits for recognition on topographic form alone.

(2) Over time sherd scatters at the surface are subject to processes of *accumulation* as a 'lag' deposit during mound erosion, *breakdown* through being trodden on, and *movement* across the surface. Of these, surface transportation is the least important.

(3) Together these processes affect surface sherd concentrations by

increasing them over the first 50-100 years after the end of occupation and thereafter decreasing them *exponentially* to comparative densities of 0.056 after 500 years and 0.0074 after 1500 years (for 4-8 cm sherds). *The concentration-time* curve (Fig. 7) is area-specific, depending mainly on climatic factors. Given local calibration, it is possible to correct sherd densities for age to give comparative values for the original sites.

(4) The concentration-time curve is increased by loss of early sherds through deposition on the site by later processes such as natural alluviation or cultural accumulation (as in a tell). The rate of attenuation of sherd concentration at the surface in these situations depends principally on the *ratio of average rate of accumulation to the depth of mixing* (through natural and cultural processes).

(5) For *alluvial* conditions similar to those measured in Oaxaca (average annual deposition of 0.25 cm) in 1000 years only 0.004 per cent of the sherds from the buried site will be lying on the present surface.

(6) For tell conditions like those in south-west Iran, sherd concentrations at the surface are more sensitive to *tell height than the proportion of old building material re-used or the number of occupation levels*. At tell heights of greater than 5 m, older periods will be represented at the present surface by less than 1 per cent of the original sherd concentration.

Taken together these results indicate a severe attenuation of site recovery for early periods. If sites are only recoverable at the surface when their surface sherds are at more than 1 per cent of the original concentration (this seems a reasonable and optimistic estimate) then sites which are more than 1,300 years old (on the basis of Fig. 7) will *only* be discovered in surface survey when they have been re-used at least once. Similarly, sites more than 2,400 years old will only be detected when they have been re-used twice, and so on. These actual figures refer to Oaxaca conditions but a comparable progression will be valid for different climatic areas.

This process of highly selective recovery means that arguments about the continuingly favoured location of certain sites over long periods must be viewed with caution where the evidence is principally from surface survey. Similarly, within any one area the numbers of sites recovered from different periods are interdependent and influenced by the overall settlement density. Ratios of sites between different archaeological periods cannot therefore be assumed to bear a simple relationship to population change or distribution through time. Errors are likely to be greatest for the lowest population densities and this, combined with progressive loss of evidence through time, means interpretation of cultural patterns for the earliest periods in any area are not only the most suspect, but may be missed altogether.

Acknowledgments

The authors undertook the fieldwork for this paper while studying the overall geographic environment as part of archaeological projects to the Valley of Oaxaca, 1966-70, led by Kent V. Flannery (University of Michigan) and supported by the Smithsonian Institution and the National Science Foundation; to the Valley of Nochixtlan, 1968-70, led by Ron Spores (Vanderbilt University) and supported by Vanderbilt University and the National Science Foundation; and to Deh Luran, 1969, led by Frank Hole (Rice University) and supported by Rice University and the National Science foundation. We would like to thank both our many colleagues in the field especially the project leaders, and the supporting institutions and foundations, without whose help this work could not have been done.

REFERENCES

Hole, F., K.V. Flannery and J.A. Neely (1969) 'Prehistory and human ecology of the Deh Luran Plain', *Memoirs of the Museum of Anthropology, Univ. of Michigan, Ann Arbor.*, 1.

Kirkby, A.V.T. (1973) 'The use of land and water resources in the past and present valley of Oaxaca, Mexico', *Memoirs of the Museum of Anthropology, Univ. of Michigan, Ann Arbor*, 5.

Kirkby, A. and M.J., Kirkby, (1974) 'Surface wash at the semi-arid break in slope', *Z. Geomorph: Supplementband* 21, 151-76.

Kirkby, M.J., (1976) 'The physical environment of the valley of Oaxaca, Mexico', *Memoirs of the Museum of Anthropology, Univ. of Michigan, Ann Arbor*, in preparation.

DISCUSSION

Mr. A. Bolt commented on the linearity of the slope process model implying that it was different from usual slope models. Dr. D. Jenkins noted that in assuming a model with a 'normal' redistribution of detritus and in plotting this distribution as a cumulative curve on probability paper, a straight line was obtained. In the example quoted, Dr. D. Jenkins thought that there seemed to be a slight tendency for deviation in a sigmoidal manner from this straight line and he wondered if this deviation could be interpreted and used to modify the initial model and the processes involved. Professor M. Kirkby said that the essence of the problem was the normality of the data and it was impossible to apply a test for normality given the undefined sample size.

D.A. Davidson
Processes of tell formation and erosion

An outstanding feature of many early settlement sites in the eastern Mediterranean and Middle East is their form. Large mounds or tells are a reflection of human occupation over a long period of time. They are best developed in Mesopotamia, but their distribution extends not only eastwards to Afghanistan and the Indus basin, but also westwards to Egypt and to the Levant including Turkey and the Balkans (Lloyd, 1963). Archaeological interpretation of these tells has depended primarily on artefacts and structures but Lloyd (1963) has stressed that there is need for detailed evaluation of stratigraphies since complete reconstructions are only possible once the processes of formation and erosion have been elucidated. In other words a geomorphological or sedimentological study ought to be an essential part of tell excavations.

In 1968 and 1969 Professor M. Gimbutas of the University of California at Los Angeles and Professor A.C. Renfrew of Southampton University organised a large project in the Plain of Drama, northeast Greece (Fig. 1). The focus of the project was excavation of the tell at Sitagroi, directed by Professor A.C. Renfrew.

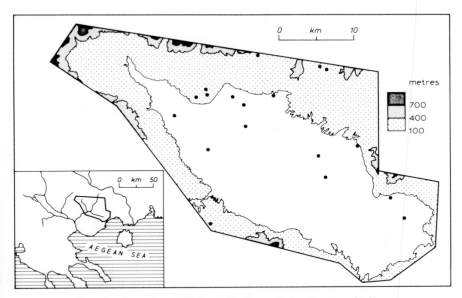

Figure 1. Location of Plain of Drama and distribution of tells.

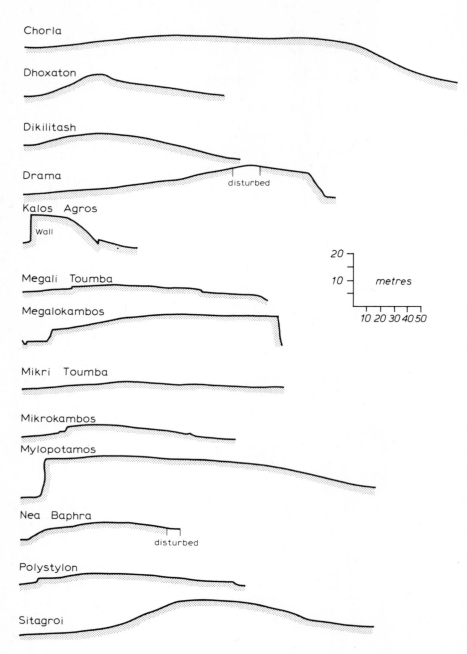

Figure 2. Profiles of selected tells.

Before attention is directed to the Sitagroi site it is interesting to consider briefly the form of other tells in the Plain of Drama. A total of 18 Neolithic and Bronze Age sites were located within the area during the project (Fig. 1) and representative cross sections of 13 of these are presented in Fig. 2. The tells are very varied in terms of present day form and magnitude. Some are extremely indistinct, such as Mikri Toumba, whilst others, such as Sitagroi and Chorla, are outstanding features in the landscape. Several are being actively eroded by rivers, of which the most spectacular example is Megalokampos, located in the neck of a meander of the river Angitis. The slightly asymmetric profiles of the tells at Chorla and Drama suggest that they have been trimmed by a river at some time in the past; the same is probably true of the Sitagroi tell. Besides modification by natural processes several sites have been influenced by various forms of construction – the best example is Kalos Agros where the site within the village has been drastically altered by fortification in historical times. But most sites

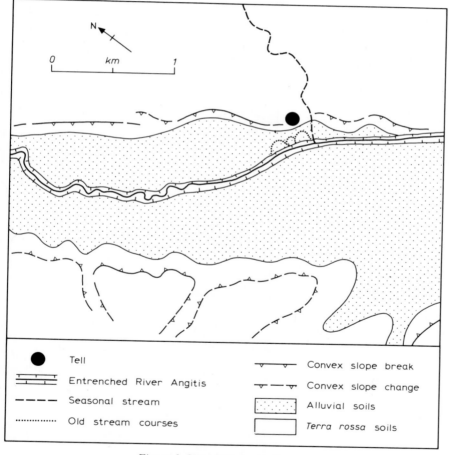

Figure 3. Site of Sitagroi tell.

are not within present-day villages; instead most are associated with alluvium (Davidson, 1971) and analysis of tell evolution must be intimately connected with the alluvial chronology, a theme which will be developed for the Sitagroi tell.

1.   The Sitagroi tell

The tell is situated at the junction of a lowland limestone area with the alluvium flanking the river Angitis (Fig. 3). The edge of the limestone area which has *terra rossa* soils with clay contents of the order of 50 per cent, is marked by a convex break or change of slope which gives way to a slope of about 3 to 5° down to the river terrace where the Angitis is incised into the alluvium to a depth of about 5.5 m. The alluvium is

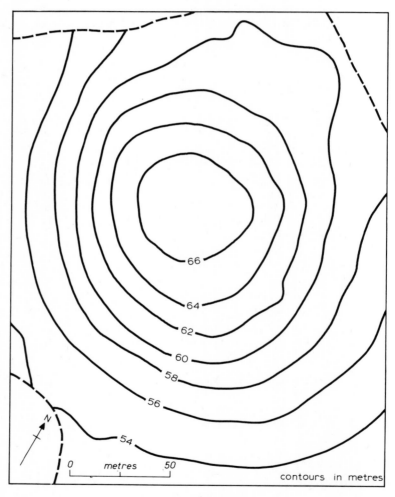

Figure 4. Form of Sitagroi tell.

varied in texture but the upper stratum is dominated by silts with a little clay. Part of the course of the Angitis has been straightened in recent times and this may account for the short abandoned channels near the site. An important point about the alluvial stratigraphy is that a stratum about 2.5 m above the river contains Classical sherds dating around 300 B.C. Thus this alluvium appears to correspond with Vita-Finzi's (1964, 1969) 'younger fill'.

The tell is about 180 m in diameter and 11 m in depth and the form of the tell is shown in Fig. 4. A slight elongation is evident in the contour pattern with the longer axis parallel to the river, suggesting perhaps that the tell has been trimmed sometime in the past by the river. The tell ranges in age from about 5400 to 2200 in calendar years B.C. and five major phases of occupation have been recognised (Renfrew 1970, 1971). In order to examine the processes of tell formation and erosion, two approaches were used: first the stratigraphy of a deep sounding through the centre of the tell was examined and second, the stratigraphy on the flanks of the mound towards the alluvium was investigated.

1.1 Tell formation

Samples were collected from apparently homogeneous layers down the 11 m deep sounding. They were analysed for particle size and total phosphate content and the methods and results have been described (Davidson, 1973) and it is only necessary to summarise the conclusions. The phosphate results confirmed what might be expected – high values indicated that the mound had evolved as a result of occupation. Phosphate contents on the whole were very variable though in certain parts of the section there were suggestions of decreasing phosphate with depth. In other words there is evidence that intensity of occupation in part increased with time.

Particle size analysis of samples of *terra rossa* soil, alluvium from the terrace of the Angitis, a wall from the final phase of occupation and from the deep sounding showed that the latter three types of material were most similar. The inference is that the tell resulted primarily from the collapse of walls which were built of alluvium derived from the nearby river. The wall sample contained 50 per cent silt and only 22 per cent clay which would have meant that walls built of this material would have required frequent rebuilding. Collapse of such walls over the long period of occupation meant that the tell gradually increased in height to give the site additional defence value as well as a commanding view over the surrounding partially wooded landscape (Wijmstra, 1969). Thus the tell seems to have been built up primarily by wall collapse in the manner envisaged by Lloyd (1963) for the Mesopotamian mounds.

Particle size analysis of 36 samples from the deep sounding gave further evidence about the occupation of the site. Particle size groups were recognised by fitting Pearson Type I curves to the size distributions and comparing the exponents of the equations. The interesting result was that a certain degree of coincidence could be recognised between particle size groups and occupation phases. The most distinct contrast in particle size curves was between occupation phases I, II and IV. A change in nature of material used for wall construction at the same time as a change in occupation phase suggests a lack of continuity of settlement, in other words an abandonment of the site for an unknown time period between occupation phases III and IV.

Through time the tell was gradually built up by wall collapse, with interruptions between certain occupation phases. It is difficult to envisage other processes, either man-induced or natural, which would have contributed substantially to tell construction. The rate of formation would have been in proportion to the number of walls and thus to the population which as already noted tended to increase through time. However, as the tell became a more outstanding feature it would have suffered increasing erosion, though it is doubtful if ever the situation arose during occupation wherein the net rate of accumulation was equalled by the net rate of erosion.

1.2 Tell erosion

In order to examine the erosional history of the tell, a series of pits was excavated along line L of the 10 m planning grid: the main area of excavation was the base of the western side of the tell. Pits IL, ZC, ZK and ZF were used to construct a section along the line L (Fig. 5a). Fig. 5b is a generalised interpretation of the stratigraphy on the flanks of the tell. One sounding was excavated on the eastern side of the tell, but time unfortunately did not permit further investigation in that locality.

Examination of the stratigraphy allows the following relative chronology to be proposed:

(i)   Faulting of the limestone – suggested from aerial photographs of the area. Weathering of limestone to a clay and formation of the palaeosol as shown in IL.
(ii)   River Angitis, fault guided, cuts broad valley.
(iii)   Deposition of sands and gravel by the Angitis (first river fill, first phase), represented in ZF and ZC.
(iv)   Continuing deposition by Angitis, but now clay (first fill, second phase), shown in ZF, ZK and ZC.
(v)   Slight incision by river, reflected in the channel between ZK and ZC. During this phase as well as in the previous one the river may have trimmed the base of the tell. The result was that the tell became separated from the river by a limestone bluff of 3 m. The river was probably flowing in a braided fashion over a broad flat marshy flood-plain.
(vi)   Erosion of tell resulting in colluvial material as in ZK and

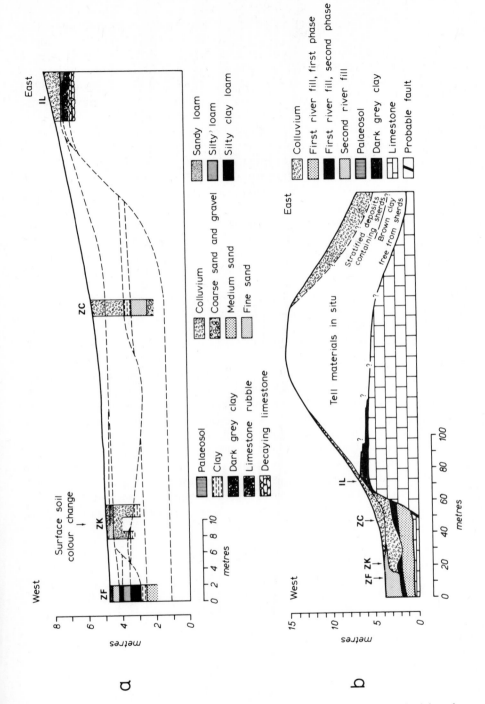

Figure 5. Stratigraphy on the lower part of Sitagroi tell with details shown in (a) and
a generalised interpretation in (b).

ZC. Sherds from all phases of tell occupation were found in this deposit.

(vii)   Incision by the river into the colluvium, but leaving a terrace between ZK and the tell.

(viii)   Aggradation in Classical times (second fill) as shown in ZF, again a certain amount of colluvium probably being removed by the river.

(ix)   Incision (about 5.5 m) by the river to the west of the section.

(x)   Erosion of tell to produce second colluvial deposit. This upper colluvial material tapers out over the second fill and this is reflected on the soil surface by a soil colour change from dark greys of the colluvium to reddish browns of the alluvium.

Thus a general sequence of events has been established by considering the stratigraphy of the lower part of the tell. The suggestion is that during its occupation the tell was adjacent to a dynamic fluvial environment. A considerable volume of tell colluvium is present in the section even though the river must have eroded this material in early phases. However, given the amount of colluvium apparent in the sections, a speculative reconstruction of the form of the tell can be made.

An immediate problem is that section line L does not pass through the top of the tell. As can be seen from Fig. 6, the 64 m contour is the highest one crossed by this line of section. The summit area (0.25 ha), almost flat, is delimited today by the 66 m contour. If the assumption is made that the summit of the tell has not decreased since final occupation, then the problem is to determine the former extent of the flat top before erosion gradually modified the tell to its present form.

The next problem is to decide if the surface of the tell has suffered from uniform erosion. In other words has an equal depth of material been lost along section line L on the tell? If it is assumed that processes of surface water erosion were dominant, then the rate of soil transport (s) can be expressed as a function of length of slope (x) and slope angle (β). Relationships of the form

$$s = k_1 \ x^m \ \tan^n \ \beta$$

have been obtained from soil erosion plots by Zingg (1940), Musgrave (1947) and Kirkby (1969) where k_1, m and n are constants. If the angle of slope is held constant, then the above equation can be simplified to

$$s = k_2 \ x^m$$

where k_2 is a constant

Thus $$\frac{ds}{dx} = k_2 \ mx^{m-1}$$

The rate of erosion per unit length of slope will be constant if m equals

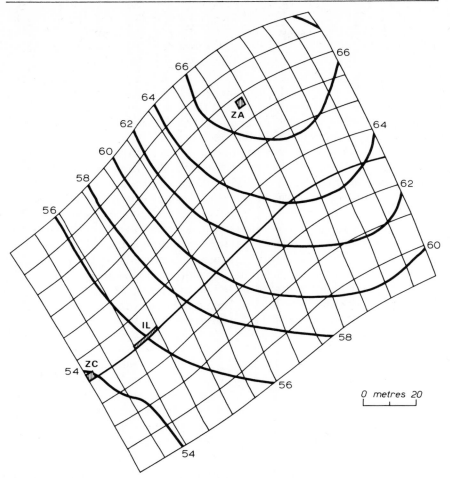

Figure 6. Axonometric view of the south-western part of Sitagroi tell showing the planning grid, contours in metres, location of certain sections and the section line for Figure 5 through ZC and IL. (Surveyed and drawn by R. Wakeham)

1. Along section line L the angle of slope of tell materials *in situ* is approximately constant almost to the crest and thus the assumption of equal depth of tell removal can be made if m equals 1. Experiments on erosion plots have produced values greater than 1 whilst Schumm (1956, 1964) for badlands in different localities obtained values of 0.77 and 1.0 (Carson and Kirkby, 1972). Given such variability in results it seems reasonable to make an assumption of an equal amount of erosion along section line L. When this is calculated for the western as well as the eastern sides of the tell, a result of 1 m is obtained in each case. This should very much be regarded as a minimum figure since, as already noted, an unknown quantity of colluvium has been removed by the river in the west; there could have been considerable

loss by deflation and nothing is known about the extent of colluvium beyond the eastern limit of the cross section (Fig. 5b). It must also be noted that section line L does not follow the line of maximum slope near the top (Fig. 6) and thus for example, the area enclosed within the 64 m contour would have lost material which has not become colluvium at the base of the section line. The important conclusion is that the top of the section line must have been at least 1 m higher than today. This figure in itself gives a clear indication of the amount of archaeological evidence which has been lost.

It is possible to speculate about the former extent of the summit area at the final abandonment of the tell. The highest point on the section line today is 64.5 m and thus at abandonment this point would have been at least 65.5 m. Given the assumptions in estimating the minimum of 1 m erosion, it can be suggested that at abandonment, the flattish summit area extended to the point corresponding to the top of section line L because the present summit is enclosed by the 66 m contour. When this former summit area is interpolated for the whole mound it can be suggested that at abandonment this area was 0.49 ha compared to 0.25 ha today. As well as a change in summit area, a change in slope would also have occurred. The maximum slope today is 15° on the western side, but this would have been about 18° during the final phase of occupation.

Conclusion

From laboratory analysis of samples taken from the deep sounding, it has been established that the mound grew as a result of house collapse and discontinuities are suggested between certain occupation phases. Such conclusions are only possible if material between occupation layers is analysed using pedological techniques. A geomorphological interpretation of the stratigraphy on the lower part of the tell has been the basis for estimating the degree of erosion. It is suggested that the extent of the flat summit at final occupation was about twice its present day area and this illustrates the amount of archaeological information which has been lost. For example, evidence of any perimeter structure on the summit of the tell has disappeared. The important conclusion which emerges from the geomorphological work in Sitagroi tell is that an earth scientist can make a particular and integral contribution to the complete interpretation of an archaeological site.

Acknowledgments

The author wishes to acknowledge the help and encouragement of Professor A.C. Renfrew. Mr. B. Thomas assisted with the description and interpretation of the stratigraphy. Mr. C. Lewis drew the diagrams. Financial support was obtained from the University of

Sheffield Research Fund, the Sir Ernest Cassel Educational Trust and from the funds of the U.C.L.A. – Sheffield University Archaeology Research Project.

REFERENCES

Carson, M.A. and M.J. Kirkby (1972) *Hillslope form and process.* Cambridge.

Davidson, D.A. (1971) 'Geomorphology and prehistoric settlement of the Plain of Drama', *Revue Géomorph. dyn.* 20, 22-6.

Davidson, D.A. (1973) 'Particle size and phosphate analysis: evidence for the evolution of a tell', *Archaeometry* 15, 143-52.

Kirkby, M.J. (1969) 'Erosion by water on hillslopes', *in* R.J. Chorley (ed.) *Water, Earth and Man.* London, 229-38.

Lloyd, S. (1963) *Mounds of the Near East.* Edinburgh.

Musgrave, G.W. (1947) 'Quantitative evaluation of factors in water erosion: a first approximation', *J. Soil Wat. Conserv.* 2, 133-8.

Renfrew, C. (1970) 'The burnt house at Sitagroi', *Antiquity* 44, 131-4.

Renfrew, C. (1971) 'Sitagroi, radiocarbon and the prehistory of south-east Europe', *Antiquity* 45, 275-82.

Schumm, S.A. (1956) 'Evolution of drainage systems and slopes on badlands at Perth Amboy, New Jersey', *Bull. geol. Soc. Am.* 67, 597-646.

Schumm, S.A. (1964) 'Seasonal variations of erosion rates and processes on hillslopes in Western Colorado', *Z. Geomorph. Supp. Band* 5, 215-38.

Vita-Finzi, C. (1964) 'Synchronous stream deposition throughout the Mediterranean area in historical times', *Nature, Lond.* 202, 1324.

Vitz-Finzi, C. (1969) *The Mediterranean valleys.* Cambridge.

Wijmstra, T.A. (1969) 'Palynology of the first 30 metres of a 120 m deep section in northern Greece', *Acta Bot. Nerrl.* 18, 511-27.

Zingg, A.W. (1940) 'Degree and length of land slope as it affects soil loss in runoff', *Agric. Engng.* 21, 59-64.

DISCUSSION

Mr. T. Wilkinson asked the speaker how much colluvium has accumulated on the side of the tell away from the river. Dr. D. Davidson could not comment on this in detail since only one small trench had been sunk under adverse conditions, but the results did suggest that the amount of colluvium on both sides of the tell was about the same. Mr. J.L. Bintliff enquired whether there was any evidence to support the view that wind-blown material had contributed to the growth of the tell. The speaker said that one would tend to expect erosion on a mound rather than deposition, but that Dr. I.W. Cornwall had suggested that the particle-size curves indicated the presence of wind-blown silt. Mr. B. Kerr asked whether the colluvium had been recognised by features other than the included pottery, and the speaker replied that although no particle-size analysis of the colluvium had been carried out, it proved an easy matter to distinguish between colluvium and alluvium in the field. Professor

A.C. Renfrew confirmed the abundance of sherds in the colluvial material.

A general discussion of slope angles then followed in response to a comment by Mr. D. Williams as to whether the mound would be stable at an angle of 18 – 20°. The general feeling was that this seemed perfectly feasible, and that silt slopes can be as steep as 25° if well drained although under bad drainage, a slope of only 15° could be supported.

J. Bintliff

Sediments and settlement in southern Greece

Six sample areas of southern Greece were selected for a study of settlement patterns from the earliest human occupation (Fig. 1). Two had been intensively surveyed for sites: the south-west Argolid Peninsular, the Agiofarango Gorge in south Crete; two had been extensively surveyed – the Sparta and Helos Plains; and two had a good number of sites discovered over a long period of investigation – the island of Melos and the Argos Plain. The location of suspected occupation sites was examined in the light of local geology, soils, geomorphology, water supplies and defensive possibilities, and where applicable, marine resources. The apparent relationship in space between sites of the same period was also studied.

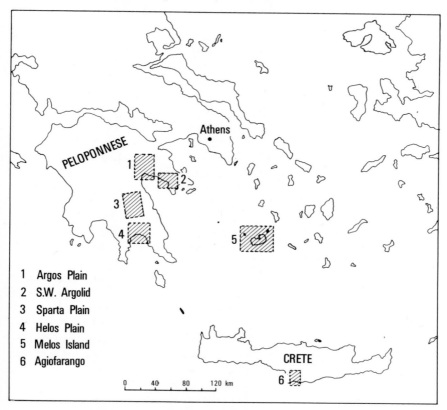

Figure 1. Map of southern Greece to illustrate the areas studied by the writer.

Hitherto the following views have prevailed amongst scholars in this field:

(i) The discontinuous development of the Mediterranean valleys proposed by the geomorphologist Vita-Finzi (1964, 1969), is neither widely applicable nor likely in its implications; Butzer (1971) and Loy (1967) criticised the scheme on grounds of general theory, but Eisma (1962, 1964) and Raphael (1973) produced studies where alluviation appeared to differ in chronology.

(ii) The most prized agricultural areas in the long-term occupance of the Greek lowlands were alluvial valley-bottomlands.

(iii) Settlements are generally placed with priority given to water and defence.

(iv) The Mediterranean landscape has been progressively run down in its farming potential ever since, and directly due to, the activity of man in the area.

(v) Settlement network analysis, based on mutual distance, is of little value to the archaeologist with his incomplete data.

1.   *Pleistocene and Holocene alluviation in Greece*

Five study areas gave evidence of at least two major phases of valley fill, identical in morphology and topography to those described by Vita-Finzi as 'older' and 'younger'; the Agiofarango, heavily eroded, contained only the more recent deposition. The older fill related to a period of low sea-level, and was associated in the south-west Argolid with Upper Pleistocene artefacts (Bintliff and Drost, 1972). It is to be interpreted as a Würmian 'periglacial' colluvium/alluvium, and further evidence for such an identification can be found in studies from other parts of Greece (Leroi-Gourhan, 1963; Dufaure, 1970; Nicod, 1963; Sordinas, 1969; Vita-Finzi, 1966).

Typical older fill features characterise, for example, the upper Argos Plain, where both the plain proper and its bordering massive piedmont fans are composed of a stony red clay that is often heavily cemented. This provides only poor to mediocre arable land for dry – culture cereals and olives. Holocene archaeological sites associated with the older fill are always on, or just below, its present surface; a good example, again from the Argos area, is the Tiryns Late Bronze Age royal tomb, excavated into a 'Red-Bed' colluvial fan. (For the location of this and other sites mentioned in the text, see Fig. 2).

After an interval of incision and thin deltaic deposition in lower reaches, renewed valley infill of a strikingly different nature began, whose cessation and subsequent incision is recent and observable today. In all study areas this latter alluvium could be dated from pottery stratified in its basal layers, and in some cases by buried structures, to late Roman and medieval times. Similar results have been obtained by Paepe (1969), Ward-Perkins (1962), Davidson

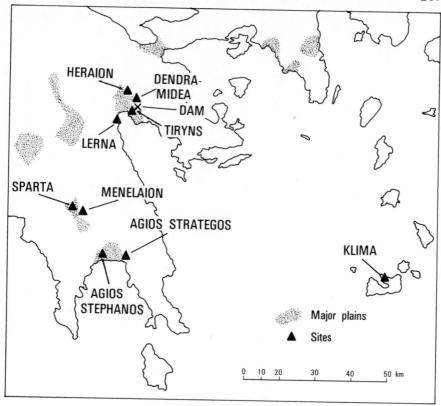

Figure 2. Map of the Peloponnese and west Aegean sea with the location of archaeological sites mentioned in the text.

(1971), and Rapp (1973). The Vita-Finzi chronology is therefore confirmed. Both Eisma (1962, 1964) and Raphael (1973) mistook the limited and agriculturally insignificant deltaic phase for the deep and extensive younger fill, and hence their slightly earlier chronology. The older and younger fills cover wide areas of southern Greece, and constitute a high proportion of its limited cultivable land.

A series of faulted basins in volcanic tuff on Melos provided clear examples of eroded remnants of older fill, while the present basin floors were composed of younger fill now undergoing incision. It is of interest that we are concerned here with closed depressions and therefore no influence of changing sea-levels on the alluvial sequence can be postulated. Both basin and valley fills on the island contained Roman pottery. Still on Melos, the ancient site of Klima lies in a small alluvial plain beside the Great Bay; the Classical and Roman city of Melos is located on the cliffs above, and throughout that time the plain was an inlet of the sea and formed the harbour for the town. By Late Roman times the port had begun to be blocked by alluvium of the second deposition, and above marine levels on a thin terrestrial bed are found the first building remains. By Early Byzantine times the space between the harbour moles (now underwater) was land and a large structure was erected there. Later alluvium covered over these walls and a rising sea reclaimed the ancient mole area (Smith, 1896).

The late Bronze Age fortress-palace of Tiryns is sited on a limestone *inselberg* in the southern and lower part of the Plain of Argos. Deep soundings show that the settlement around the citadel hill rests on buried older fill, while the fortress is now surrounded superficially by younger fill, which has covered over the prehistoric lower town and now occupies the area between it and the sea. The older fill forms the present-day surface higher up the Plain, and if the gradient of this exposed surface is extrapolated to the site, there is good agreement with the level of the old land-surface. Even allowing for sea-level rise, the coast would still in prehistoric times have been much closer and Tiryns would have functioned as a port; and this would explain the atypical location on older fill. (Lehmann, 1937; Gercke, 1971, 1973). To the east of Tiryns and in a higher part of the plain there is a Mycenaean dam (Karo, 1930; Verdelis, 1963; Gercke, 1973). In this area two westward-flowing torrents incised into the older fill to a depth of over ten metres during the prehistoric period. The more northerly of the two became a threat to the stability of the Mycenaean town at Tiryns, and therefore its course was blocked off upstream by a stone and earth dam, and the flow diverted southwards along an artificial canal to the other torrent. The abandoned torrent ceased to develop and no further incision took place in it; it consists only of older fill, and this is important as it demonstrates that upstream sources control both incision and aggradation phases. The canal, on the other hand, which was most probably dug about 1250 B.C., experienced later incision followed by the second alluvial period and finally the present phase of renewed incision.

On the other side of the Argos Plain, at Lerna, a prehistoric town mound has its earliest foundations on a regular older fill land-surface, partly below present sea-level, and now cut off from the nearby coast by a narrow fringe of dark younger fill. As with Tiryns, extrapolation of the gradient plane despite necessary sea-level adjustments, would put the sea at the very edge of the mound. Again, as with Tiryns, the lower fertility of the older fill soil forming the hinterland to the site, compared with other soils available in the Plain, is amply compensated for by the potential of fishing and commerce.

A valuable local fixed point for the alluvial sequence in the Sparta Plain can be found in the riverside quarter of the ancient city. Immediately beside the present bed of the river Evrotas runs a modern and artificial levee, protecting behind it a low strip of recently reclaimed riverbed. Behind this in turn is a raised alluvial terrace, deposited against a higher plateau of Tertiary marls and sands. Beneath the alluvial terrace, and resting on a red clay older fill landsurface, were excavated ancient walls, roads, wells and altars

(Dickins and Wace, 1906; Dawkins, 1909). All these remains were covered by several metres of post-Roman younger fill.

In the Helos Plain, the Late Bronze Age centre of Agios Strategos (Waterhouse and Hope Simpson, 1960) is another site separated from the coast by a badly drained band of second fill and lagoonal deposits. There is good evidence for identifying the location with the Homeric city of Helos 'on the shore'. A parallel situation is found on the opposite (western) edge of the same plain – the contemporary centre of Agios Stephanos (Waterhouse and Hope Simpson, 1960; Lord William Taylour, 1972). The settlement lies on a ridge about two kilometres from the sea, but the intervening land is still below sea-level and is superficially composed of second fill. Recent research indicates that under this fill lie deep marine deposits of Holocene date (Rapp, 1973). The site is unique in the region in lacking a fertile arable base, and would have been a port in the prehistoric era.

2. *Priority arable land*

Over 150 sites were examined in the field, comprising all known prehistoric and a fair sample of the historic sites in each area. With the exception of a few, interpreted as basically fishing and herding locations, almost all postulated pre-Roman settlements lay amid light, nutrient and moisture rich soils, which were eroded *in situ* from Pliocene marls, flysch (in the Argos Plain), and serpentine (in the south-west Argolid).

A good example of flysch preference is the significant exposure that forms the basis of the palace centre at Dendra-Midea (Argos area). Flysch zones are also diagnostic for prehistoric settlements throughout Switzerland (Gates, 1972). Not far from Dendra the major early centre at the Heraion is placed neither on the limestone hills of the plain edge, nor the older fill of the plain proper, but locates on the fine soils of the bright yellow Neogen marls at an intermediate elevation between them. The rendzina soils found on the marls are often the key to early settlement elsewhere, for example the Italan Tavoliere (Jarman, 1972), northern France (Bender, 1971), and southern England (Evans, 1971). Research into early mining techniques suggests that prehistoric prospectors relied on simple characteristics such as colour and feel in differentiating between minerals. The striking soft sandiness and bright yellow and orange hues of the marls and flysch can often today single out an ancient site before its location is found from a map, and once early farmers became aware of the remarkably favourable properties of these soils, they would encounter no problems in searching out similar exposures. An interesting feature of the Helos Plain is a number of inconspicuous low Neogene hillocks that rise above the dark alluvium of the basin, several of which are prehistoric occupation sites. The alluvium is historic fill, and if this later deposit is discounted, the prehistoric

landscape around these settlements can be reconstructed as a low relief
marl hilland.

On Melos, quite individual in geology, there was a locational
preference partly on marls but chiefly in closed or inadequately
drained depressions, where chemical breakdown, of little effectiveness
normally in Greek soils, produced highly fertile basins of infill derived
from the infertile glassy sand blanketing the isle. There is an obvious
contrast between the normal Melian groundsurface of tuff sand and
eruptive scree, with little or no soil, and the fine soils that have
accumulated in the closed basins. It is the latter location that is
usually favoured by the prehistoric communities of the island.

Close exploitation of the younger alluvial fill is predictably post-
Roman in date, while settlement on the older fill begins generally with
the recent introduction of irrigation from deep wells and mechanical
ploughing, which vastly transform the dry, stony, heavy soil.

3.   The role of fresh water and defence

Sources of water and defensive requirements appear only locally and
temporarily to have significantly determined settlement situation,
and in nearly all such examples there exists an adjacent zone of
notably favourable arable land. A good example of this is the site, now
under excavation, of what may be the Mycenaean palace centre for
the whole of the Sparta Plain – the Menelaion. The immediate
environs of the ancient site are dissected and rugged, appropriate for
defence but inhospitable for farming. However the wider view from
the site demonstrates how it dominates a very broad and fertile
lowland of marls and sands, which were later occupied by the
Classical and modern town of Sparta. To take another apparently
'defensive' location, consideration can be given to the imposing
mountain of crystalline limestone that forms the Mycenaean palace-
citadel of Midea (Argos Plain). It rises on the edge of the older fill of
the main plain, and neither the latter nor the mountain itself are of
particular arable fertility. However recent work by Hägg (1962)
favours the suggestion that the main settlement was in fact at the
mountain foot amid a localised pocket of flysch and marls, precisely
where the modern village of Dendra-Manessi is found.

4.   The degeneration of the landscape

The arrival of the historic fill may have considerably increased the
agricultural and maritime depression of the early medieval
Mediterranean, already suffering from invasions and pestilence, by
creating swamps on a large scale along all stages of the rivercourses
and especially in the plains, which were often malarial. The
complexities of coastal change, involving probably absolute movement
in both sea-level and bedrock, combined with sediment sinking and

the actual deposition, have yet to be unravelled and were inadequately tackled by Vita-Finzi (1969). However the vast bulk of recent *coastal plain* fill can be correlated with the early medieval *valley* fill, and together they constitute the most fertile and profitable soils of the Mediterranean. In contrast to the older fill they are loose, fine-textured, and at all times are very close to the watertable. Significantly, their exploitation was only on a small scale until the arrival of political stability, associated drainage and irrigation works, and citrus farming. The natural change in stream regime from aggradation to downcutting, which took place in the same late medieval period, aided the human efforts at reclamation and these soils became open to the farmer.

The nature and dating of the younger fill precludes direct human causation, and has ultimately in all the areas examined brought improved agricultural production in comparison to the potential of the prehistoric landscape. A similar conclusion was reached by Hutchinson in north-west Greece (Hutchinson, 1969). The most probable explanation for this deposition is climatic change. The studies of H.H. Lamb (1966) on recent climatic changes provides a sequence of striking reversals over the period during which clear discontinuities in river regimes have been noted. His postulated southward movement of depression tracks into the Mediterranean would be contemporaneous with the historic fill there. He cites evidence for unusually cold and wet conditions in south Europe at this time, while north Europe underwent a notably warm and dry period.

Human erosive activity paralleling alluviation is improbable considering the alternation of high and low population, intensive and extensive agriculture during this period (Russell, 1972), and the evidence of morphology and vegetation remains from the fill itself (Vita-Finzi, 1969, 1971). It should be noted that the last century has witnessed truly massive deforestation and human interference in Greece, perhaps more than ever before, but no aggradation has ensued.

5. *Settlement network analysis*

Finally, though settlements were closely tied to priority arable resources, in at least three study areas during the Late Bronze Age, settlement hierarchies developed based on mutual distance within an isotropic surface. Sampling problems were obviated by selecting the higher order communities, which are very well represented in archaeological surveys, for the basis of the network, and a period when exact contemporaneity could be plausibly claimed from pottery studies.

Acknowledgments

The fieldwork from which this paper has been drawn is to form the original core of a Ph.D. thesis, 'Human settlement and natural environment in prehistoric southern Greece', to be submitted at the University of Cambridge during 1975. Fieldwork costs in Greece were financed by the Department of Education, Corpus Christi College and the Greek State Scholarship Foundation.

REFERENCES

Bender, B. (1971) Lecture given in Cambridge on 'The French Neolithic'.

Bintliff, J.L. and B. Drost (1972) 'Preliminary report on the palaeoecology of the south-west Argolid', cyclostyled.

Butzer, K.W. (1971) *Environment and Archaeology*. Chicago and New York.

Chorley, R.J. and P. Haggett (1968) *Socio-economic Models in Geography*. London.

Davidson, D. (1971) 'Geomorphology and prehistoric settlement of the Plain of Drama', *Revue Géomorph. dyn.* 20, 22-26.

Dawkins, R.M. (1909) 'Excavations at Sparta', *Annual of the British School at Athens* 15, 6-11.

Dickins, G. and A.J.B. Wace (1906) 'Excavations at Sparta', *Annual of the British School at Athens* 12, 284-302.

Defaure, J-J. (1970) 'Niveaux d'abrasion marine Quaternaires autour du Péloponnese', *Annls. Géogr.* 79, 325-342.

Eisma, D. (1962) 'Beach ridges near Seljuk, Turkey', *Tijdschr. K.* 79, 23446.

Eisma, D. (1964) 'Stream deposition in the Mediterranean area in historical times', *Nature, Lond.* 23, 1061.

Evans, J.G. (1971) *in* D.D.A. Simpson (ed.) *Economy and Settlement in Neolithic and Early Bronze Age Britain and Europe, Leicester, 11-73.*

Gates, T. (1972) personal communication.

Gercke, P. and G. Hiesel (1971) 'Gräbungen in der Unterstadt von Tiryns von 1889 bis 1929', *Tiryns. 5*, Deutsches Archäolog. Institut Athen. Mainz.

Hägg, R. (1962) 'Research at Dendra', *Opuscula Atheniensia* 4, 79.

Hutchinson, Sir J. (1969) 'Erosion and land use: the influence of agriculture on the Epirus region of Greece', *Agric. Hist. Rev.* 17, 85-90.

Jarman, M. and D. Webley (1972) Unpublished report on the early farmers of the Tavoliere Plain, South Italy.

Karo, G. (1930) 'Archaeology in Greece', *Archäologischer Anzeiger*, Jahrbuch des Deutschen Archäologischen Instituts Bd. 45, 112-3.

Lamb, H.H. (1966) *The Changing Climate*. London.

Lehmann, H. (1937) *Argolis 1*. Athen, Deutsches Archäologisches Institut.

Leroi-Gourhan, A., and J. and N. Chavaillon (1963) 'Paléolithique du Péloponnese', *Bull. Soc. Préhist. fr.* 49, 249-265.

Loy, W.G. (1967) *The Land of Nestor*. Office of Naval Research Report 34, National Research Council (U.S.A.)

Nicod, J. (1963) 'Problèmes de morphologie karstique en Grèce', *Méditerrannée* 4, 15-25.

Paepe, R. (1969) 'Geomorphic surfaces and Quarternary deposits of the Adami area (south-east Attica)', *Thorikos 4*, 7-52.

Raphael, C.N. (1973) 'Late Quaternary changes in coastal Elis, Greece', *Geogrl. Rev.* 63, 73-89.

Rapp, G. (1973) personal communication.

Russell, J.C. (1972) 'Population in Europe 500-1500', chapter 2 *in* C.M. Cipolla (ed.) *The Fontana economic history of Europe: the Middle Ages.* London.

Smith, C. (1896) 'Excavations on Melos', *Annual of the British School at Athens*, 2, 63-9.

Sordinas, A. (1969) 'Investigations of the prehistory of Corfu during 1964-66', *Balkan Studies* 10, 393-424.

Verdelis, N. (1963) 'Neue Geometrische Gräber in Tiryns', *Mitteilungen des Deutschen Archäologischen Instituts*, Athenische Abt. Bd. 78, 5.

Vita-Finzi, C. (1964) *Nature, Lond.* 202, 1324.

Vita-Finzi, C. and E.S. Higgs (1966) 'The climate, environment and industries of Stone Age Greece, Part 2', *Proc. prehist. Soc.* 32, 1-29.

Vita-Finzi, C. (1969) *The Mediterranean Valleys.* Cambridge.

Vita-Finzi, C. and G.W. Dimbleby (1971) 'Medieval pollen from Jordan', *Pollen et Spores* 13, 415-20.

Ward-Perkins, J.B. (1962) 'The historical geography of South Etruria', *Geogrl. J.* 128, 389-405.

Waterhouse, H. and R. Hope Simpson (1960) 'Prehistoric Laconia, Part 1', Annual of the British School at Athens 55, 67-107.

DISCUSSION

Attention was focussed on the nature and origin of the younger fill. Mr. J. Wagstaff gave supplementary evidence to support a post-Roman and medieval date for the material. Professor A.C. Renfrew commented that a long period of relative stability is implied before the formation of the younger fill and he wondered if one possible explanation for the formation of the younger fill was an increase in deforestation during Classical times. He also noted that extensive deposition in certain areas must have been paralleled by erosion in others to infer a substantial change in relative land quality. The speaker replied that literary evidence for deforestation during Classical times was sparse and unreliable. Miss C. Delano Smith made the point that changes in sea level could also have initiated aggradation.

T.J. Wilkinson

Soil and sediment structures as an aid to archaeological interpretation: Sediments at Dibsi Faraj, Syria

During four seasons in 1972 and 1973 archaeological rescue excavations were conducted on the Late Roman/Byzantine site of Dibsi Faraj, Syria (Fig. 1). For two of these seasons the writer was employed to examine site sediments and the regional geomorphology in order to elucidate some of the stratigraphic and archaeological problems. This paper outlines how the study of sediment and soil structures assisted in the interpretation of the site stratigraphy.

The citadel of Dibsi Faraj is situated on low bluffs and alluvial fans on the right bank of the River Euphrates approximately 18 km downstream of the modern town of Meskene (Fig. 2). It is a large site

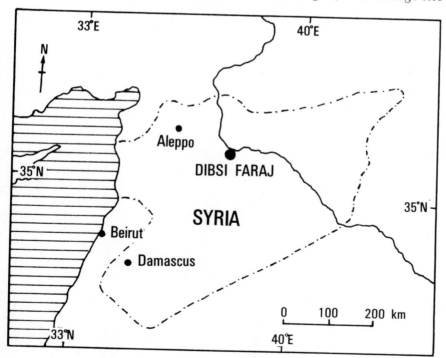

Figure 1. Location of Dibsi Faraj.

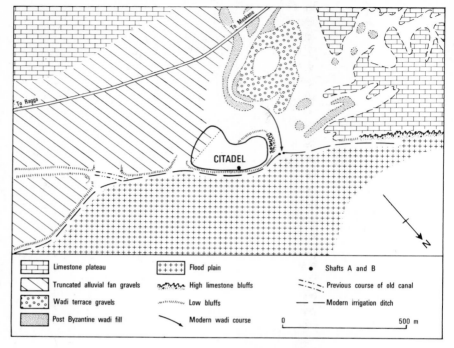

Figure 2. The physical setting of Dibsi Faraj.

occupying some 5.5 ha and was constructed of large limestone and conglomerate blocks as well as baked and mud-bricks. Later occupation, extending into the Islamic period, used more mud-brick, and because of the confining defensive walls sediments from such constructions along with copious supplies of town refuse accumulated to depths of up to four metres.

In any suite of occupational deposits four major characteristics can be described namely colour, composition, texture of these components and, their structure. In natural soils structure represents the degree and nature of aggregation and the distribution of voids, whilst in sediments it is the arrangement of sediment particles as a result of processes of deposition or post-depositional disturbance. Such sediment and soil structures lead to an understanding of the mode of deposition of the sediment, the direction of movement of the depositing medium, the rate of deposition (e.g. rapid, slow accretion, stable surface, continuous or cyclic deposition), the character of the post-depositional environment and the relationship of one stratum or feature to another.

At Dibsi Faraj very great 'topographic' variations and multiple sediments sources resulted in complex deposits; some resembled natural mineral sediments, but many deposits were hybrid soil/sediment/artificial materials. All deposits were examined in site sections with the aid of a hand lens and Munsell colour charts and detailed descriptions were then made. Eventually the environment of deposition of certain soils could be diagnosed and significant structures classified. The latter fell into two major categories:

(i) Primary structures, those produced during or just before the deposition of the sediment.

(ii) Secondary structures, those produced after deposition and which represent a re-arrangement or transformation of the organic/mineral particles by an applied force, the deposition of clay or calcium carbonate within the original fabric or, biological/human disturbances. Time periods for structure formation overlap, but usually primary structures are produced more rapidly than secondary, and, theoretically, over long time periods soil horizons should form.

1.   Sedimentary processes and resultant structures

1.1 Primary structures

Water-lain sediments. On sites possessing large areas of impermeable ground surfaces, storm runoff is rapid and when urban debris is available large amounts of water-lain sediments can accumulate in depressions. Storm runoff forms sheet flow over large areas, but where flow is confined or runs across pre-existing sediments, channel flow may result. Because of the short transport distance at Dibsi Faraj, sediment rounding was rare.[1] Gravels were usually clean, except in areas of illuviation, and sorting varied. In areas of sheet flow, well sorted lag gravels and sands occurred, but in depositional 'sink' areas graded couplets of clay-silts-medium gravels predominated. Couplet thickness varied from 2-20 cm and although some occurred in, or were cut by, channel deposits; most formed gently dipping beds with sediments. Flow-directional structures such as cross laminations and couplets or still-water silts and clays. These beds, formed by flash floods, superficially resemble basal turbidite beds rather than alluvial sediments. Flow-directional structures such as cross laminations and pebble imbrication were rare and reliance had to be placed on flow geometry, gradient and downstream textural changes. Transverse and longitudinal sections aided these studies and tentative conclusions concerning flow direction, flow power and material source could be made (see section 2). Flash flood sediments appeared to predominate after at least part of the site had been abandoned and when large areas of impermeable ground surface still remained. Such sediments were typical of early urban decay.

Wind-blown sediments. These occurred over most of the site. Most

1.   Rounded sediments did occur but were usually transported by man from the flood plain or excavated from terrace gravels beneath the site.

frequent were wind-blown topsoils, massive, soft, loess-like soils of low (estimated) bulk density, permeated by fine root holes, plant impressions and insect burrows. They usually post-dated the flash flood deposits and apparently formed during the final urban decay period when runoff was diminished by infiltration into surface sediments. Such additions increased the overall infiltration capacity of the site, thus further inhibiting runoff.

Of more direct archaeological interest were the accumulations of charcoal and ash deposits of very low bulk density found in several situations near walls. In one case the sediments were stratified and possessed up to 14 black/brown[2] layers forming accumulations up to 1.2 m in thickness. Strata boundaries were moderately distinct and 'flame structures' were common together with one normal fault. These 'flame structures' were triangular features intruding into the layer above and they appeared to represent deposition from turbulent, lee-side eddy currents.[3] The strata in question did not represent a series of discrete periods of deposition, but appeared to be formed by particles of different densities deposited by one major wind of fluctuating strength and direction.

The faulting observed could be traced by the displacement of strata, and it was apparent that immediately after faulting little time had elapsed for the further slumping of the deposits and subsequently deposition was continuous across the faulted area. As a supplement to this, observations showed that contamination by adventitious sediments and insect burrows were rare, for in this case they had been excluded by the rapid rate of sedimentation.

It was tentatively concluded that deposition was from an ash and charcoal laden wind in a turbulent lee side environment. The time period of accumulation was probably short (several hours-days) and the material must have been derived from a large, neighbouring burnt area.

1.2 Primary-secondary structures

Carbonate precipitates. Calcium and magnesium carbonates are frequently precipitated from clean water flows and flow-stones occur as pipe linings, coatings on walls due to seepage from pipes and also where carbonate-rich waters (flowing, for example, through areas of collapsed mortar) have filtered through the soil. Pipe linings occur as thin carbonate laminations precipitated and aligned as minor bed forms with the palaeo-current flow. Flow direction can sometimes be

2.   Brown varied from pale brown 10YR 6/3 (dry) to light brownish grey 2.5Y 6/2 (dry).

3.   The mechanism of formation of flame structures remains controversial, but has been ascribed to the drag effect of a sediment laden current on the underlying sediments (Conybeare and Crook, 1968) or differential deposition caused by differential settling rates (Dzulynski and Walton, 1965). In the case in question the latter suggestion is favoured, but deformation of the underlying strata by drag cannot be discounted.

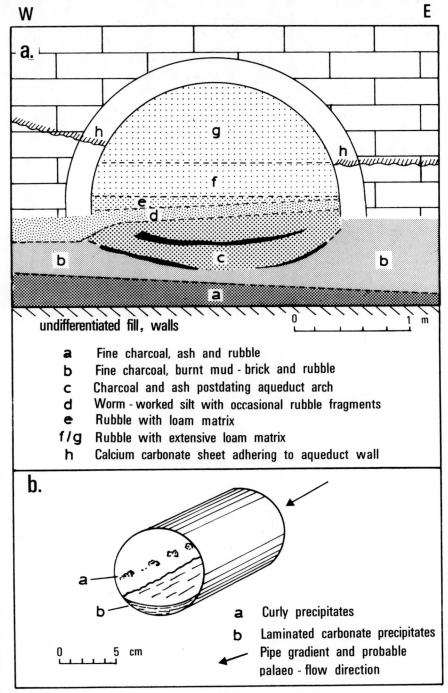

Figure 3. (a) Stratigraphic sequence associated with an aqueduct. Note on stratigraphic correlations: *c-h* all post-date the construction of the aqueduct although *d* is non-conformable due to biological mixing. *d*, and its topsoil *e*, can be tentatively correlated with *h* by virtue of their similar moist depositional environment. However, correlation by height equates *h* with *f* or *g*. Clearly, unless either correlation can be shown to be unjustified, a compromise must be sought. (b) Idealised sketch of carbonate precipitates in a water supply pipe.

established by bed morphology (if it is not a plane bed) and it appears possible that depth of flow may be estimated from the deposits adhering to the pipe walls (Fig. 3b). If the pipe was not full, the overall pipe gradient can be used to check the direction gauged from the bed forms. Where pipes run through buildings and along walls thin carbonate sheets sometimes adhere to the walls below the pipe course; therefore even if the relevant pipe is absent, its course can be traced by studying the distribution of such sheets. Some sheets were permeated by fine charcoal flecks, indicating that deposition was in a dirty, charcoal-rich environment, and it could be concluded that the pipe was in operation during or after the nearby stokeholes.

Particle orientation. Particle and fabric orientation is one of the fundamental principles used in understanding the rate of deposition of gravity deposited sediments, and it is intuitively used by most archaeologists. Rapid collapse leads to a random orientation of masonry and rock fragments which are permeated by large interstitial voids or, less frequently, a sparse fines matrix. Slower accretion leads to the orientation of particles as they align parallel to the slope. In the accretion range, fall-sorting zones can still be distinguished, but where deposition is very slow, the larger fragments tend to lie in a poorly sorted matrix of wind-blown and slowly accreted fines. Below walls built of readily weathered blocks, small rock fragments flake off the wall and accumulate slowly to form a fine, well-bedded deposit grading to sheet or puddle deposited fines down slope.

Of greater diagnostic significance are silts and clays which, due to an applied stress from above, possess a secondary platy structure aligned at right angles to the stress direction. These structures usually occurred as remnants of pre-existing mud floors or external courtyard or ground surfaces, but care had to be taken to differentiate between compressional structures and other oriented fabrics. Platy structure reinforces the old rule-of-thumb criterion of sediment hardness for the identification of mud floors, and when combined with the evidence of plant impressions and insect holes it enables a tentative diagnosis of the internal or external nature of the surface to be made.

1.3 Secondary structures
Illuvial deposits. Illuviation occurs when waters rich in silts and clays percolate through pre-existing permeable materials. The transported sediments are progressively deposited as coatings on stones, cracks and soil aggregates (peds) until silt/clay bridges occur and voids are infilled. A similar process also occurs where carbonate saturated waters flow through archaeological sediments. Illuvial accumulations frequently occur in the vicinity of drains, conduits or depositional 'sinks'. They enable drains and conduits to be traced even if the feature itself is not encountered and because of their secondary

4.   Usually described as flute casts, moulds or marks depending upon their morphology, (Conybeare and Crook, 1968).

nature, they can aid in the relative dating of specific features (Fig. 4, shading d).

Biogenic structures. The Dibsi Faraj site sediments exhibited varying numbers of insect larval holes throughout the profile for if the surface remains stable for several years, insect activity becomes concentrated in the upper layers and an incipient A_1 horizon forms, permeated by abundant small (2-5 mm) tubes. Such A_1 horizons characterise the local steppe soils. Because of the rapid rate of sedimentation and soil disturbance, true soil profile development was inhibited and these biogenic horizons were used in the place of paleosols as an indication of earlier stable ground surfaces.[5] Not all insect passages, however, are formed near the surface; ants nests, horizontal lenticular voids filled with illuvial clay and pillow-shaped faecal material, penetrate well below the ground surface and are a poor indication of the position of the old topsoil. In general though floral and faunal pores are an indication of at least temporary stability, but should be treated with caution unless they are abundant or form a distinct horizon.

In arid lands abundant evidence of worm action indicated the previous existence of moist soils in the area. This evidence is variable and not always clear; in homogenous silts and clays worm action was found to produce a firm deposit with common, irregular voids, which readily broke into a sub-rounded blocky structure.[6] Hand lens examination showed ped faces to be curved with occasional small, spherical, silt deposits adhering to them. Identification of worm action becomes clearer in heterogenous sediments where one sediment is injected into the next, while soft, wet rocks can be permeated by small worm holes about 2-3 mm in diameter. If a worm-worked structure is demonstrated, this can indicate the previous existence of a damp zone or the possibility of abundant horizon mixing (pottery chronologies of uncertain use, see Fig. 3a).

Voids. Voids are fundamental to the definition of soil structure, and in archaeological sediments at Dibsi Faraj they took three main forms:

(i) Interstitial voids in collapsed materials. Discussed in section 1.2 under 'particle orientation'.

(ii) Large interconnected chambers forming cavernous deposits in organo-mineral sediments. Decay of the contained organic matter takes place, leaving the inert mineral materials as a skeleton. Such structures are fragile and liable to collapse and the best developed cavernous structures were found in rubbish deposits permeated by calcareous illuvial materials which, when indurated, provide a rigid skeleton for the sediment. In this form they can greatly hinder sieving.

---

5.   These are generally referred to as 'biopores' and when present in large numbers they can indicate the presence of buried A_1 horizons in natural soils (de Meester and van Schuylenburgh, 1970).
6.   Described by E.A. FitzPatrick (1971) as 'vermicular'.

(iii) The most distinctive voids in the Dibsi Faraj soils occurred in vesicular silts, soils resembling the surface horizons of certain irrigated and arid and arctic soils. These silts were permeated by abundant small (0.5-2.0 mm) circular vesicles, which according to Miller (1971) are produced by wetting and drying cycles within the soil. On the site, in a layer of fine, accreted limestone debris, common vesicles indicated the previous operation of wetting and drying cycles, whilst laminated illuvial silts infilling cracks in the same deposit enabled the number of cycles to be estimated. In this case the cycles were caused by a nearby irrigation ditch, but in other cases features not previously associated with water accumulation, could be demonstrated to have emptied and filled cyclically.

Voids similar to vesicles are common in some illuvial sediments, but they can normally be differentiated from vesicles by their irregular walls possessing multiple centres of curvature.

2. Stratigraphic interpretation from two shafts

Behind the northern defensive wall two deep shafts were partially excavated (Figs. 2 and 4). The internal shaft (A) was cut in the local soft, white limestone, and near its excavated base two tunnels or chambers were found, but time only permitted the partial excavation of these. The external shaft B flanked the flood plain and was excavated until the water table was reached. The excavation of the shafts was a lengthy process taking three men approximately four weeks for each shaft and at the end of that period it was apparent that much work remained. Time was at a premium but by examining the deposits, the local geomorphology and nearby water supply systems an attempt could be made to unravel its geometry and possible use.

The sediments contained within the two shafts are summarised in Fig. 4. Strata *a*, *b* and *c* were probably laid down by flash floods draining from the main site; flow power declined from *a* to *c* with the sink for deposits *a* being in *b* and *c* and the sink for deposits *b* being in *c*. This suggests that sediments entered the system from at or below street level and that the chambers (a) and (b) are connected to form a continuous flow system linking the street drainage with the shaft. Above the upper chamber (a) in A and throughout shaft B, flash flood deposits were absent and instead, collapsed and accretion debris predominated (Fig. 4, deposits *f*, *g* and *h*). In A this change in the nature of deposition could be ascribed to a halt in flash flood sedimentation due to the blockage of the system and after this, gravity fall accumulation became predominant.

Syperficial examination of the deposit *c* could have led to its diagnosis as a graded sequence deposited in a well, but its association with deposits *b* and *c* and the evidence from illuviation of water flow *through* the fabric indicated the existence of a different sedimentary

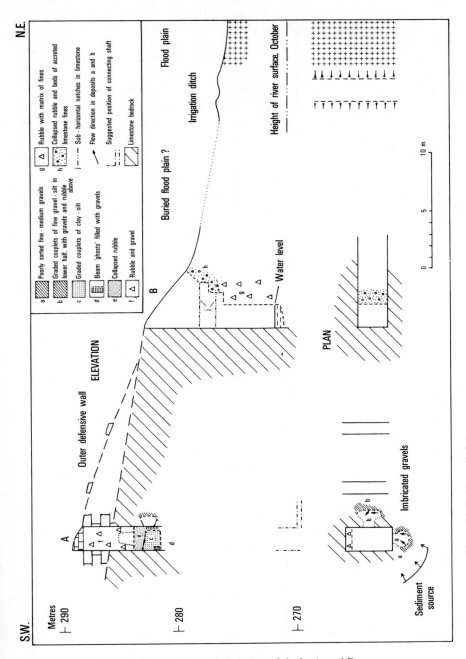

Figure 4. Plan and elevation of shafts A and B.

environment. The conclusion that the deposits were a product of flash floods during urban decay was reinforced by the occurrence of post-Byzantine sherds throughout the deposits.

The position of shaft B adjacent to the flood plain suggests its possible role as an intake from a canal which once crossed the flood plain in the vicinity. Traces of such a canal occur in sheltered positions downstream and excavations along with the study of the canals hydraulic characteristics suggest that it once penetrated at least 8 km upstream, only to be subsequently removed by the migrating Euphrates.[7] However, no canal silts or water marks exist at the required level in the shaft and an alternative hypothesis was sought. Instead it is suggested that shaft B, being partly excavated into flood plain gravels, acted as a well, receiving clean, rapidly recharged water from the flood plain. From this shaft, water could then be led by a tunnel to shaft A which being cut entirely in sparsely jointed or massive limestones, would have received only slow ground water recharge. This clean water could then be fed into pipes for domestic consumption and bath systems, access being via the chambers (a) and (b). Fig. 4 shows a tentative reconstruction of the possible link tunnel.

Conclusions

During the period of maximum occupation of the Dibsi Faraj citadel, building materials and refuse were either cleared away, removed by floods and drains from the site or dumped in waste areas. Upon the decline of the site, drains were blocked and characteristic flash flood deposits accumulated in depressions, especially along the low northern wall. The accumulation of water-lain and other sediments increased the site infiltration capacity and consequently wind-blown sediments became more significant. These eventually covered the water-lain sediments and today blanket the entire site. Within this very general scheme, details of sedimentation lent themselves to the following archaeological interpretations:

(i) Establishing the nature of the environment contemporary with specific features.

(ii) Tracing building structures and establishing the generalised pattern of partially excavated buildings.

(iii) Understanding water conveyance systems.

(iv) Examining minor details of chronology; for example, the length of time for the deposition of certain sediments or the relationship of one feature or stratum to another. The latter can sometimes be established by 'genetic equivalence', in other words by tentatively

---

7.   Geomorphological and archaeological evidence has demonstrated that in the Dibsi Faraj area the brunt of fluvial erosion has been born by the south bank of the Euphrates valley, (T.J. Wilkinson, 1975).

correlating deposits of a similar depositional environment, but different stratigraphic position.

(v) Finding old mud-floor levels and external ground surfaces.

(vi) Demonstrating the existence of biologically disturbed strata.

This selection of structures in archaeological sediments is by no means exhaustive and only applies to certain urban sites in semi-arid areas. Sediments from sites which have only used mud-brick for building, are more homogenous and less readily differentiated, but even on sites exhibiting a wide range of sediment types, the structures which have been described in this paper are not always common and when they do occur, interpretations must remain tentative until further 'process' or comparative studies have been undertaken.

Acknowledgements

This work was a contribution to the excavations at Dibsi Faraj, Syria, jointly financed by Dumbarton Oaks, Washington and the University of Michigan. The author would especially like to thank Richard Harper (director of excavations), Yvonne Harper, Ali Taha (Palmyra Museum), and all members of the Dibsi Faraj excavations for their help and encouragement during the two seasons of 1972 and 1973.

REFERENCES

Conybeare, C.E.B. and K.A.W. Crook (1968) *Manual of Sedimentary Structures*, Dept. of National Development, Bureau of Mineral Resources (geology and geophysics). Bulletin no. 102.

Dzulynski, S. and E.K. Walton (1965) *Sedimentary Features of Flysch and Greywackes*. Oxford.

FitzPatrick, E.A. (1971) *Pedology*. Edinburgh.

de Meester, T. and T. van Schuylenburgh (1966) 'Genesis and morphology of reddishbrown soils of the great.Konya basin,'*in* T. de Meester (ed.) *Soils of the Great Konya Basin, Turkey*, Wageningen, Netherlands, 221-7.

Miller, D.E. (1971) 'Formation of vesicular structure in soil', *Proc. Soil Sci. Soc. Am.* 135, 635-7.

Wilkinson, T.J. (1975) 'Erosion and sedimentation along the Euphrates valley, Syria: interpretations from archaeological sites', *in* W. Brice (ed.) *Historical Geography of the Middle East*. London.

A.J. Tankard
and F.R. Schweitzer

Textural analysis of cave sediments: Die Kelders, Cape Province, South Africa

This paper presents an analysis of the sediments from a coastal cave which bears both Middle Stone Age and Late Stone Age cultural material. The aim of this study is threefold: to present a quantitative and qualitative description of the cave sediments; to determine the environment both within the cave and in the neighbourhood during the time of sediment accumulation; and to examine the change that has taken place subsequent to deposition.

For many years sedimentologists have used grain size parameters, with varying success, to determine sedimentary environment. Sorting (standard deviation or disperson), skewness and kurtosis have been found to be environment-sensitive parameters. It has been found that on the phi scale dune sands are generally positively skewed and beach sands negatively skewed (Folk and Ward, 1957; Mason and Folk, 1958; Friedman, 1961; Duane, 1964). One of the problems of interpretation of grain size distribution is that the same processes occurring within a number of environments result in similar textural responses (Visher, 1969). In the last decade a powerful new tool, the scanning electron microscope, has been used to complement conventional grading analysis. During transportation, sedimentation and compaction detrital grains may be mechanically abraded and chemically altered (Krinsley and Takahashi, 1962). Krinsley and Margolis (1969) found it possible to distinguish between aeolian and littoral grains by means of surface textures.

Textural analysis of cave sediments may be invaluable in palaeo-environmental interpretation but is, unfortunately, too seldom used. Lais (1941) explored the application of textural analysis to Alpine cave deposits while Shackley (1972) presents a more recent discussion. In South Africa Brain (1958) has very successfully used sediment analyses in examining the Sterkfontein cave fills, while Butzer (1973) has carried out a detailed examination of the sediments of a coastal cave.

1.   Location and geological setting

Die Kelders I is a north facing cave situated on Walker Bay (Fig. 1).

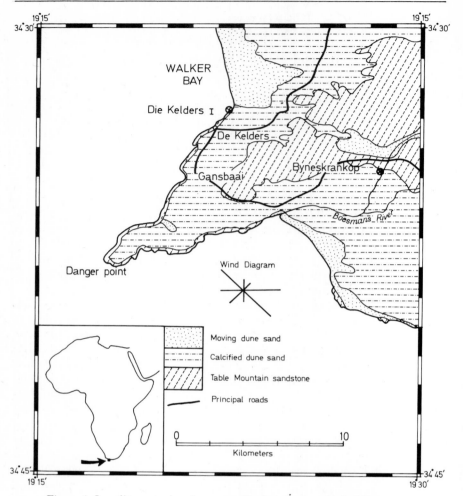

Figure 1. Locality map showing generalised geology and wind diagram.

Coordinates are 34° 32.8′ S and 19° 22.3′ E. It is one of a complex of caves situated 160 km south-east of Cape Town. The cave is about 8 m above mean sea level (a.s.l.) and 10 m inland from the high-watermark (Fig. 2). Excavation was carried out intermittently by one of us (F.R.S.) at the eastern side of the cave from mid-1969 to early 1973.

The oldest rock unit encountered is folded quartzite of Palaeozoic age, the Table Mountain sandstone. In the area of the cave the quartzite dips 40° to the north and has been locally planed at 7-8 m a.s.l. The planed surface rises gently to the southeast. Lying unconformably upon the quartzite is the Late Tertiary Bredasdorp Formation, a marine limestone. The basal part of the limestone is an angular to subangular breccioid horizon, passing upward through

Figure 2. Die Kelders I showing erosion of cave along interface between Bredasdorp limestone (B) and Table Mountain sandstone (Tm). Overlying the Bredasdorp limestone is Late Pleistocene aeolianite (A).

horizontally bedded fine quartzose sands with much comminuted shell, and finally into massive limestone.

The youngest sediments in this area are Late Pleistocene aeolianites (Fig. 2) of Würm age and superficial modern dune sands. The aeolianites exhibit high angle crossbedding dipping predominantly in a north to north-westerly direction. Just north of the cave complex the aeolianites contain artefacts similar to those from the Middle Stone Age horizons in Die Kelders I. (The Middle Stone Age is thought to range from 100000 to about 35000 B.P.; see Klein, 1974). The artefacts occur in a poorly sorted outwash deposit interbedded with the aeolianite. The Middle Stone Age artefacts prove the aeolianite to be no older than the Middle Stone Age. Furthermore, the occurrence of some rubble of this aeolianite within the Die Kelders I stratigraphic succession suggests that lithification must have occurred prior to about 6000 B.P. Locally on promontories and stacks of Palaeozoic quartzite there are outliers of aeolianite which show a more seaward extension of the dune field during the lowered sea levels of the last glacial age. Northwards along Walker Bay the aeolianites extend to below beach level and are being eroded back by the present sea. Along the length of Walker Bay, these easily eroded aeolianites show no evidence of having been planed by any sea levels higher than present.

The Die Kelders cave complex is eroded along the interface between the Palaeozoic and Tertiary rocks. The backs of the caves reach 7-10 m a.s.l. The frequent occurrence of caves at this altitude along the southern Cape coast suggests that they were formed during a period or periods of higher than present sea level. Die Kelders I is separated from the sea by a beach rampart of rounded quartzite boulders which rises to 2 m a.s.l. Erosion along the strike of the steeply dipping quartzite has formed deep passages. The Middle Stone Age deposits in Die Kelders I are preserved in the extension of one of these passages.

2.   General features of the cave deposits

2.1 Stratigraphy

Excavation at Die Kelders I has proved the existence of more than 7 m of sediment. The stratigraphy of the cave deposits lying between a bar of Table Mountain sandstone and the back of the cave can be summarised as follows (Figs. 3 and 4).

(1) An incompressible horizon of quartzite 'beach' boulders and coarse poorly-sorted interstitial quartz sand containing weathered echinoid spines (layers 16 and 17).
(2) Middle Stone Age sequence comprising alternating occupation and non-occupation layers. The occupation layers are compacted and characterised by cultural material and faunal remains. The

Figure 3. Die Kelders I excavation. Numbers refer to layers. The extensive tabular boulders of roof rock (R) in layer 6 and talus (T) deposit in layer 2 are shown.

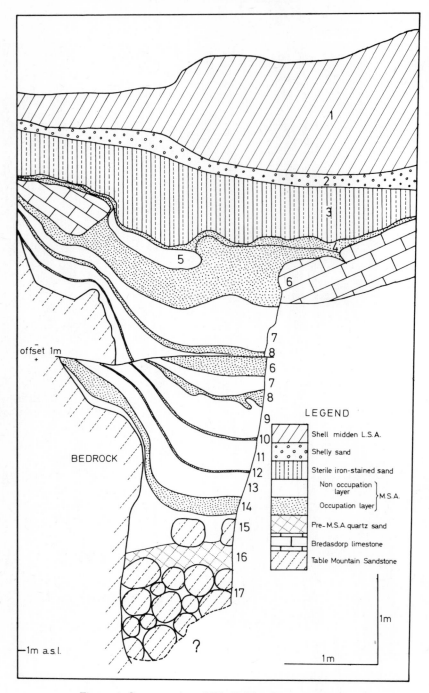

offset 1m

BEDROCK

—1m a.s.l.

?

1

2

3

4

5

6

7
8
6
7
8
9
10
11
12
13
14
15
16
17

LEGEND

Shell midden L.S.A.

Shelly sand

Sterile iron-stained sand

Non occupation layer } M.S.A.

Occupation layer

Pre-M.S.A. quartz sand

Bredasdorp limestone

Table Mountain Sandstone

1m

1m

Figure 4. Cross-section of Die Kelders I stratigraphy.

non-occupation layers consist of fine to medium quartz sands and micro-fauna (layers 4 through 15).

(3) Sterile yellow sands, fine to medium grained, iron-stained (layer 3).

(4) Shelly quartz sand, fine to medium grained, echinoid spines, foraminifers (layer 2).

(5) Late Stone Age shell middens with intercalated lenses of fine to medium calcareous sand (layer 1). Radiocarbon dates on charcoal samples range from 2020 B.P. to 1465 B.P. (Schweitzer, 1970).

The history of Die Kelders I deposits begins with accumulation of round 'beach' boulders and cobbles at a time when the cave was being scoured by the sea. With the onset of glaciation in the high latitudes and the consequent lowering of sea level, the cave became available for occupation and accumulation of the Middle Stone Age succession.

The Middle Stone Age quartzose sediments (layers 4 through 15) have been modified by the addition of cultural debris (artefacts and bone), small rodent bones probably dropped by predators such as owls, spalled fragments of roof rock and humus. The mud-humus context reaches a maximum in occupation layers 6 and 14. The Middle Stone Age is represented in all by six occupations (layers 4, 6, 8, 10, 12 and 14) which are separated by sterile non-occupation layers (5, 7, 9, 11 and 13). The occupation layers are generally darker coloured than the non-occupation layers due to the higher mud and humus content, and characterised by artefacts and vertebrate remains. The occupation layers have a range of carbon values from 0.16 to 0.36 per cent while the range in the non-occupation layers is 0.08 to 0.12 per cent (carbon determined by oxidation and titration). (Carbon in soils can be as much as 1 to 2 per cent.) There is a total absence of marine shells suggesting that the shoreline was at that time some distance from the cave. But a few seal and penguin bones do suggest that the sea was still exploited.

Occupation layer 6 near the top of the Middle Stone Age sequence is the most conspicuous, being 0.7 m thick. It is a dark yellowish brown colour (10 YR 3.2-4.2: Rock-Color Chart of the Rock-Color Chart Committee, 1970), but mottled red by haematite and yellow by limonite. The quartz grains in this horizon are extensively iron-stained. The underlying and older occupation layers are less distinct in terms of coloration and thickness, and considerable compaction has taken place due to leaching of the mud and humus. In layers 4, 10 and 12 leaching and compaction have resulted in the horizons being represented largely by artefacts and bone. In this respect it is envisaged that the gully between the quartzite wall rock and the back of the cave, which was wave-scoured, acted as a natural channel for drainage of ground-water and elutriation during accumulation of Middle Stone Age sediments.

Overlying the Middle Stone Age succession are the yellow iron-stained sands (layer 3) (10 YR 7.4-6.4, 7.6). In contrast to the underlying strata the yellow sands are sterile in that they contain no lithic component or large vertebrate bones, indicating a period of non-occupation of the cave. The period of non-occupation started at 35000 to 40000 B.P. and ended in the Late Stone Age occupation at about 2000 B.P. The virtual absence of even rodent bones in these sediments suggests that the cave environment was hostile to occupation at that time. The yellow sands are entirely quartzose and contain no shell fragments suggesting that the shoreline was still distant.

Overlying the yellow sands are the shelly quartzose sands which contain echinoid spines and foraminifers. Complete echinoid spines suggest that the shoreline had returned close to the cave. The brief period of accumulation of the shelly sands is followed by the shell middens of Later Stone Age man.

An important constituent of the cave sediments is roof rock debris in layers 5, 6, 13 and 14. In the Middle Stone Age layers 13 and 14 the inclusions occur as angular, granule and pebble-size, spalled flakes. Microscopic examination shows that they are fresh, unweathered and with sharp edges, suggesting mechanical rather than chemical release from the roof. Whereas the loam and iron-staining of the Middle Stone Age sands suggest a wetter or damper microclimate, it also seems feasible that these spalled flakes can be attributed to frost action. In layer 6 there are fewer flakes of roof rock, but an almost impenetrable cover of unweathered large tabular boulders of roof rock occurs, the uppermost of which show solutional weathering. So extensive and massive are the tabular roof rock boulders that pneumatic drilling was required to remove them. It is thought that such an extensive collapse of roof rock could be explained only by a catastrophic earthquake shock. Frost action is an unlikely explanation because of the limited amount of small cryoclastic debris. But it should be noted that recent earthquakes (1969) have collapsed cliff rock along the present shore.

2.2 Structure

Layers 16 and 17 at the base of the succession are incompressible sands and 'beach' boulders. Fig. 4 clearly illustrates how the Middle Stone Age succession consists of alternating occupation and non-occupation strata which dip towards the centre of the cave. On the east face (Fig. 4) dips range from about 35° to nearly vertical against the quartzite wall rock, but flatten out to a subhorizontal attitude away from the wall rock. Sections of the south wall show that the general dip is 20° to 25° to the centre of the cave. It is most significant that the dips of 35° to 90° against the wall rock are greater than the angle-of-repose for clean, dry quartz sands.

Occupation layer 14 is more loamy in content than either the sediments above or below. But microscopic examination shows that the quartz grains from underlying layer 15 are humus-coated. The vertical dip in the central part of layer 14 could only have arisen by compaction or removal of part of layer 15. Since it is difficult to account for a massive removal of quartz sand, it must be assumed that layer 15 at one stage contained an appreciable mud-humus content. If it is also correct to assume that, at the time of formation of layer 14, the occupants lived on an almost horizontal surface, then the present configuration of the base of this layer suggests that the compaction of the underlying sediment (layer 15), by leaching and dewatering, has been of the order of 75 to 80 per cent. Convergence of the laminae of layer 14 over the protruding quartzite (Fig. 4) is the direct result of this compaction. Occupation layers 4, 10 and 12 have been so condensed that they are discernible only by their content of artefacts and bones. Layer 4 is 60 to 80 per cent compacted.

The progressive decrease in dip with decreasing age of the occupation layers and the thickening of non-occupation sediments towards the centre of the cave shows that subsidence due to compaction has occurred from the earliest phase of deposition to the final phase of Middle Stone Age occupation. Whereas the laminae in the sediments overlying the Middle Stone Age horizons are always sharply defined, the laminae in the Middle Stone Age succession are wavy and considerably less distinct, a natural response to compaction. Compaction has also resulted in the formation of numerous microfaults in the sediments.

The laminae of the overlying yellow sands are always sharply defined (Fig. 3) showing a minimum effect of compaction of the underlying strata. The maximum dip is 20 to 25° towards the back of the cave (southward) while the sediments are horizontally-bedded on the south wall. There is a general expansion of the laminae towards the back of the cave, illustrating clearly that the top of the Middle Stone Age succession formed a basin at the time of deposition.

The general expansion of the laminae and beds (terminology after Ingram, 1954) toward the back of the cave and the overall subhorizontal attitude of the layer 3 sediments is clearly indicative of deposition into standing water. Ponding of the water took place between the back of the cave and the quartzite bar at the cave entrance as the result of compaction of the Middle Stone Age strata. The total absence of roof rock debris suggests rapid deposition. Throughout layer 3 there is a cyclic repetition of thin-bedded to medium-bedded limonite-stained sands and laminated haematite-stained sands. The haematite-stained laminae, which also contain the only animal remains (dune moles), indicate drier phases in the sedimentation cycle. Numerous microfaults and micro-thrustfaults in layer 3 are the direct result of compaction due to dewatering.

There is some evidence that the first few centimetres of the overlying shelly quartz sand (layer 2) were deposited in water. Everywhere the layer 2 sediments are subhorizontally bedded. The lowest 50 cm of layer 2 includes much poorly-sorted and poorly-bedded gravel and boulder-size Pleistocene aeolianite which has formed as a talus deposit (Fig. 3). The talus material originated by weathering of the Pleistocene aeolianite capping the Tertiary limestone roof rock. The weathered material gravitated down the flank of a dune which had accumulated in the front of the cave. The surface of the talus slope formed a basin in which sedimentation once again took place briefly into standing water. This bed includes some lime mud.

3.   Grain size distribution

3.1 Laboratory methods

Random sampling from the various levels of Die Kelders I sediments was carried out by one of us (F.R.S.) during the four years of excavation. The sampling was supplemented by several visits to the cave during 1973.

Grain-size determinations were made on the sand fraction ($>63\mu$) by conventional dry-sieving methods using half-phi intervals. The use of the formulae of Folk and Ward (1957) for computing the statistical parameters necessitated the use of settling tube methods on the few samples where the sand fraction was less than 95 per cent. The statistical parameters were determined from graphs on arithmetic probability paper with an arithmetic gradescale:

$$\text{Phi Mean diameter } (M_z) \quad = \quad \frac{\phi16 + \phi50 + \phi84}{3}$$

$$\text{Standard deviation } (\sigma) \quad = \quad \frac{\phi84 - \phi16}{4} + \frac{\phi95 - \phi5}{6.6}$$

$$\text{Skewness (Sk)} \quad = \quad \frac{\phi16 + \phi84 - 2\phi50}{2(\phi84 - \phi16)} + \frac{\phi5 + \phi95 - 2\phi50}{2(\phi95 - \phi5)}$$

$$\text{Kurtosis (Kg)} \quad = \quad \frac{\phi95 - \phi5}{2.44\,(\phi75 - \phi25)}$$

For simplification the phi scale has been used in all grain size determinations. Shackley (1972) gives a table relating phi units to millimetres.

3.2 Mineralogy and grain shape

The major component of the cave sediments is quartz, with a heavy mineral fraction of less than one per cent consisting of ilmenite with minor amounts of garnet. In the basal marine horizon (layer 17) there is a trace of well-rounded pelletal phosphorite in the 1.5 to 2.5ϕ range. The sediments in the Middle Stone Age horizons tend to have a significant humus content. Iron-staining of the quartz grain surfaces increases upwards through the stratigraphic column to the top of the yellow sands (layer 3).

The sediments overlying the yellow sands are composed of quartz sand and shell fragments. The $CaCO_3$ content of this bed (layer 2) ranges upward from 53 per cent. The shell midden (layer 1) is composed very largely of whole and fragmented marine mollusc shells with ash and a very minor fraction of quartz sand.

The quartz grains range from angular to round with the majority sub-round. The sediments low in the stratigraphic column tend to be sub-angular to sub-round, with the very coarsest grains being angular. Generally the sphericity and degree of rounding increase with diminishing grain size. In layers 1 and 2 the quartz grains are better rounded than those of the underlying sediments. The quartz sands contain some euhedral quartz crystals, generally with well defined and polished facets, although frosting has occurred in some instances.

From the basal marine unit upwards through layer 3 the grain surfaces are lightly polished, with occasional grains displaying better polish or frosting. In layers 1 and 2 with the improvement in rounding there is also a significant increase in the degree of frosting, probably due to post-depositional diagenesis.

3.3 Cumulative size frequency distribution curve

The cumulative curve (Fig. 5) depicts graphically the grain size variation in the Die Kelders I sediments. The textural classes suggested by Cornwall (1958) have been adopted. The cumulative curve is semi-quantitative with respect to 'layers 6 and 17 because in both these levels the exposure allowed only an estimate of boulder size. The blanket of collapsed roof rock at the top of layer 6 is not shown.

In layer 17 the 'beach' boulders and cobbles constitute about 75 per

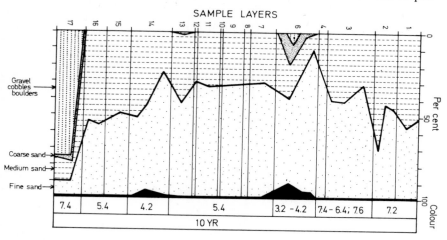

Figure 5. Cumulative curve showing the broad constitution of the cave sediments.

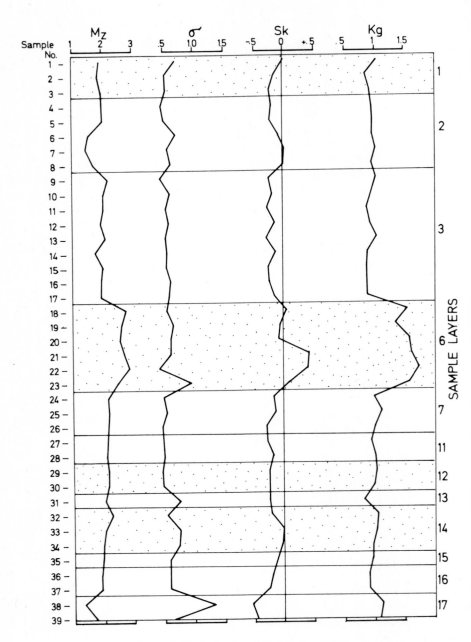

Figure 6. Variation in grain size parameters.

cent of the material. The 'beach' boulders are a waterworn round to sub-round quartzite. Again in layer 6 there is a high proportion of rubble and large tabular boulders of collapsed roof rock. The mud content throughout the column is generally less than 1 per cent, with two notable exceptions. Occupation layers 6 and 14 have a high mud-humus content, and these correspond to dark coloration (10 YR 3.2-4.2).

3.4 Character of the sand fraction

The sediments of Die Kelders I have so far been considered in totality. The sediments transported into the cave came from some outside source and have subsequently been modified by roof rock debris and local loam formation. In this section it is proposed to examine quantitatively the sand fraction with a view to understanding the source of the sediment and, of course, the environment outside the cave at the time of sediment accumulation. The sieved fractions of the sediments were examined microscopically and macroscopically and all cryoclastic, lithic and bone material removed. The shell fraction in layers 1 and 2 has not been removed since the shell particles have, like the quartz sand, been transported into the cave from an outside source.

(i) *Mean.* The mean sediment size virtually throughout the sediment column is close to 2ϕ (Fig. 6). The range of mean grain size measurements is 1.38 to 2.79ϕ with an average value of 2.06ϕ and a standard deviation of the mean size 0.32ϕ. The sand is thus classified as fine sand, but close to the medium sand-fine sand boundary.

The coarsest sediment is the interstitial component from the 'beach' boulders at the base of the succession. If coarseness of the sediment is a reflection of current strength, then a substantial current is indicated. This would be expected from the locality of the marine horizon where waves probably surged into the cave.

Coarser sand (ca. 1.5ϕ) is again encountered in layer 2 due to the addition of much platey shell debris which occurs in the coarser grades. Decrease in grain size (average 2.61ϕ) in the Middle Stone Age succession is the result of admixture of a mud-humus fraction. In occupation layer 12 the mud fraction has been almost totally removed by leaching, with subsequent compaction, and the grain size is now marginally coarser than the sediment as a whole.

(ii) *Standard deviation.* To describe the sorting of the cave sediments, the classification of Folk (1966) has been used:

Very well sorted	less than 0.35
Well sorted	0.35-0.50
Moderately well sorted	0.50-0.71
Moderately sorted	0.71-1.00
Poorly sorted	1.00-2.00
Very poorly sorted	2.00-4.00
Extremely poorly sorted	more than 4.00

The plot of standard deviation (Fig. 6) shows that the sediments average moderately well sorted (average 0.61; standard deviation of the standard deviation 0.15; distribution unimodal). In the basal marine unit the interstitial sand is moderately to very poorly sorted, suggesting an inefficient transporting medium. Sorting is somewhat variable throughout the Middle Stone Age succession, largely due to admixture of 'fines'. The yellow iron-stained sands (layer 3) contrast markedly with the Middle Stone Age sediments. Here the spread of sorting values is small (mean 0.57; standard deviation 0.07), where two-thirds of the samples have standard deviations between 0.50 and 0.64.

(iii) *Skewness.* Skewness measures the non-normality of a population. The Die Kelders I sediments are nearly all slightly negatively skewed, implying that the size frequency distribution has a tail towards the coarser grades. Generally skewness is between -0.10 and -0.25 (Fig. 6), but with a range +0.42 to -0.53. The interstitial sands from the 'beach' boulders (layer 17) are markedly negatively skewed (-0.43 and -0.53). This is undoubtedly the result of surges of sea water flushing out the 'fines'. Admixture of mud and humus in occupation layer 6 has produced positive skewness. While the negative skewness of layer 3 is due mainly to the absence of fine material, in layers 1 and 2 the negative skewness has been produced by admixture of platey shell material which occurs in the coarse grades.

The significance of the skewness values in these cave sediments is difficult to interpret. Folk and Ward (1957), Mason and Folk (1958), Friedman (1961) and Duane (1964) agree that beach sands are generally negatively skewed and dune sands positively skewed. In this respect the interstitial sand from the boulders at the bottom of the sedimentary column can safely be interpreted as marine. Friedman (1961) explains the negative skewness of beach sands as due to the operation of two opposing forces, the incoming wave and the outgoing wash, which effectively winnow out the fine material. Positive skewness of dune sands he attributes to unidirectional flow.

(iv) *Kurtosis.* Kurtosis compares the sorting in the 'tails' with the sorting in the central part. Normal curves have Kg equal to one. The Die Kelders I sediments are predominantly platykurtic to mesokurtic (platykurtic = 0.67-0.90; mesokurtic = 0.90-1.11) with values close to one. The range of kurtosis values is 0.81-1.74. The largest deviation from the normal curve is in layer 6 where the sediment is leptokurtic (leptokurtic = 1.10-1.50); that is, better sorted in the central part than the tails.

(v) *Scatter diagrams.* By themselves the individual grain size parameters do not reveal much about the origin of the sediment, although they do define the sediments quantitatively. By plotting the grain size parameters against each other in 6 two-variant scatter diagrams their geological significance is revealed.

In Fig. 7 the mean grain size for the Die Kelders I sediments is

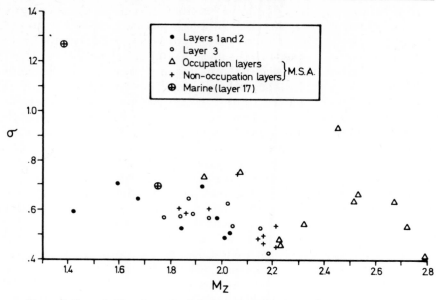

Figure 7. Scatter plot of standard deviation (sorting) against mean grain size.

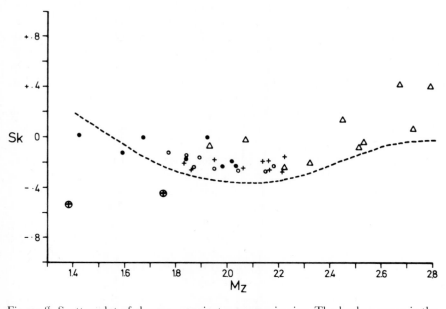

Figure 8. Scatter plot of skewness against mean grain size. The broken curve is the dividing line Friedman (1961) found to separate dune from beach sands.

plotted against standard deviation (sorting). The sorting is practically independent of mean grain size, although sorting is poorest with the coarse marine sand and the fine loamy sand from occupation layer 6. Best sorting is associated with the most prominent sediment mode.

In Fig. 8 mean grain size is plotted against skewness. Apart from the two marine samples, the path followed by all other size-skewness values is arcuate. Coarser than 2.4ϕ the sediments are generally negatively skewed. The decrease in grain size below 2.4ϕ due to addition of a mud-humus mode to the sand mode is related to an increase in positive skewness.

It was mentioned earlier in a discussion of skewness that many authors have found dune sands to be generally negatively skewed and beach sands positively skewed. It was suggested that in the case of the Die Kelders I sediments the sign of skewness was perhaps not necessarily diagnostic. If the sands had originated from dune accumulation, it is likely that they had been dumped over the lip of the cave, from the higher ground above, and rapid burial would not favour efficient sorting. Friedman (1961) plotted mean grain size against skewness and found an almost complete separation of the dune sand and beach sand parameter plots. The dividing line between these two fields computed by Friedman is shown in Fig. 8. It is observed that the two undisputed marine sediments lie well within the beach sand area, while only two of the other 37 samples lie just within the beach sand realm. It is argued that a dune origin for these sands is highly likely. In a cave environment sorting would be negligible and the skewness value would be expected to be inherited. Postive skewness of the samples finer than 2.4ϕ is not inherited; it is due to the addition of small amounts (3-10 per cent) of mud-humus in the primary sand mode. It was also shown earlier that the Middle Stone Age succession has been compacted considerably, probably due to leaching of humus and elutriation of the silt-clay fraction. Negative skewness of many of the samples could possibly have arisen in such a fashion.

In Fig. 9 mean grain size is plotted against kurtosis. It will be seen that two separate fields occur. For the bulk of the samples kurtosis values range from 0.80 to 1.05, showing that the sand mode gives a near normal curve. In the occupation layers the admixture of 3 to 10 per cent fine material has resulted in not only a decrease in mean grain size, but also a concomitant increase in kurtosis. This addition of a fine mode results in a poorer sorting in the tails when compared with the central part. The occupation layer sediments tend toward leptokurtosity ($Kg > 1.5$). Of most significance is the fact that with progressive leaching and elutriation of the humus-mud fraction, the grain size parameters approach the field of the average 'clean' sediment with kurtosis less than 1.05. In other words, a common origin for the sand fraction of these sediments is indicated.

The same trend is revealed in the other scatter diagrams; skewness

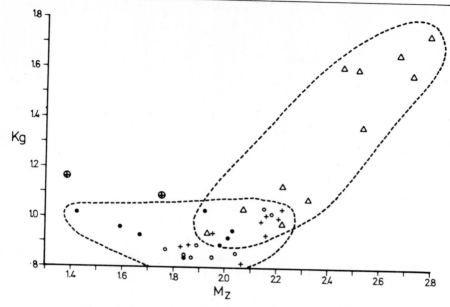

Figure 9. Scatter plot of kurtosis against mean grain size.

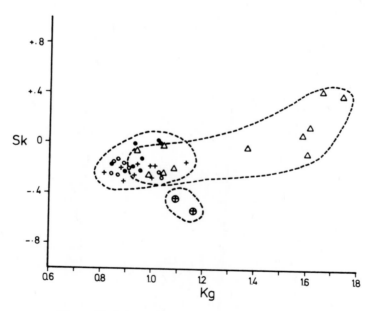

Figure 10. Scatter plot of skewness against kurtosis.

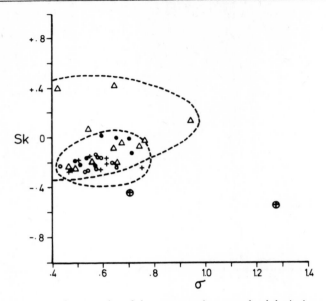

Figure 11. Scatter plot of skewness against standard deviation.

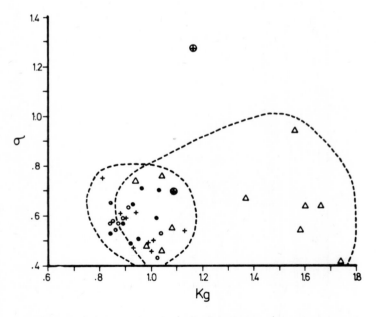

Figure 12. Scatter plot of standard deviation against kurtosis.

versus kurtosis (Fig. 10), skewness versus standard deviation (Fig. 11), and standard deviation versus kurtosis (Fig. 12). In all of these scatter plots it is observed that the marine sands at the base of the Die Kelders I succession are separated from the other sediment size parameters. For most of the sands the plots are very concentrated, suggesting a common origin. The only wide scatter of points occurs in the occupation layers where the fine mode ranges from 3 to 10 per cent. With progressive post-depositional removal of this fine mode, the sand fraction approaches the other tight cluster of points. In Fig. 10 in particular high positive skewness due to admixture of the fine mode in the cave sediment correlates with increasing leptokurtosity as the sorting in the tails becomes progressively poorer compared with the central part.

3.5 Summary

In general, then, it is seen that the grain size parameters for the sand fractions of all, save the marine sands, are very similar. Even sand from layers 1 and 2 which was blown into the cave from the nearby shoreline, conforms to the general pattern. The similar grain size parameters indicate a similar source for the sediments. Modification due to addition of a coarse shell mode or a fine mud-humus mode affects the skewness and kurtosis values most markedly.

Grading analyses show that the Die Kelders I sediments are uni-modally distributed, and the degree of rounding supports a single origin. Any sub-populations with significantly different rounding in any one sample occur to only a minor extent. Furthermore, the sub-round character of most of the sediment samples from layers 3 to 15 and their light polish are suggestive of an aeolian origin.

It was shown earlier that after the sea withdrew from the cave, the shoreline remained far distant for a long period. Its return is reflected in layers 1 and 2. Dune-bedded aeolianites along this coast show that the dominant sand movement was from the south-east. The cave is north facing, and moving dunes above the cave would dump sand down their advancing front to the foot of the cliff where there would be little opportunity for sorting as the sediment would be rapidly buried and re-working would be negligible. It was argued that the sign of skewness did not necessarily negate an aeolian origin. A plot of mean grain size against skewness confirmed the aeolian nature of the sands.

4.   Electron microscopy

With a view to supporting the conclusions reached by conventional grading analysis, an extensive examination of the quartz grain surface textures was carried out by means of scanning electron microscopy. Krinsley and Takahashi (1962) and Krinsley and Margolis (1969) have found it possible to distinguish between aeolian and littoral grains by means of surface textures. They attributed the different

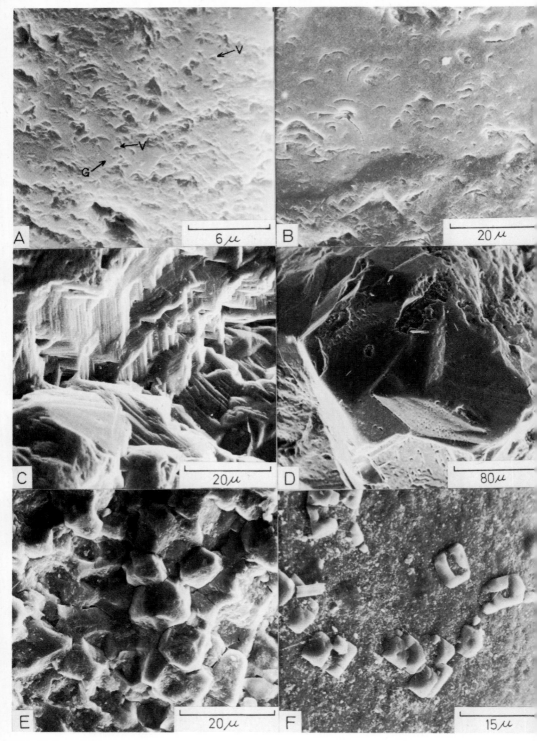

Figure 13. Scanning electron micrographs. (A) Grain from layer 2 showing mechanical V-shaped patterns (V) and grooves (G). (B) V-shaped markings the result of chemical etching. (C) Chemical etching. (D) Reprecipitation forming new crystal faces. (E) Incipient crystal growth. (F) Calcite crystal growth on the surface of a shell grain.

textures to differences in the velocities of the impingeing grains and the viscosity of the transporting media. The littoral environment is characterised by small pyramidal indentations (V-shaped patterns) on the rounded edges of grains, and blocky conchoidal breakage patterns. Aeolian textures, on the other hand, are characterised by meandering ridges, graded arcs, flat pitted areas, and parallel, upturned series of plates which follow the cleavage direction(s) in quartz. These various mechanical features may be obliterated by later diagenesis.

4.1 Laboratory method

The quartz grains were treated with dilute hydrochloric acid for 30 minutes to remove surface impurities. The grains were then spread on a stub and coated with a gold-palladium alloy, vacuum evaporated. Coating was done from high and low angles and the samples rotated to ensure an even coating.

The scanning electron microscope used in this study is a JEOL JSM U3 at Rhodes University, Grahamstown.

4.2 Results

Fig. 13a illustrates the type of modification a quartz grain repeatedly reworked in a littoral environment may undergo. The V-shaped patterns and grooves are considered to be diagnostic of a beach environment. The V-shaped indentations are the most common of mechanical features observed on littoral grains. Randomly oriented V-notches such as those illustrated in Fig. 13a are thought by Krinsley and Donahue (1968) to be caused by gouging in a high-energy beach environment. According to Krinsley and Donahue the grooves are generally associated with non-oriented V-notches and are used as a criterion for wave action. Throughout the Die Kelders I stratigraphic succession the only occurrence of such littoral textures is on sand grains from layer 17 (high sea level deposits of the last interglacial age) and layers 2 and 3 (Holocene rise of sea level to its present position). Sand grains from layers 16 through 3 have few mechanical surface textures preserved, but the most commonly preserved are parallel upturned plates typical of dune sands.

In contrast Fig. 13b shows the effect of chemical etching along the trigonal cleavage of the quartz to produce large-scale oriented V-shaped petterns. Chemically etched V-shaped patterns are attributed to slow solution at high pH (Krinsley & Donahue, 1968). The *en echelon* orientation of the Vs should be noted. Chemical etching controlled by the internal symmetry of the quartz is further illustrated in Fig. 13c.

Re-precipitation features, or overgrowths, are common in the Die Kelders I sediments. In Fig. 13d re-precipitation has produced crystal faces in optical continuity with the quartz grain. Overgrowth may also result in smaller crystals which impart a blocky appearance to the grain surface (Fig. 13e).

Sand size shell grains are a major component of layers 1 and 2. Although the compound microscope shows them to be well-rounded and polished, scanning electron micrographs show the grain surfaces to be characterised by re-precipitation of calcite (Fig. 13f).

Post-depositional diagenesis of the quartz and shell grain surfaces is common and reflects the highly active chemical environment of the Die Kelders I sediments (pH range 7.9 to 8.9).

5. *Synthesis*

The Die Kelders cave complex is eroded along the interface at 7 m a.s.l. between Palaeozoic sandstone and the Late Tertiary Bredasdorp Formation. The cave complex was probably formed as a result of wave-cutting by a Late Pleistocene sea level at 7-10 m above present sea level, and subsequently enlarged by sub-aerial erosion. In Die Kelders I, sediments of marine origin are restricted to spherical and water-worn quartzite boulders and poorly sorted interstitial sand at the base of the succession. The height of the top of these 'beach' boulders above sea level, 2 m, coincides with the height of the beach rampart in front of the cave and beach boulder floors in presently exposed grottoes in the quartzite. The Middle Stone Age industry overlying the marine horizon (layers 16 and 17) dated the 'beach' zone as pre-Würm, while the thinness of the sediments separating the marine deposit from the first occupation suggests the sea had not long vacated the cave.

In Fig. 14 the history of Die Kelders I from the time of formation of the basal marine horizon to the final Late Stone Age occupation is illustrated.

Withdrawal of the sea from the cave complex was followed by several periods of occupation, each being separated by sterile quartz sands. The Middle Stone Age occupation is characterised by typical artefacts and an abundance of well preserved bones. (Klein, 1974, suggests that the Middle Stone Age in South Africa ranged from about 100000 to about 35000 B.P.). The Middle Stone Age sediments are devoid of shell remains, largely because the shoreline had retreated with lowering of sea level. During the Würm-maximum lowering, the shoreline was approximately 20 km distant, assuming a lowering of 120 to 140 m, but less than 10 km during the interstadial (-25 m). That exploitation of the more distant shoreline took place is evident from the occasional seal and penguin remains in the deposit. The top of the Middle Stone Age deposits is characterised by massive, tabular boulders of collapsed roof rock which was the result of a catastrophic earthquake shock. Solutional weathering of the top layer of boulders suggests a long period of non-deposition in the cave, although a condensed occupation horizon (layer 4) does overlie these boulders.

Overlying the Middle Stone Age sediments are sterile iron-stained

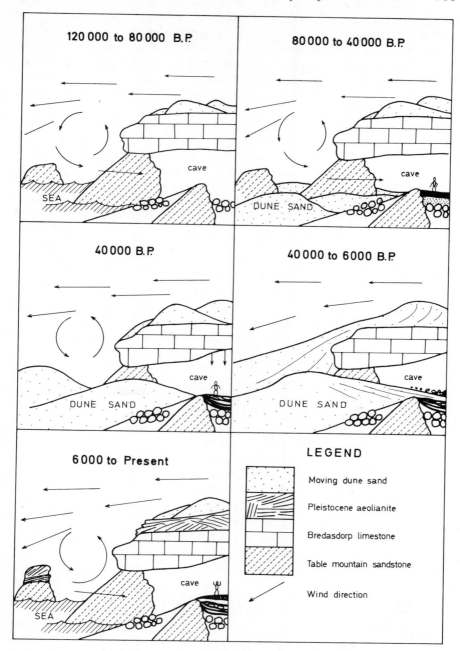

Figure 14. Cartoon summarising the evolution of Die Kelders I cave sediments from time of occupation of the cave by sea to the Late Stone Age midden deposit. Note how the southerly wind moves dune sand over the brink of the cliff, which results in near burial of the cave from about 40000 to about 6000 B.P. With the Flandrian transgression the dune sand field was eroded away, leaving only outliers of aeolianite on promontories and on stacks, besides that on high terrain. Note also the collapse of roof rock at about 40000 B.P.

quartz sands which are negatively skewed, moderately well sorted and have a silt content always less than 0.7 per cent. These sterile sands bear little sign of occupation of the cave. Even small rodent bones are scarce. This period of non-occupation spanned from about 40000 B.P. to probably 6000 B.P. when shelly sands and littoral textures on the quartz grains show that the shoreline had once again returned near to its present position.

Initially the most perplexing aspect of this study was the apparent non-occupation of the cave for some 35,000 years, from about 40000 to about 6000 B.P. In this respect grading analyses and the use of scatter diagrams to present these results have been of inestimable value. It was shown that the sediments from layers 1 to 16 are fine grained, but bordering on medium sand size. They are largely moderately well sorted and negatively skewed. If the sign of skewness was diagnostic the sediments should be interpreted as inherited from a beach environment, dune sands being generally positively skewed. Neither was kurtosis, by itself, found to be significant. But plotting all the grain size parameters as six two-variant scatter diagrams proved more illuminating. Essentially, the parameters separate into three broad groups in all the scatter diagrams. One group, the interstitial marine sands (layer 17) are not of interest. The other two groups are overlapping and consist of samples from occupation and non-occupation layers. In the non-occupation samples the grain size parameters are tightly clustered in most of the scatter plots, suggesting a common origin. The differences encountered in the occupation samples have been shown to be due entirely to admixture of a fine mode – mud (silt and clay) and humus. With progressive leaching and elutriation of the humus and mud respectively, the occupation layer grain size parameters approach those of the non-occupation parameters more closely. Where the humus-mud mode has fallen to less than 1 per cent the parameters are the same.

Comparing mean grain size with skewness, Friedman (1961) found an almost complete separation between beach and dune sands. In plotting these two parameters (Fig. 8) we have also plotted Friedman's dividing line between dune and beach environments. Fig. 8 shows that all but two of the sediments above the marine horizon fall on the dune side of Friedman's boundary, suggesting that all the sediments above layer 17 are of aeolian origin. Although these sediments had an aeolian origin the iron-stained sand of layer 3 was clearly deposited in ponded water. But sedimentation was rhythmical and fairly rapid. Narrow haematite-stained laminae with microfaunal remains suggest that the pond periodically dried out. Layer 2 was in part deposited in standing water. The lower part of this layer is a talus deposit comprising shelly sands and aeolianite rubble that gravitated down the flank of a high dune that stood at the cave entrance.

There is also other evidence to suggest a seaward extension of the dune belt during this period of lower sea level. Overlying promontories of quartzite along the coast and, in one case, occurring on a small stack, are aeolianites. All along Walker Bay these aeolianites extend to below present sea level and are presently being eroded back by the sea. More specifically, above the cave the aeolianites extend to the edge of the cliff, and are found to contain Middle Stone Age artefacts very similar to those in Die Kelders I. A panoramic view of the aeolianite shows not only the original extent of the dune field, but also strongly suggests that the cave complex may have been completely buried at some stage. Microscopic examination shows that the quartz grains of Die Kelders I average subround and are very lightly polished. The grains from the aeolianite are also subround and have the same degree of polish. In contrast sand grains from the beach in front of the cave are all well rounded and polished.

Non-occupation of the cave from about 40000 B.P. to 6000 B.P. can be attributed to two basic causes. Firstly, for much of this period dampness of the cave probably discouraged occupation. The layer 3 sediments even demonstrate the presence of standing water. Secondly, regression of the sea during the Würm left behind vast quantities of sand exposed on the continental shelf. With increased wind activity associated with glacial conditions in the high latitudes, a dune field built up above and in front of the cave. Not only would an active dune field discourage occupation, but dumping of sand over the lip of the cave would have built up a high dune there, and probably even have completely buried the cave at some stage. Sand falling over the brink of a cliff would enter a sand shadow where further transportation and sorting would be ineffective.

It may be argued that for the 35,000 years in question Die Kelders I was not necessarily isolated from occupation, but that for some reason this entire coastal plain was uninhabited. But a test-pit excavated by one of us (F.R.S.) at Byneskranskop (Fig. 1), which is only 10 km east-south-east of Die Kelders I, has shown that the coastlands were indeed occupied for at least the last 15,000 years. Byneskranskop cave faces south-east, and is situated about 60 metres a.s.l. Burial beneath moving dune sands from the south-east would be unlikely there, while the absence of a bar across the cave entrance may imply somewhat better drainage. At Byneskranskop cave the Late Stone Age spans about 10,000 years from about 15000 to about 5000 B.P. The upper age limit is inferred from the lack of pottery, which occurs in Die Kelders I at 2000 B.P. (Schweitzer, 1970). The lower limit is based on the fauna, artefacts and a single radiocarbon date of 10570±85 B.P. (UW-287, charcoal). These conclusions are based on a 2.4 m deep test-pit, the charcoal coming from a depth of 1.6 m.

Any arguments about microclimate must necessarily be somewhat subjective. It has been shown how the structure of the bedrock in the cave forms a natural channel through which groundwater movement could leach out the humus and elutriate the sediment. The lower members of the Middle Stone Age succession have been compacted by

75 to 80 per cent. The presence of standing water in which layer 3 sediments and a small part of the layer 2 sediments accumulated, and the iron-staining of the Middle Stone Age sands, suggest a wetter climate. Furthermore, these wetter conditions are also recorded in the poorly-sorted outwash deposits containing Middle Stone Age artefacts interbedded with the aeolianite above the cave. Cryoclastic debris has been found in the Middle Stone Age succession which probably corresponds to a similar cold phase recorded by Butzer (1973) from the Nelson Bay Cave 380 km to the east. Winter temperatures at Die Kelders are today not low enough for frost action on the roof rock. The top layers of the Middle Stone Age succession are blanketed by large roof rock boulders, but with a smaller amount of cryoclastic debris than is found in the lower layers. Solutional weathering of the uppermost of these boulders also testifies to wetter conditions. The Die Kelders I sediments, which accumulated largely during the Würm lowering of sea level, indicate not only very active dune conditions but also a wetter and considerably colder climate.

A detailed examination of the quartz grain surface textures by electron microscopy has shown that most mechanical textures have been obliterated by post-depositional diagenesis. Layer 17 sand grains at the base of the succession do still have littoral textures preserved, while layer 3 sand grains possess aeolian textures. The sudden return of the sea which removed the dune field in front of the cave is recorded by the occurrence of beach textures on sand grains in layer 2. These sands were blown from the beach which was probably only a short distance from the cave. The beach is now only 10m from the cave excavation. The high degree of chemical energy which the diagenetic textures suggest is significant when discussing radiocarbon dates. The pH range for the Die Kelders I sediments is 7.9 to 8.9, which is to be expected in a cave beneath a mass of limestone. Sand size shell grains in layers 1 and 2 commonly bear calcite crystal overgrowths on their surfaces. Radiocarbon dates from such a cave should be considered as minima. In the cave complex a stalactite formation with embedded Middle Stone Age artefacts shows an early period of carbonate activity. But later stalactite formation overlies the Late Stone Age horizons, and appears to be continuing to the present day.

Acknowledgments

The writers wish to express their thanks to Mr. R. Cross of Rhodes University for taking the electron photomicrographs and to Mr. V. Branco for making the illustrations. F.R. Schweitzer has been supported in his work by a grant from the Human Sciences Research Council while A.J. Tankard acknowledges a grant-in-aid from the Council for Scientific and Industrial Research.

The authors are most grateful to Drs. A.B.A. Brink, D.K. Hobday and T.C. Partridge for stimulating discussions. Mr. S. Fourie of Fedmis (Pty.) Ltd kindly performed chemical analyses.

REFERENCES

Brain, C.K. (1958) 'The Transvaal ape-man-bearing cave deposits', *Transv. Mus. Mem.* 11, 1-131.

Butzer, K.W. (1973) 'Geology of Nelson Bay Cave, South Africa', *S. Afr. archaeol. Bull.* 28, 97-110.

Cornwall, I.W. (1958) *Soils for the archaeologist* (3rd ed.) London.

Duane, D.B. (1964) 'Significance of skewness in recent sediments, western Pamlico Sound, North Carolina', *J. sedim. Petrol.* 34, 864-74.

Folk, R.L. (1966) 'A review of grain-size parameters', *Sedimentology* 6, 73-93.

Folk, R.L. and W.C. Ward (1957) 'Brazos River Bar: a study in the significance of grain-size parameters', *J. sedim. Petrol.* 27, 3-26.

Friedman, G.M. (1961) 'Distinction between dune, beach and river sands from their textural characteristics', *J. sedim. Petrol.* 31, 514-29.

Ingram, R.L. (1954) 'Terminology for the thickness of stratification and parting units in sedimentary rocks', *Bull. geol. Soc. Am.* 65, 937-8.

Klein, R.G. (1974) 'Environment and subsistence of prehistoric man in the southern Cape Province, South Africa', *World Archaeol.* 5, 249-83.

Krinsley, D.H. and J. Donahue (1968) 'Environmental interpretation of sand grain surface textures by electron microscopy', *Bull. geol. Soc. Am.* 79, 743-8.

Krinsley, D.H. and S.V. Margolis (1969) 'A study of quartz sand grain surface textures with the scanning electron microscope', *Trans. N.Y. Acad. Sci.* 31, 457-77.

Krinsley, D.H. and T. Takahashi (1962) 'Applications of electron microscopy to geology', *Trans. N.Y. Acad. Sci.* 25, 3-22.

Lais, R. (1941) 'Uber Höhlensedimente', *Quartär* 3, 56-108.

Mason, C.C. and R.L. Folk (1958) 'Differentiation of beach, dune and aeolian flat environments by size analysis, Mustang Island, Texas', *J. sedim. Petrol.* 28, 211-26.

Shackley, M.L. (1972) 'The use of textural parameters in the analysis of cave sediments', *Archaeometry* 14, 133-45.

Schweitzer, F.R. (1970) 'A preliminary report of excavations of a cave at Die Kelders', *S. Afr. achaeol. Bull.* 25, 136-8.

Visher, G.S. (1969) 'Grain size distribution and depositional processes', *J. sedim. Petrol.* 39, 1076-106.

DISCUSSION

Dr. E. Tratman asked the speaker whether there could be any climatic changes involved in the abrupt transitions visible in the sequences and Mr. F. Schweitzer replied that a change to colder conditions could be postulated. Mr. A. Fleet asked for further information on the application of the scanning electron microscope to the study especially since the speaker had mentioned that the obliteration of mechanical features had been affected by a highly energetic post-depositional chemical environment. The speaker commented that the grain surface textures did indeed include high-energy chemical features, for example the surface reprecipitation of silica.

Mr. J. Bintliff asked where the approximate position of sea level would have been at the time of occupation. The speaker replied that 18,000 years ago, during the last major regression, the sea would have been some 120 m lower and that a large area of Walker Bay would have been land.

A. Straw

Sediments, fossils and geomorphology a Lincolnshire situation

The Lincolnshire Wolds west and south-west of Louth comprise an area of rolling chalk-land, drained by a group of valleys which converge on the town from the west and south (Fig. 1). The interfluves and the valleys are occupied in part by glacial drifts that consist of tills and outwash materials which were emplaced during the Wolstonian (Older Drift) and Devensian (Newer Drift) glaciations (Straw, 1957). The more recent Devensian drifts lie within the lower ends of the valleys and over spurs to a maximum height of 113 m O.D., but westward dissected Wolstonian glacial materials persist only on certain interfluves and form part of a sheet of drift that once overspread the whole of the central Wolds and much of the central vale of Lincolnshire.

Four and a half km west-north-west of Louth, at Welton-le-Wold (TF 282884), large quarries belonging to Stephen Toulson and Sons, Ltd. have been worked for many years in sand and flint gravel deposits that are overlain by thick grey-brown and creamy-brown tills. These sediments, because of post-depositional erosion of the Welton valley, now form a spur between this valley and a tributary on its northern side.

In September 1969, some thirty years after strip excavation northward into the spur had commenced, mammalian remains were discovered by a local geologist, Mr. C. Alabaster, along a 40 m stretch of the flint gravels beneath the till in the eastern part of the main section. Through 1970 and 1971, further fossils and four flint hand-axes were recovered, and in 1970 and again in 1972 after it had been worked back a further 6 m, the section was surveyed by the author and stratigraphic details recorded by him and Mr. Alabaster. The main section of the quarry, over 600 m long, was last worked in December 1970, and is deteriorating rapidly.

1. General stratigraphy and fossils

The section showed some 6 to 13 m of grey-brown and creamy-brown tills containing much chalk and flint and some erratic igneous and Carboniferous rocks, overlying at least 8 m of interbedded flint gravels, sands and fine-grained sediments. The section is summarised in Fig. 2 which illustrates the presence of a weak unconformity within

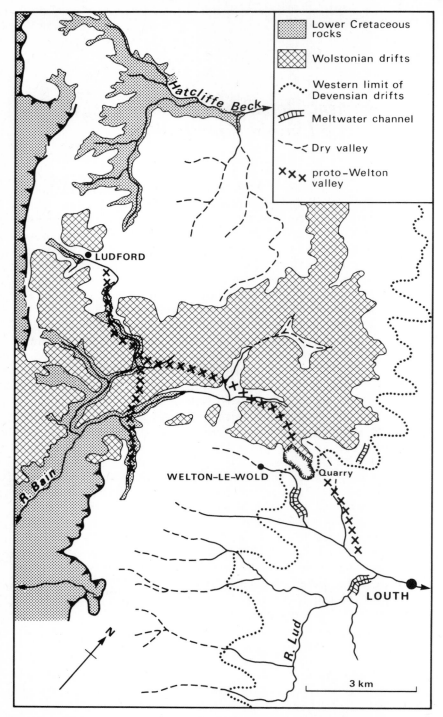

Legend:
- Lower Cretaceous rocks
- Wolstonian drifts
- Western limit of Devensian drifts
- Meltwater channel
- Dry valley
- proto–Welton valley

Hatcliffe Beck

● LUDFORD

R. Bain

WELTON–LE–WOLD ●

Quarry

R. Lud

● LOUTH

N

3 km

Figure 1. Geological and geomorphological features of the central Lincolnshire Wolds.

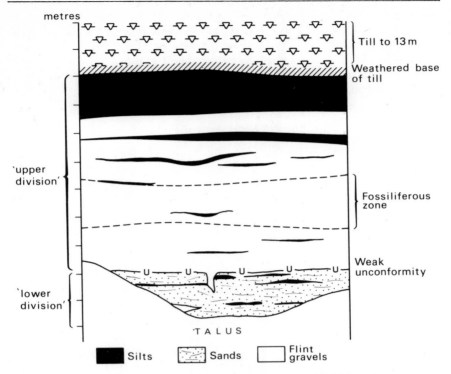

Figure 2. Diagrammatic section of the Welton-le-Wold deposits.

the bedded materials and the location of the 1.5 to 2 m thick zone from which most of the mammalian remains and artefacts were taken. This zone lies within the gravel-dominated upper division of the sub-till deposits which averages 6.5 m in thickness. The lower division, probably of similar thickness but mostly obscured by talus, has a higher proportion of sands than gravels, and is apparently unfossiliferous.

The mammalian remains include a complete tusk, 2.2 m long, a partial tusk and three molars of straight-tusked elephant (*Palaeoloxodon antiquus*), part of an antler of red deer (*Cervus cf. elephas*), two anklebones of a larger deer, possibly the Irish Giant deer (? *Megaloceros sp.*), and cheek teeth fragments of horse (*Euus sp.*). These identifications were kindly made for the author by Mr. P.J. Boylan, Director of Leicester Museum and Art Gallery, who also described the artefacts as somewhat crude but typical Acheulean hand-axes, comparable with Middle and Upper Swanscombe palaeoliths. In spite of its meagre character, he regards the assemblage as temperate and interglacial, and, as the presence of human artefacts renders a Cromerian age unlikely and *Equus* is rare in the main Ipswichian fauna, the assemblage is probably of Hoxnian date.

Questions arise that relate to the restricted occurrence of the finds (though this is of least concern in this paper), and to the nature of the fine-grained materials, sands and gravels. The relations of these to the overlying till, to the contained fossils and artefacts, and to the pre-Glacial relief of the locality therefore need elaboration, in order to provide a better understanding of the environmental circumstances of the deposition of these materials.

2.   *Field observations*

Field examination showed that the upper division of the bedded deposits was dominated by thick seams of poorly-sorted, indistinctly bedded and loosely-packed flint gravels, and it is in this component that the fossils and artefacts were preserved. The flint fragments were mostly angular though not razor-sharp, and the majority were between 2 and 7 cm in length along the longest axis, with occasional pieces to 15 cm. Chalk was absent and the flints lay at all angles within a loose matrix of medium and coarse sands, which were composed, curiously, of angular chips of flint and rounded quartz grains. The sand grains and flint fragments alike were coated lightly or heavily with iron and manganese oxides which were precipitated from percolating groundwaters after deposition of the gravels. Whilst the absence of chalk is remarkable because chalk slopes rise above the deposit to north, south, and west, the angular character of the flint gravels is consistent with a dominance of physical weathering, which, discounting a hot arid environment, presumably involved frequent freeze-thaw oscillations.

The flint gravels were interspersed with seams of stone-free fine-grained material particularly near the upper part of the section, and incorporated thin lenses and discontinuous layers of similar material and coarser sand.

The lower division, where exposed, contained a higher proportion of sands, the seams displaying current-bedding, but seams of flint gravel and thin lenses of fine material also occurred. The weak unconformity was marked by occasional structures that are interpreted as ice-wedge pseudo-morphs as well as by the clearer evidence of stratification in the lower deposits. The sands at least appeared to be water-deposited, but neither ferro-manganese coatings to sand grains and flints, nor fragments of chalk were observed.

The field evidence suggests that the gravelly members of the deposit were transported and aggraded essentially as soliflual deposits under cold climate conditions, an environment that would contrast sharply with the temperate woodland character of the mammalian remains.

The till rested sharply on the underlying materials though with an undulating base. A compact sandy clay, it contained many boulders and smaller pieces of chalk and flint, and fewer fragments of hard grey and yellow sandstones, dolerite and Scottish igneous rocks. Although a few rounded flint and quartzite pebbles occurred, most fragments were sub-angular to sub-rounded and carried faceted surfaces and striations. The flints however were commonly fractured, retained some cortex, and were mostly unpatinated and grey in colour.

The absence of rolled chalk and flint fragments and of erratic non-Cretaceous rocks in the underlying gravels militates against their interpretation as outwash from the ice that laid down the till.

3. Laboratory analyses

In order to extend or confirm the provisional conclusions reached on field evidence concerning the depositional environment of the sub-till materials, and the implication that the fossils and artefacts are derived rather than contemporaneous, samples of the various components were collected for laboratory analysis. To date only particle-size analysis, and microscopic examination of the coarse silt and sand fractions with reference to grain shape and mineralogical composition have been attempted, but they have yielded data which help to resolve the problems.

3.1 Particle-size analysis

Several samples of the flint gravels and fine-grained materials have been collected from different levels within the upper division of the sub-till deposits, and of sands and fine-grained materials from the lower division. On completion of particle-size analyses, sand grade fractions in particular were retained for microscope examination.

Particle-size analysis was based on British Standard sieving and

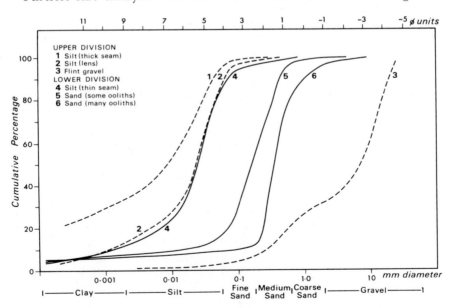

Figure 3. Particle-size distribution of Welton-le-Wold sediments.

Table 1. Mean sorting indices of Welton-le-Wold deposits.

Upper Division

1	Silt (thick seam)	4.75
2	Silt (lens)	2.65
3	Flint gravel	2.15

Lower Division

4	Silt (thin seam)	1.93
5	Sand (some ooliths)	1.40
6	Sand (many ooliths)	0.83

(Sorting index $\sigma\ \phi = \frac{1}{2}\,(\phi 16 - \phi 84)$

hydrometer methods (B.S. 1377: 1967) and included pretreatment, determination of specific gravity, moisture content, and dispersing agent and meniscus corrections. Calculations based on the observed weights and corrections were carried out by computer using a program developed by Dr. R.A. Cullingford, to produce the particle-size distribution of the sample in the form of cumulative weight per cent against phi diameter and mm diameter.

The results of some of the particle-size analyses are illustrated in Fig. 3 in the form of smoothed curves for the flint gravel, sands, and fine-grained materials, and in Table 1.

The gravel is bimodal, poorly sorted (high index), and is composed mainly of material coarser than medium sand. The fine-grained materials are also poorly sorted but from whatever seam, lens or division, they consist predominantly of silt, and the curves are consistent with an interpretation of the material as of wind-blown origin. The better sorting (low index) of the sands of the lower division supports the field evidence of fluvial deposition.

3.2 Microscope analysis

The sand and coarse silt fractions of the various deposits were examined by petrological microscope in order to determine grain shape and mineral composition.

Of the flint gravel of the upper division, approximately 60 per cent of the coarse sand fraction consisted of angular chips of flint and hard cortex, while most of the remainder comprised sub-rounded shiny quartz grains. In the medium sand fraction, these two components were present in equal proportions, but only a fifth of the fine sand fraction consisted of flint chips, and the coarse silt fraction was composed almost entirely of angular and sub-angular quartz grains.

The silt layers of the upper division contained some sands but these fractions constituted less than 15 per cent of the deposits. Again the proportion of flint chips to quartz grains decreased and quartz grains became more angular toward the finest fraction. Some ferro-manganese concretions were present in fractions of both the gravels and the silts.

The angularity of the flint chips and of the quartz grains in the finer fractions supports the belief that both the flint gravels and silts have been subjected mainly to mechanical comminution and to deposition by agencies other than running water.

The silts of the lower division contained no particles of manganese oxide, and angular flint chips comprised only 3 to 4 per cent of each of the sand fractions. The bulk of the deposit therefore was composed of quartz grains, angular and sub-angular in the finer fractions, sub-rounded in the coarser fracton. Up to 4 per cent of the medium and fine sand fractions consisted of small rounded darkbrown grains and hemispherical shells of ooliths of hydrated ferric oxide (limonite).

The lower division sands consisted largely of rounded and sub-rounded clear, cloudy and yellow-stained quartz grains. Angular chips of flint occurred in the coarser fractions in minute proportions, but a more noticeable constituent was the limonite ooliths. In some of the sands they were so frequent that separation of this component in bromoform was feasible, revealing that some 30 per cent of the medium sand fraction and 45 per cent of the fine sand fraction consisted of limonite ooliths, some whole but most chipped and fractured.

The main characteristics of the lower division sands were the higher proportion of rounded and sub-rounded quartz grains, the better sorting (Table 1), and the presence of oolitic limonite. The sorting rather than the shape, for reasons elaborated below, supports the field evidence that running water was a major agency of deposition. Although the silts included no manganese oxide and contained some limonite oolith fragments, in particle-size distribution they are remarkably similar to the silts of the upper division (Fig. 3), and a wind-blown origin would again seem very likely.

It is of interest at this point to comment on the possible sources of the constituents of the gravels, sands and silts. The Upper Cretaceous chalk is clearly the source of the flints, but the quartz and limonite ooliths must have another provenance. The very fact that these two components alone form the bulk of the deposit excluding the flint would seem to preclude their derivation from glacial drifts that may have been expected to have provided a wider range of minerals. Although Tertiary deposits, not hitherto discovered in Lincolnshire, might have provided them, it is much more probable that the rounded quartz grains and the ooliths were eroded from marine Lower Creceous rocks. Two members of these rocks, the Spilsby Sandstone and Carstone, are composed largely of rounded quartz grains, and certain clayey beds (Claxby Beds and Fulletby Beds) of the intervening Tealby Series are rich enough in oolitic limonite to have been quarried formerly as iron-ore (Swinnerton and Kent, 1949).

If largescale derivation of ooliths and quartz grains has occurred then the depositional environment of the Welton deposits is indicated more by the angularity of the chips of flint and the chipping and fracturing of some of the quartz and ooliths than by the roundness of the other quartz grains and ooliths.

4. *The depositional environment*

It is now possible to postulate the general environmetal circumstances concerning the aggradation of the gravels, sands and silts at Welton-le-Wold.

On the basis of field evidence, supported by certain results from laboratory analyses, it is suggested that the flint gravels were laid down as a sludge or soliflual deposit, perhaps redistributed in part during periods of snow-melt and affected periodically by shallow cryoturbation. The sands owe their character more to fluvial deposition, but the silts, both in their grain-size characteristics and in the manner of their stratification, are probably wind-blown.

A repetitive seasonal pattern of silt, sand and gravel aggradation may be envisaged, in that wind action, winnowing silt and fine sand, may have been prevalent on a cold dry winter surface, while mass transport and meltwater action brought the gravels and sands over a partially thawed summer surface.

The deposits as a whole therefore appear to be part of a sheet of material which, as chalk slopes rise to the north and south, can be regarded as a valley-floor accumulation aggraded by alternating soliflual, niveo-fluvial and aeolian processes under 'periglacial' conditions that presumably preceded advance of Wolstonian ice into the locality.

It follows that the contained mammalian remains and hominid artefacts of temperate interglacial character must be derived elements, and taking into consideration for example the relative concentration and well-preserved appearance of the elephant remains, it is suspected that all the remains and artefacts were eroded from some proximate but as yet undiscovered interglacial deposit.

5. *Geomorphological consequences*

The presence of distinctive rounded quartz grains and limonite ooliths in the Welton sediments has a geomorphological significance that should not pass unmentioned.

Evidence has been presented that the sediments are a periglacial valley-floor accumulation, and there is every indication that the former valley would have drained eastward down the chalk dip-slope. Although Lower Cretaceous rocks are known to outcrop at about present sea-level beneath Devensian drifts at Louth, it is very unlikely that the quartz and ooliths were introduced from this locality. Lower Cretaceous rocks do however outcrop extensively in the western Wolds along the scrap face and in inliers in the heads of the major Wold valleys (Fig. 1).

The present Welton-le-Wold valley is rather short and does not include any such inlier. However on its northern side, west-north-west of the quarry, the interfluve with the upper part of the south-draining Bain system is overspread by thick Wolstonian glacial drifts. It has been demonstrated elsewhere that the Bain system, which has widely exposed the Lower Cretaceous rocks, is essentially a post-Wolstonian development (Straw, 1966), and in 1961 a pre-glacial east-draining valley system, such as shown in Fig. 1, was postulated following a detailed examination of drainage evolution and planation surfaces of the whole of the Lincolnshire Wolds (Straw, 1961). It is particularly satisfying that material support for such a proto-Welton valley, that could have had an overall gradient of about 1:350, has emerged from examination of the Welton sediments.

Conclusion

This paper has reported on a local but fascinating geological, geomorphological and archaeological situation. The role of laboratory sediment analysis has not been great, but the data acquired have supported conclusions based on field evidence regarding the depositional environment of the Welton sediments and in consequence the derived character of the fossils and artefacts. The analyses also revealed the presence of mineral constituents that are firmly believed to be of Lower Cretaceous provenance; the geomorphological significance accrues from the support that this gives to demonstrating the former existence of a long pre-glacial valley system heading in the western Wolds. That such a valley would have been part of the Hoxnian land surface occupied by the temperate mammals, including hominids, lends support to the view that the reconstruction of former landscapes and environments through comprehensive physico-geographical studies is a viable, and one may even say necessary, contribution to our knowledge of Quaternary events and of the environment of early man.

Acknowledgment

The author is particularly grateful to Mr. C. Alabaster, who not only recovered most of the fossil material but brought it to the author's attention, and also provided some of the stratigraphic details. He wishes to thank Mr. P.J. Boylan, Director of Leicester Museum and Art Gallery, for identifying the mammalian remains and artefacts, and Stephen Toulson and Sons, Ltd. for unrestricted access to the quarry on many occasions since 1955.

REFERENCES

Straw, A. (1957) 'Some glacial features of east Lincolnshire', *E. Midld. Geogr.* 1, 41-8.

Straw, A. (1961) 'The erosion surfaces of east Lincolnshire', *Proc. Yorks. Geol. Soc.* 33, 149-72.

Straw, A. (1966) 'The development of the middle and lower Bain valley, Lincolnshire', *Trans. Inst. Br. Geogr.* 40, 145-54.

Swinnerton, H.H. and P.E. Kent (1949) *The geology of Lincolnshire*, Natural History Brochure No. 1, Lincolnshire Naturalists' Union, Lincoln.

DISCUSSION

Mr. J.J. Wymer made some comments on the Hoxne stratigraphy and its relationship to this Lincolnshire site. He asked Professor A. Straw if the deposits had been examined for pollen. The speaker replied that a pollen analysis had not been carried out, but that he was doubtful whether pollen would be preserved. In answer to a question from Miss C. Burek about evidence for the direction of movement of the till above the gravels, Professor A. Straw said that he had looked at the distribution of erratics and the general relief as well as at the orientation of particles in the till fabric. These were almost exclusively aligned north to south, and some east coast sedimentary rocks occurred in the deposit. Dr. J. Catt expressed interest in the iron and manganese coatings of the flints, and asked the speaker when these were deposited. Professor A. Straw said that the cause was lateral movement of ground-water through the gravels so permitting post-till deposition of secondary minerals. Dr. A. Weir asked if the particle-size distributions of the silts had been examined in greater detail in order to determine comparative transportation distances. The speaker replied that the gravels seemed to have been aggraded cyclically and that the wind would have blown silts along the gravel surfaces to create fine-grained bands within the lower deposit. The higher silts were thicker and finer-grained, and were possibly produced by wind picking up material derived from the ice advancing from the north.

J.J. Wymer
The interpretation of Palaeolithic cultural and faunal material found in Pleistocene sediments

The majority of Lower Palaeolithic cultural and faunal material is found in sediments that are either fluviatile, lacustrine or aeolian. Flint artefacts and the fragmentary bones or teeth of large mammals comprise most of the evidence, but there can be various other material in the same sediments: molluscs, beetles, ostracods, micro-fauna, macroscopic plant remains, pollen, etc. The archaeological problem is to assess how the cultural material relates to all the associated evidence and the sediments themselves. This has not always been done as critically as it should have been, and is one of the major factors in our misunderstanding of many events in the Pleistocene. For example, a fluviatile river gravel may contain hand-axes and the bones and teeth of large mammals constituting a 'warm fauna'. This does not necessarily imply that the river deposited this gravel under warm conditions, for the artefacts and bones could just as likely have been derived from an earlier sediment destroyed by the processes which formed the gravel. However, similar assemblages of bones and artefacts within lacustrine clay-muds or fine flood silts will be contemporary with the deposition of the sediments, for natural forces under such conditions of sedimentation would have been insufficient to transport them to their place of rest. The first matter, therefore, is to consider the status of the material. Is it

    (i) in a primary context,
    (ii) derived from a primary source, or
    (iii) derived from a secondary source?

Derived material is not valueless, but there are obvious limitations to its use. Only material in a primary context is suitable for assessing correct associations: the age of the sediment will date it, and much can be learnt from a study of the spatial distribution of the material within the sediment.

As mentioned, material in river gravels is, from the very nature of the sediment, almost certain to have been derived from pre-existing surfaces or deposits. Solifluction, head or till are also deposits in which it would not be expected to find material in a primary context. Conversely, it is unlikely that material in fine lacustrine, fluviatile or aeolian sediments will be derived. Such is obvious, but it is rare in the field for the evidence to be so clear-cut. There can be no standard maxims to cover all eventualities and most sites need to be considered in their own right. A few examples are outlined below, from sites within the experience of the writer.

1. *Fluviatile*

Artefacts in river gravel constitute some 95 per cent of the evidence for Lower Palaeolithic human activity in Britain, and over most of the northern hemisphere where the drainage systems have had to react to the drastic changes in regime caused by different phases of the Ice Age. Unless there is any evidence to the contrary, such artefacts should be regarded as derived and, as far as dating goes, are contemporary with or, much more likely, earlier than the final deposition of the gravel. The study of the archaeological contents of the gravels of such rivers as the Thames and the Somme has done little to elucidate more than the broadest general sequence of stone industries in those areas. The complexity of terrace formation has possible been underestimated; local correlations may be valid, but those over a wide area are suspect. Altitude of bench or surface level of terraces above the present rivers, or sea level, is not a reliable method of correlating deposits, for it is now evident that there have been numerous rises and falls of the sea during the Pleistocene, connected with glacial episodes, so that opportunities have existed for sediments to accumulate at the same level at quite different periods. However, it does still seem to hold good that the oldest terraces of the Thames are at the top (Wooldridge's Stage I) and the most recent are at the bottom (Buried Channels and flood plains of Wooldridge's Stage III) (Wooldridge, 1957), but it does not follow that the intervening ones are in a perfectly neat order of descent through time.

The horizontal distribution of artefacts in river gravels, mapped over a whole drainage system, imply two things: either the concentration along the old river courses reflects that the river valleys were favoured regions for the Lower Palaeolithic hunters, or that the denudation of the higher ground was so great at various stages of the Pleistocene, that upland surfaces were swept clean of their archaeological litter, which came to rest as constituents of the gravel sediments in the valleys. Both implications are probably true. The gravels of Savernake, for instance, contain numerous hand-axes which could well have been swept off the Wiltshire Downs. Lacaille (1971) has drawn attention to high-level finds of palaeoliths on the Wiltshire Downs. These may have escaped the erosive forces because of their exceptional height and distance from the heads of the valley, even allowing for the radical changes in topography since the Middle Pleistocene on the chalklands of Britain. Similar high-level finds have been made on the Hampshire and Surrey chalk downs and several other situations could be cited, all of which could represent a distribution that once went well beyond the river valleys. However, there are many palaeolithic sites along the river valleys where the material, although derived, is unlikely to have travelled a few metres from its original source. Such sites as Caversham, Furze Platt and Burnham in the Thames Valley (Wymer, 1968) have large concentrations of artefacts, the majority in a sharp or just slightly rolled condition. As Treacher observed, it looks as if the river had just 'turned the palaeoliths over'.

Mapping of the find-spots of palaeoliths in sediments along the Thames Valley shows a concentration of major sites at the confluences of major tributaries, as in the Reading-Sonning area where the Kennet and Loddon-Blackwater meet the Thames. Evidence is accumulating throughout Africa and elsewhere that palaeolithic hunters favoured areas with wide sheets of water. These distributions along the Thames Valley may reflect this.

The degree of abrasion on an artefact in a fluviatile deposit will be connected with the distance it has travelled from its primary source. This is something that can be measured objectively and there are possibilities for experimentation and quantitative work (Shackley, 1974). A suggested method for recording the condition of an artefact is to examine the divisions between the flake beds on the face of the artefact, not the edges, as use or secondary working could confuse the result. As an artefact is subjects to natural rolling or abrasion a facet will develop on these divisions and it will obviously become wider the greater the rolling. Suggested categories are:-

mint -	as unbraided as when freshly struck; an extremely rare condition for an artefact from a river gravel
sharp -	the divisions are dulled but no macroscopically measureable facet has been produced
slightly rolled -	a facet has been produced but is nowhere more than 1 mm wide
rolled -	the facets are more than 1 mm wide but nowwhere more than 2 mm
very rolled -	the facets are more than 2 mm wide

An application of the analysis of artefacts by these categories is shown in the diagram for Clacton (Fig. 1) (Singer *et al.*, 1973). At this site, artefacts from an area excavation throughout up to 70 cm of coarse gravel were found at all levels in various conditions. The possibility was that the slightly rolled, rolled and very rolled categories of artefacts represented derived material in the gravel, and that the mint and sharp artefacts were from actual occupation on the gravel when it was in a wet and plastic state. The majority of those in a primary context would thus be expected to occur near the top, and this is borne out in the diagram.

The horizontal distribution of artefacts in river gravels is unlikely to do more than record natural scatters, but even this is informative. The

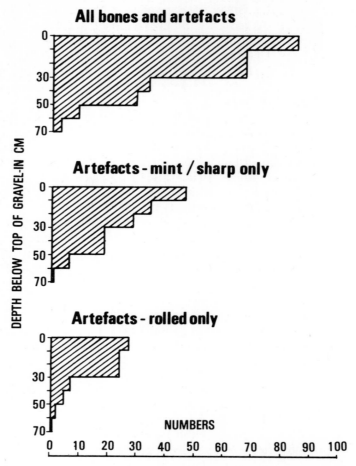

Figure 1. Vertical distribution through gravel at Clacton-on-Sea. Comparison of fresh and rolled artefacts: the former are regarded as in a primary context from occupation of the surface of the gravel when in a wet and semi-plastic state; the latter as derived from earlier deposits or surfaces.

Swanscombe Skull occurred in the lower part of a stratum referred to as the Upper Middle Gravel. It was possible to follow this particular bedding plane in the deposit as excavation took place and the plan (Wymer, 1964) shows a dispersal pattern with its point of origin towards the position where the three human bone fragments were found. The interpretation is that both the artefacts and the skull were derived from a primary source nearby.

It is unfortunate that the rare sites in the Thames Valley where palaeoliths were found in concentrations covered by or within fine fluviatile silts or loams (brickearths), such as at Crayford and Ealing, were never subjected to area excavation, for the material (as recognised by the excavators), was in a primary context. Such sites appear to have been riverside settlements, even if temporary ones, that were gently covered by successive layers of annual flood silt. The material is contemporary with the sediment, the relative positions of artefacts and associated remains have been preserved, and conditions are at an optimum for obtaining the maximum information on activity patterns, contemporary environments and dating. Similar conditions fortunately exist in one of the levels currently being investigated at Hoxne. Material is within a fine silt, interpreted as a flood plain sediment. It is in sharp condition, having only been subjected to the gentle flow of water, and must be contemporary with the sediment. It is particularly informative as both evolved flake tools and hand-axes are present, proving their contemporaneity.

2. *Lacustrine*

Lakesides have been greatly favoured places for hunting groups throughout the Pleistocene period, from the earliest known sites in East Africa. The formation of lacustrine clay-muds, coupled with oscillations in the water table, provide highly suitable conditions for the preservation of the evidence in a primary context. Several lakes appear to have existed in Britain on the edge of the receding ice-sheets during the middle or latter part of the Pleistocene, which slowly silted up during the succeeding interglacial periods. An important group around Caddington in Bedfordshire is still imperfectly understood. The famous Hoxne lake, type site of the interglacial, is currently being examined by the writer on behalf of the University of Chicago. Most of the Hoxnian lake muds are archaeologically sterile, but a fall in the water level during the Hoxnian zone III late-temperate stage produced an alder swamp around the edges with Acheulian working sites on the dryer spots. One, at least, has been perfectly preserved on this temporary surface, covered by lake-muds of the following zone IV early-glacial stage. Palaeolithic material is thus in a primary context and, in this instance, is associated with faunal remains, wood, molluscs and beetles. Sites of this nature require area excavation over as big an area as possible.

Another aspect of lake-muds, as now demonstrated at Hoxne by Drs. Aitken and Thompson, is that the remanent magnetism is retained. The oscillations of the earth's magnetic field at this period of the Pleistocene are similar to those of the Flandrian Lake Windermere (Molyneux *et al.*, 1972) and, to judge by the similar sequence of pollen zones, a similar time-span is implied (Thompson *et al.*, 1974). It is hoped that further work on the palaeomagnetism of other lake sediments will allow comparisons to be made between them. The technique promises to be a useful archaeological tool, especially when coupled with analyses of the pollen generally found in lake sediments.

3.   Aeolian

The classic aeolian sites of Europe are those of the loess belt. This fine, wind-blown dust has covered numerous palaeolithic sites of Late Pleistocene date. The best known are those on the valley of the Somme, where commercial exploitation of gravel and brickearth has exposed them. Palaeosols are sometimes contained within the loess and palaeolithic artefacts have been associated with them and, unless there has been later subsidence or movements due to compression, solifluction or cryoturbation, they are in a primary context. Such sites warrant large-scale horizontal area excavation but, unfortunately, most of the prolific sites appeared in the nineteenth century before such techniques were considered. The site of Cagny-la-Garenne shows a tripartite division of the loess and emphasises the possibility of obtaining important sequences from a combined pedological, palaeo-botanical and stratigraphical study of them.

In spite of the calcareous nature of the loess in western Europe, a modification of the Frenzel method has enabled pollen to be extracted from it (Bastin, 1964).

Dune sands constitute an aspect of aeolian sedimentation that often has archaeological associations. It is usually very difficult to assess whether levels of archaeological material under or within dune sands are *in situ* or are merely an assortment of material brought down to a common level by deflation. There is a well-known site in the Eastern Cape of South Africa called Geelhoutboom where about 2 km^2 of the local veldt has been stripped of its vegetation and loose mantle of sand, exposing the surface of the Table Mountain Sandstone. Palaeolithic artefacts of quartzite abound on its surface and also within the body of the sand. The process is currently active and the accumulation of material by deflation can be witnessed.

A similar problem had to be contended with at Elandsfontein, the site of the Saldanha Skull, also in the Cape. Dune erosion here may have been the result of indiscriminate grazing in the nineteenth century, and some 4 km^2 of sandveldt have been transformed into a miniature Sahara, with barchan dunes established from one side to the other. Complex soil changes have taken place within some of the sand bodies which must date to different and probably much earlier phases of sand movement, for they are parallel to the prevailing winds instead of being at right angles to them. Calcretes and silcretes have formed within these ancient sand bodies, now eroding away into grotesque shapes. Within hollows, scooped down to harder, calcareous formations, lie prodigious quantities of fossil animal bones including extinct species. Stone artefacts, including hand-axes, are also present, and the Saldanha Skull was found under these conditions. Other complexities are so-called ferricrete ridges, interpreted by Butzer as the 'casts' of fossil water courses. A ferruginous soil horizon had also formed within the modern sandveldt, about a metre below the present surface (Singer and Wymer, 1968).

At one place it was possible to demonstrate that the hand-axes and faunal remains in one concentration were in a primary context and not there as a result of deflation. They occurred beneath about 2 m of sand with a ferruginous horizon half way down the profile, in a small area that had escaped erosion. Further artefacts and faunal remains lay exposed around the edges of this raised area, and the first task of the investigation by the University of Chicago in 1965 was to excavate and see whether the material really did continue at the same level under the raised, uneroded area. This it did, so the inference was that the material around the edge had, until recently, also been in a primary context, but had now been exposed by recent deflation and moved slightly. Advancing dune fronts would eventually cover it up again. The possibility had to be considered that this process of exposure and burying had occurred several times either recently, or during the Pleistocene, but an analysis of the condition of the artefacts showed conclusively that this was not so. The artefacts were made mainly of various forms of silcrete, some of which was quite soft. Those exposed by recent deflation had their edges and divisions between the flake beds smoothed to a lesser or greater degree, whereas those excavated *in situ* were as fresh as when first made. The dulling of artefacts by sand abrasion is a rapid process and it would have been impossible for material to have accumulated gradually on a common level by deflation and still retained its sharpness.

The conclusion is that where there is reason to think that archaeological material, in particular sediments, is in a primary context, it is imperative that controlled area excavation be made, and particular attention given to the condition of the artefacts. Only then is it possible to make meaningful assumptions from associated material or the sediments themselves.

REFERENCES

Bastin, B. (1964) 'Essais d'analyse pollinique des loess en Belgique, selon la methode de Frenzel', *Extrait de Agriculture*, 12, 703-6.

Lacaille, A.D. (1971) 'Some Wiltshire palaeoliths', in G. de G. Sieveking, ed., *Prehistoric and Roman Studies*, British Museum, London.

Molyneux, L., R. Thompson, F. Oldfield, and M.E. McCallan (1972) 'Rapid measurement of the remanent magnetization of long cores of sediment', *Nature, Lond.* 237, 42.

Shackley, M.L. (1974) 'Stream abrasion of flint implements', *Nature, Lond.* 248, 501-2.

Singer, R. and J.J. Wymer (1968) 'Archaeological investigations at the Saldanha Skull site in South Africa', *S. Afr. Archaeol. Bull.* 23, 63-74.

Singer, R., J. Wymer, B.G. Gladfelter, and R.G. Wolff (1973) 'Excavation of the Clactonian industry at the Golf Course, Clacton-on-Sea, Essex', *Proc. Prehist. Soc.* 39, 6-74.

Thompson, R., M.J. Aitken, P. Gibbard, and J.J. Wymer (1974) 'Palaeomagnetic Study of Hoxian lacustrine sediments', *Archaeometry* 16, 233-7.

Wooldridge, S.W. (1957) 'Some aspects of the physiography of the Thames Valley in relation to the ice age and early man', *Proc. Prehist. Soc.* 23, 1-19.

Wymer, J.J. (1964) 'Excavations at Barnfield Pit, 1955-60', *in* C.D. Ovey (ed.) The Swanscombe Skull, *Roy. Anthropol. Inst. Occ. Paper*, 20, 19-60.

Wymer, J. (1968) *Lower Palaeolithic Archaeology in Britain As Represented by the Thames Valley*. London.

DISCUSSION

Dr. R.A. Webster enquired whether stone artefacts were susceptible to experimental investigation to determine their amount of wear. The speaker said that he had not done such work himself, but referred the question to Dr. M.L. Shackley who briefly outlined the recent work which she had done on the experimental abrading of implements (Shackley, 1974). Mr. P.C. Buckland then asked the speaker whether the alignments of implements on the sites of Hoxne and Clacton had been plotted and interpreted, and Mr. J.J. Wymer replied that the orientation of all reasonably-sized artefacts had been recorded. However, there was no evidence at Hoxne of any water movement to cause alignment, and at Clacton the people seemed to have been occupying the surface of the gravel when it was in a semi-plastic state. Under these conditions, alignment of implements is not likely to be related to water action.

G.H. Cheetham
Palaeohydrological investigations of river terrace gravels

As the archaeologist probes further and further back into prehistory, he becomes increasingly dependent upon the principles of stratigraphy and the analysis of sediments to gain an understanding of the environment of prehistoric cultures. When considering the Palaeolithic archaeology of the Thames Valley, for example, one must of necessity refer to alluvial terraces. Although there are considerable problems with regard to the derivation of artefacts, one has to be careful not to develop unnecessarily deterministic reasoning by implying desertion during certain periods, especially during the deposition of cold climate terraces, since it is often extremely difficult to distinguish between derivation from previous alluvial phases for which no sedimentological evidence may have survived, and derivation from contemporaneous occupation. Whatever the situation may be, it is of interest to the archaeologist and the Quarternary geologist to examine the sediments of alluvial terraces to gain an understanding of the depositional environment and how it changes through time, and to see how this relates to the archaeological evidence.

The object of this short discussion is to examine river terrace gravels from a palaeohydrological point of view, with reference to the Kennet Valley in Berkshire, initially with regard to general aspects of the fluvial environment, and then to consider a specific sedimentological problem and its archaeological implications.

Palaeohydrology, to follow the definition provided by Schumm (1965), is the study of fluvial processes which have operated in the part and their hydrologic implications. Differing approaches have been adopted in the literature (compare, for example, Schumm, 1968; Dury, 1964; Jopling, 1966; Cotter, 1971).

A specific study of the palaeohydrology of the Thames Valley was undertaken by Kimball (1948) in which he utilised the Manning equation, that is:

Dimensions

$$V = \frac{1.49}{n} R^{\frac{2}{3}} S^{\frac{1}{2}}$$

where V = velocity   L/T

R = hydraulic radius   L

S = slope of the water surface   L/L

n = Manning coefficient of roughness   L

and the flow continuity equation, that is:

Dimensions

$$Q = A\bar{V}$$

where Q = water discharge L^3/T
       A = cross-sectional
           channel area L^2
       $\bar{V}$ = mean flow velocity L/T

to show that towards the close of the Atlantic Period, using land snail evidence to estimate precipitation, the Thames probably had a mean annual discharge four times greater, and channel dimensions greater than the present-day Thames in the following proportions: width x2; depth x1.5; flow velocity x1.25.

However, when considering the environment of prehistoric man and their dynamic interaction, one must also consider the important question of channel pattern, in the sense that a series of smaller braided channels presents a totally different picture archaeologically with regard to site location, site catchment and cross-channel mobility than does one outsize channel.

The discriminant function between meandering and braiding, established by Leopold and Wolman (1957), may be expressed as:

$$S_b = 0.012 \, Q_b^{-0.44}$$

where S_b = channel slope
       Q_b = bankfull discharge in
              $m^3/sec.$

In sedimentological terms, meandering and braided streams are characterised by quite different bedding sequences both vertically and laterally (Allen, 1965a). The sediments of meandering streams display an 'upward-fining' sequence (Allen, 1965b) from coarse bedload, through cross-bedded sands, to fine textured overbank sediments, corresponding to the lateral shift of a point bar and meander (see Visher, 1965), whereas the sediments of braided streams display a complex of channels and bars, with fine-grained sediments being comparatively rare (McDonald and Banerjee, 1971; Williams and Rust, 1969). As far as coarse gravel deposits are concerned, such a complex is frequently reflected by what may be called 'matrix-bedding', that is the lateral and vertical alternation within a body of gravel of relatively uniform grain size, of texturally different matrix infills, which may be sand, silt, clay, a mixture of these, or in fact no matrix at all. The alluvial terraces of the Kennet Valley currently being investigated display this sort of sequence (compare Church, 1972).

Having made these few general remarks of archaeological application, it is now proposed to examine a specific sedimentological problem by way of illustration. On the surface of the Floodplain

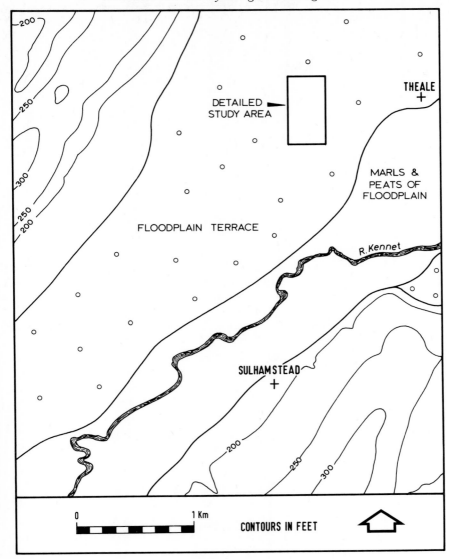

Figure 1. Eastern end of the Kennet Valley showing location of detailed study area.

Terrace of the Kennet Valley (Fig. 1), a tributary of the Thames, a complex series of braided palaeochannels is preserved west of Theale in Berkshire (grid squares SU 6270, 6370, 6271, 6371). Wymer (1968) draws attention to others further west in grid squares SU 6169 and 6269. The Floodplain Terrace, whose exact age is uncertain but thought to be Devensian, is a low morphological feature situated 3-4 m above the contemporary floodplain.

The palaeohydrological significance of meandering palaeochannels has been studied widely by Schumm (1968, 1972). The interpretation

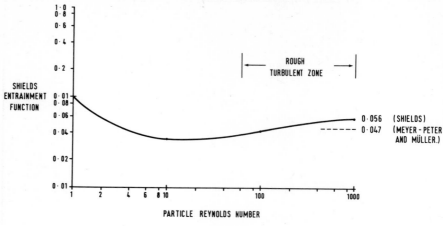

Figure 2. Shields' diagram.

of braided palaeochannels, however, presents greater difficulties. The approach adopted here is to examine the palaeo-flow characteristics associated with the incipient motion of the palaeochannel traction load, which is composed largely of gravel-sized material. First of all, it is necessary to outline the relevant literature on the hydraulics of gravel-based streams.

The initiation of motion, or entrainment of sediment, can be studied by comparing fluid forces acting to create motion with the resisting forces of the particles. When the two are equal, a situation of incipient motion, or the threshold condition, is reached. One of the most frequently quoted studies is the work of Shields (1936), as illustrated by Fig. 2, in which an entrainment function, θ, defined as:

$$\theta = \frac{\tau_O}{\gamma'_s\, D_m} \quad \text{in which} \quad \begin{aligned} \tau_O &= \gamma dS = \rho V_*^2 \\ \gamma &= \rho g \\ \gamma'_s &= \gamma_s - \gamma \end{aligned}$$

Dimensions

where
θ = a dimensionless entrainment function
τ_O = bed shear stress .... M/LT^2
ρ = fluid density .... M/L^3
g = acceleration due to gravity .... L/T^2
γ = specific weight of water .... $M/L^2 T^2$
γ_s = specific weight of sediment .... $M/L^2 T^2$
γ'_s = submerged specific weight
　　 of sediment .... $M/L^2 T^2$
d = depth of flow L
S = slope .... L/L
D_m = particle diameter L
V_* = shear velocity .... L/T

was plotted against particle Reynolds number, defined as

$$Re = \frac{D_m V_*}{\upsilon} \qquad \text{in which } \upsilon = \frac{\mu}{\rho}$$

Dimensions

where υ = kinematic viscosity L^2/T

μ = fluid viscosity M/LT

In the rough turbulent zone,[1] Shield's data shows a rise in the values of the entrainment function up to around 0.06 within the range of particle Reynolds numbers corresponding to flow over gravel beds. A re-examination of the entrainment conditions of coarse bedload by Neill (1968a, 1968b), however, has demonstrated that entrainment occurs at values lower than this, in fact as low as 0.03 for uniform materials, and that this entrainment function becomes a constant value above particle Reynolds numbers of about 70. Although there seems to be general agreement that the entrainment function becomes a constant value in the rough turbulent zone (Neill, 1968b), the value of this function appears to be related to sediment sorting, particle shape and packing. The relatively low values reported by Neill were obtained from experiments on uniform materials. Experiments on grain size mixtures and irregular particle shapes tend to show slightly higher values, implying the resisting moments to be greater. Consider, for example, the Meyer-Peter and Müller bedload equation (1948) which was derived quite independently of Shields and was based on both uniform materials and mixtures. Their data satisfies the following equation:

$$G = 8 \sqrt{\frac{g}{\gamma}} \; (\gamma dS - 0.047 \; \gamma'_s \, D_m)^{3/2}$$

where G = weight of sediment transport per unit time and width, and D_m = effective particle size of a grain size mixture.

$$\text{i.e. } \frac{\Sigma d_i \; \Delta p_i}{100}$$

where Δp_i is the % of particle size d_i

If G = 0 (i.e. at the threshold of motion), it follows that:

$$\frac{\gamma dS}{\gamma'_s \, D_m} = 0.047$$

which is simply restating the value of θ, the Shields entrainment

---

1. In hydraulics, three types of flow are commonly recognised, i.e. laminar flow, smooth turbulent in which turbulent flow is developed over a laminar sub-layer, and rough turbulent in which grain sizes are sufficiently large to create turbulence around the grains.

function, implying that entrainment will occur at values of around 0.047 as far as grain sizes greater than 2 mm are concerned (for the specifications of the Meyer-Peter and Müller equation, see Yalin, 1972, p. 112). The slope term, S, can be interpreted as either bed-slope or water-surface slope since they are normally regarded to a large extent as interchangeable values.

Others have devised more complicated entrainment equations, either experimentally or semi-theoretically. For example, Helley (1969) developed a complex relationship involving bed velocity, drag coefficient, particle axes, and particle turning arms. Thompson (1965) considers fluctuating lift forces associated with eddies arising from boundary roughness; an extension of the work of Prandtl (1952) who remarked on the relatively low velocities required to entrain pebbles, in contrast to the over-simplified Hjulström curve (1935) which considers velocity alone, divorced from other important hydraulic parameters.

The Theale palaeochannels, which are surface forms infilled with fine sediment (see Fig. 3), are considered here with respect to the Shields entrainment function and the Meyer-Peter and Müller bedload equation. A detailed study area was set úp around a pipeline section exposing the palaeochannels in August, 1973. Within the area, shown on Figs. 1 and 3, the palaeochannels were accurately located by a detailed network of augering, although the general distribution of palaeochannels over a much wider area is known from aerial photographs. The section enabled bedload to be sampled at maximum channel depth. The well-preserved network of palaeochannels would suggest an almost undisturbed alluvial surface, in which case maximum channel depth can be interpreted as maximum bankfull depth. Channel slope was measured by levelling maximum channel depth on either side of the section for a variable distance depending on

Table 1. Summary of Theale palaeochannel data.

Palaeochannel No.	D_m (mm)	d_{max} (cm)	$\dfrac{\tau_o}{\gamma'_s D_m}$	$\dfrac{\tau'_o}{\gamma'_s D_m}$
1	22	91	0.056	0.053
2	18	80	0.060	0.057
3	13	60	0.064	0.060
4	16	76	0.070	0.066
5	15	62	0.059	0.056
6	19	84	0.065	0.061
7	17	76	0.066	0.062
8	14	61	0.062	0.059
9	16	75	0.068	0.064

τ'_o = corrected value of bed shear stress (see text)

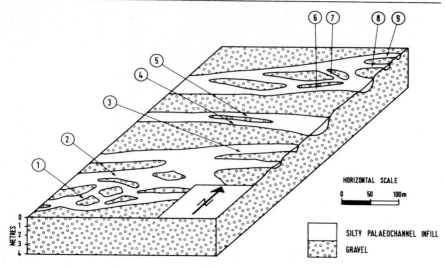

Figure 3. Braided palaeochannels of detailed study area. Numbers refer to palaeochannels listed on summary table.

whether the channel was a major or minor one. Because of the inadequacies of using augering depths, the gradients used in the subsequent calculations were only taken to three decimal places, giving consistent values of 0.002 for all the channels studied.

τ_0 was calculated as the 'depth-slope product' as it is frequently called, which is essentially valid for the central part of a wide channel (Neill, 1968a). Rubey (1938), however, has shown that 'bed-velocity' is more significant than the 'depth-slope product' as far as the threshold conditions of particles larger than 2.5 mm are concerned. An examination of Neill's experimental data (1968a) shows that τ_{0_0} calculated from velocity profiles tends to be less than measurements of the 'depth-slope product' and for his calculations, he took the mean value between the two. This being the case, the calculated values of Shields entrainment function for the Theale palaeochannels are likely to be slightly overestimated. For this reason, the calculated values were corrected by a constant factor obtained from Neill's experimental data by regressing his values of γdS and ρV_*^2, and using the regression equation as a correction factor. From the results, as shown in the final column of Table 1, one may conclude that the bedload of the palaeochannels would be around or slightly above the threshold condition at bankfull discharge, with very low transport rates calculated from the Meyer-Peter and Müller equation, assuming a critical threshold value of 0.047. At half bankfull, further calculations showed no sediment transport and Shields values well below the threshold condition. Any appreciable movement of the terrace gravels, therefore, would be associated with overbank discharges, a point which is reinforced when one considers that the calculations are based on maximum depth and take no account of shallower depths within the channel and flow over channel bars. Relatively high gravel transport rates associated with the more catastrophic events, however, lead to the formation of gravel dunes, as observed both in the field and in the laboratory (Thiel, 1932; Meyer-Peter and Müller, 1948). The sediments of the Floodplain Terrace of the Kennet Valley, however, do not show such features, but display rudimentary bedding indicative of moderately low transport rates. One must also bear in mind the frequent shifting and abandonment of channels so characteristic of braided streams and the high quantity of sediment comprising the Floodplain Terrace.

A consideration of all the above points would seem to suggest a distinct type of hydrologic régime, characterised by overbank discharges of a relatively frequent and persistent recurrence, not a surprising conclusion but one that does have important archaeological implications in that it emphasises dynamic aspects of the environment. It may be that such discharges were of a seasonal nature, perhaps in the form of a niveo-fluvial stream régime characterised by seasonal flood peaks in response to snow-melt. Such a cold climate process is in accordance with the Devensian age tentatively ascribed to the Floodplain Terrace on the basis of stratigraphic and morphological evidence. Further insight into the palaeohydrology of the Kennet catchment is provided by a close examination of the relationship between dry valleys and the Kennet Valley terraces, an aspect which is currently being investigated, particularly with respect to drainage density and discharge (Gregory, 1971).

The value of palaeohydrology in archaeological investigations is essentially twofold: it provides background information on fluvial palaeo-environments during the Quaternary, as illustrated by the Theale palaeochannel analysis, especially in areas where few *in situ* riverine sites are preserved and where one is dealing with derived artefacts whose exact relationship with the sediments is frequently uncertain, by the very nature of the deposits and their formative processes. Secondly, it enables riverine sites that have been preserved to be understood more clearly in terms of environmental processes. Archaeologically, what is important when considering gravel deposits and braided streams is the relatively frequent environmental transformation from a series of small channels to one sheet of water. This clearly has important implications with regard to the dynamic relationship between man and the environment, whether one is considering site location, site catchment analysis (Higgs and Vita-Finzi, 1972), or terrain adjustment and prehistoric communities (Davidson, 1972). The application of fluvial hydraulics to riverine sites offers much potential to the archaeologist in gaining a clearer understanding of the fluvial environment and its dynamic processes during the time of occupation.

Acknowledgments

The author wishes to thank Professor John R.L. Allen, Sedimentology Research Laboratory, Department of Geology, University of Reading, for kindly commenting on the manuscript, and Dr. Peter Worsley, Department of Geography, University of Reading, and Dr. Claudio Vita-Finzi, Department of Geography, University College, London, for their encouragement.

REFERENCES

Allen, J.R.L. (1965a) 'A review of the origin and characteristics of recent alluvial sediments', *Sedimentology* 5, 89-191.

Allen, J.R.L. (1965b) 'Fining upwards cycles in alluvial successions', *Geol. J.*, 4, 229-46.

Church, M. (1972) 'Baffin Island Sandurs: a study of Arctic fluvial processes', *Geol. Surv. Canada*, Bulletin 216.

Cotter, E. (1971) 'Palaeoflow characteristics of a large Cretaceous river in Utah from analysis of sedimentary structures in the Ferron Sandstone', *J. sedim. Petrol.* 41, 129-38.

Davidson, D.A. (1972) 'Terrain adjustment and prehistoric communities', in P.J. Ucko *et al.* (eds.), *Man, Settlement and Urbanism*, London, 17-22.

Dury, G.H. (1964) 'Principles of underfit streams', *Prof. Pap. U.S. geol. Surv.* 452 A, B and C.

Gregory, K.J. (1971) 'Drainage density changes in south-west England', in K.J. Gregory and W.L.D. Ravenhill (eds.), *Exeter essays in Geography*, Exeter, 33-53.

Helley, E.J. (1969) 'Field measurement of the initiation of large bed particle motion in Blue Creek, near Klamath, California', *Prof. Pap. U.S. geol. Surv.* 562 G.

Higgs, E.S. and C. Vita-Finzi (1972) 'Prehistoric economies: a territorial approach', *in* E.S. Higgs (ed.), *Papers in Economic Prehistory*, Cambridge, 27-36.

Hjulström, F. (1935) 'Studies of the morphological activity of rivers as illustrated by the River Fyris', *Bull. geol. Instn Univ. Uppsala*, 25, 221-527.

Jopling, A.V. (1966) 'Some principles and techniques used in reconstructing the hydraulic parameters of a palaeo-flow régime', *J. sedim. Petrol.* 36, 5-49.

Kimball, D. (1948) 'Denudation chronology: the dynamics of river action', *Occ. Pap. Univ. Lond. Inst. Arch.* 8, 21pp.

Leopold, L.B. and M.G. Wolman (1957) 'River channel patterns – braided, meandering and straight', *Prof. Pap. U.S. Geol. Surv.* 282-B.

McDonald, B.C. and I. Banerjee (1971) 'Sediments and bedforms on a braided outwash plain', *Can. J. Earth Sci.* 8, 1282-301.

Meyer-Peter, E. and R. Müller (1948) 'Formulas for bed load transport', *Proc. 2nd Congress I.A.H.S.R.*, 39-64, Stockholm, June 1948.

Neill, C.R. (1968a) 'A re-examination of the beginning of movement for coarse granular bed materials', *Int. Report 68, Hydraulics Research Station*, Wallingford.

Neill, C.R. (1968b) 'Note on initial movement of coarse uniform bed material', *J. Hydraulics Research* 6, 173-6.

Prandtl, L. (1952) *The essentials of fluid dynamics*. Glasgow.

Rubey, W.W. (1938) 'The force required to move particles on a stream bed', *Prof. Pap. U.S. geol. Surv.* 189 E, 121-40.

Schumm, S.A. (1965) 'Quaternary palaeohydrology', *in* H.W. Wright and D.G. Frey (eds.), *The Quarternary of the United States*, Princeton, New Jersey, 783-93.

Schumm, S.A. (1968) 'River adjustment to altered hydrologic regimen – Murrumbidgee River and palaeochannels, Australia', *Prof. Pap. U.S. geol. Surv.* 598.

Schumm, S.A. (1972) 'Fluvial palaeochannels', *in* J.K. Rigby and W.K. Hamblin (ed.), *Recognition of Ancient Sedimentary Environments*, S.E.P.M. Spec. publn. 16., 98-107.

Shields, A. (1936) 'Anwendung der Ahnlichkeitsmechanik und der Turbulenzforschung auf die Geschiebewegung', *Mitteilungen der Preuss. Versuchsanstalt für Wasserbau und Schiffbau*, Berlin, Heft 26. English translation by W.P. Ott and J.C. van Uchelen, *U.S. Dept. Agric., Soil Conservation ser., Coop. Lab., Calif. Inst. Tech.*

Thiel, G.H. (1932) 'Giant current ripples in coarse fluvial gravel', *J. Geol.* 40, 452-8.

Thompson, S.M. (1965) 'The transport of gravel by rivers', *Proc. 2nd Australian Congress on Hydraulics and Fluid Mechanics*, Auckland, N.Z., A259-74.

Visher, G.S. (1965) 'Fluvial processes as interpreted from ancient and recent fluvial deposits', *in* G.V. Middleton (ed.), *Primary Sedimentary Structures and Their Hydrodynamic Interpretation*, S.E.P.M. spec. publn, 12, 11632.

Williams, P.F. and B.R. Rust (1969) 'The sedimentology of a braided river', *J. sedim. Petrol.* 39, 649-79.

Wymer, J. (1968) *Lower Palaeolithic Archaeology in Britain As Represented by the Thames Valley*. London.

Yalin, M.S. (1972) *Mechanics of Sediment Transport*. Oxford.

Section 4
Biological Sediments

G.W. Dimbleby

A review of pollen analysis of archaeological deposits

There are still some archaeologists who regard pollen analysis as a dating method. The fact that as such it has now been largely superseded by radiocarbon dating means that to them pollen analysis must have little relevance in the excavation of archaeological sites. Conversely, there are those who believe that a pollen analysis is a desirable adjunct to any excavation and proffer humus-rich samples with little understanding of where their pollen content, if any, may have come from. There are also many archaeologists today who do see the value of pollen analysis as a means of answering specific questions, but regrettably they are often frustrated by the inability to find anyone to carry out the analyses for them. In an attempt to rationalise the various concepts of how pollen analysis may be used, and thereby to make the most efficient use of the limited specialist resources, a brief review of the possibilities seems called for. It may also serve to encourage the development of specialist services in a way which could help excavators to extract the maximum of information from their sites.

In this context the discussion is about the pollen analysis of sub-aerial mineral deposits, a milieu very different from the materials such as peat and lake sediments which are the medium of more orthodox pollen studies. Consideration has to be given as to how the pollen got into these deposits and how it became distributed through them. There is no value in a pollen analysis which cannot be interpreted; if pollen of different ages and from different sources is present in a way which makes it impossible to separate the different elements, the analysis is practically useless. In many cases, however, pollen of mixed origin is present but can be distinguished numerically, so that it is not necessary to insist upon the absolute stratification sought in conventional pollen analysis.

The most significant type of information which can emerge from pollen analysis of archaeological sites is ecological. It may be possible to tell what the contemporary environment was like and how it had changed, perhaps under human influence, up to the time the monument was built. For this purpose it is necessary to find some part of the site which was an exposed surface at the time and which was collecting pollen from the contemporary pollen rain. Normally this was the old land surface so often preserved beneath earthworks

(Waterbolk, 1954; Dimbleby, 1954). Given an understanding of how pollen is preserved and distributed in a soil it is often possible also to make other deductions – non-ecological – about the nature of the monument. But first it is necessary to understand the pollen content of a soil, so the principles will be briefly re-stated (for greater detail see Dimbleby, 1957, 1961a).

Every soil is receiving the pollen rain of its locality. What happens to this depends mainly on the biological activity of the soil, and this in turn is closely connected with the chemical and physical conditions of the soil. For convenience we speak of pollen preservation, but the real concern is with the different rates of decay of pollen. These are highest in soils with high bacterial and actinomycete activity; in the moist temperate region these are soils whose pH lies above about 5.5. Consequently pollen is not well preserved in circumneutral soils, though it should be added that where other factors such as aridity control microbiological activity this correlation may not apply (Martin, 1963). As will be shown, pollen may be present in small amounts in biologically active soils, but if so it is likely to be distributed through the profile by earthworms and other soil fauna. On the other hand, if the soil is acid, pollen only decays very slowly and some types apparently persist for several millennia, though differential decomposition of different pollen types occurs and has to be allowed for (Havinga, 1971).

In moist temperate regions, therefore, pollen analysis can most usefully be applied to archaeological sites in regions of base-poor soils. Furthermore, the acidity also excludes the soil-mixing fauna, so that a pattern of depth-distribution of pollen may develop, with older pollen lying deep in the soil, giving way to progressively younger pollen nearer the surface. It is not apparent how the pollen moves in the soil, but it is clear that it is not as discrete grains. It seems probable that the slow movement (perhaps 15 cm in 5000 years) is connected with the downward movement of humus, for in preparations the pollen is not released as free grains until the humus is destroyed by acetolysis. Whatever the mechanism, there is no sorting of the grains by size as might occur if they were moving as separate grains.

The longer the pollen lies in the soil the more chance it has of being destroyed. Consequently the greatest amount of pollen is found near the surface and there is a steep falling off of the absolute frequency with depth. A surface can therefore be recognised by the high frequency and the pollen types which were most abundant in the contemporary vegetation will stand out clearly in the pollen spectrum.

To illustrate how these principles may be used in the service of archaeology, consider a round barrow consisting of a mound with a turf-core and a perimeter ditch (Fig. 1). This simple model contains most of the conditions met with in much more elaborate structures and can be used as a guide to the application of pollen analysis to such structures.

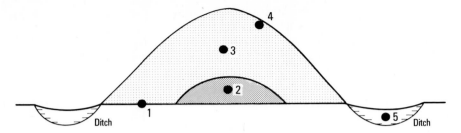

Figure 1. Sketch of a round barrow with a turf-core and perimeter ditch. Five
   sampling localities for pollen analysis are indicated.

There are five points at which pollen analysis could be used, either
to give information about the contemporary environment or to shed
light on the structure of the mound. These will be considered in turn.

1.   The old land surface

This is by far the most informative part of the site. If the soil profile is
complete it will reveal the contemporary vegetation from which may
be inferred the agricultural, pastoral or other activities of man. It will
often show the changes which have taken place in the vegetation for
some time prior to burial. Correlation between the soil profile itself
and the pollen profile may amplify this information.

Very often a buried soil has been truncated; this can be recognised
by the absence of a pollen-rich surface layer in a material which is
acid enough to preserve pollen.

For these purposes it is necessary to take a series of contiguous
samples through the soil profile, though a simple sample from the
actual surface may serve to give a picture of the contemporary
landscape.

2.   The turf stack

If the old land surface is truncated the turves of a turf stack may serve
to provide a pollen spectrum of the time. It is reasonable to assume
that most turves were cut near the site (though exceptions do occur)
and from an analysis of several of them the proportions of tree, shrub
and herbaceous pollen may be used to give information about the
setting of the barrow: for example, whether it was in a clearing.

A series of samples through the turves will quickly establish where
the turves were inverted or not; this is not always apparent from the
section. (Dimbleby, 1962: Burton Howes; Winter Hill).

3.   The body of the mound

Mounds may be composed of material from various sources. They may be made of scraped-up topsoil, of subsoil quarried from a ditch, or possibly of material brought to the site from elsewhere. Pollen analysis will distinguish straightaway between pollen-rich topsoil and pollen-deficient subsoil material, and series of samples through the mound may reveal a complex stratification of both types of material. Occasionally it can be shown that material alien to the site has been introduced, as for instance if material containing bog pollen is found in an earthwork on freely-draining soils (Dimbleby, 1962: Burley Barrow).

It should be mentioned that this type of investigation is very time-consuming and unless the information so produced is likely to be of vital importance, it is not an exercise that will be welcomed by the specialist pollen-analyst.

4.   Surface of the mound

Just as the old land surface can contain a vegetation record going back for centuries or longer, so the surface of the barrow may contain a sequence subsequent to the construction of the barrow. Erosion of the barrow surface may take time to stabilise, but after that a continuous record in broad terms may be obtained (Thompson and Ashbee, 1957). In topsoil mounds this pollen profile, perhaps 20-40 cm deep, may merge into the contained pollen of the mound. Nevertheless, the two sources can usually be distinguished (Ashbee and Dimbleby, 1959).

One particularly valuable application is to the primary mound surface of a barrow which has been heightened subsequently. With the old land surface, the present-day surface of the mound, and intermediate surface, it is sometimes possible to see the course of prehistoric deforestation by comparative pollen analysis of the surfaces of different age (Dimbleby, 1962: Burton Howes).

5.   Ditch fillings

Samples of ditch fillings are frequently submitted for pollen analysis, apparently because their humic nature and relatively damp condition are believed to favour pollen preservation. Generally, however, the pollen in their deposits is so complex in origin that no interpretation is possible. The Overton Down and Wareham earthwork experiments (Jewell and Dimbleby, 1966; Evans and Limbrey, 1974) demonstrated that the filling of a ditch is a complex process of the weathering of the walls of the ditch and the collapse of the topsoil. Consequently, in a soil in which pollen is well preserved, the ditch fill will contain pollen derived from the old topsoil. Some of this pollen may be much older than the ditch itself. In addition, material may wash into the ditch from the adjacent earthwork,-and wind-blown dust and organic matter will collect here. All these may be pollen-rich. Finally, there is the annual pollen rain adding contemporary pollen to the contained pollen. There is no means of sorting all these different elements out if the ditch is progressively filling. If, however, there is a long pause phase, during which some soil development takes place, then a limited amount of information may be obtainable from this soil surface.

For these reasons ditch fillings are rarely worth analysing; they can produce nonsense values which in the mind of the archaeologist cast doubt on the reliability of the method.

Other sources of error

There are other sources of possible error, some of minor importance, and others of greater significance in certain circumstances. To conclude this review, several of the latter category will be considered briefly.

(a) Disturbance of the old land surface by natural or semi-natural processes

Reference has already been made to truncation of the soil profile and how it shows up in the pollen record. This may occur by erosion or by deliberate action such as sod-paring, and unless there are turves in the make-up of the earthwork, there is no means left of finding out what the contemporary vegetation was.

On sloping sites soil creep may have taken place. This leaves its record in the pollen profile – successive samples may show widely different pollen spectra. Here, too, it may be impossible to interpret the pollen analyses in terms of contemporary vegetation, but at least it is established that there has been disturbance.

A third circumstance in this category is where there is aggradation on an old land surface. This may be a single episode, where an old pollen-rich surface is buried by blown sand or soil (Dimbleby, 1961b), and as such it is readily recognisable. Less easy to detect is the gradual accumulation, perhaps of top-soil dust. This would produce a continuous series of pollen curves with no sharp break. Suspicions that this may have occurred usually arise when the high pollen concentrations characteristic of a surface are found over a depth of 10 to 20 cm instead of the more usual 2 to 5 cm or so. Where aggradation of this sort takes place it is difficult to separate pollen of the contemporary pollen rain from that which was contained in the blown dust itself.

(b) Cultivation

Continued cultivation by any means will in time produce a complete mixing of the topsoil, effectively destroying any stratification which may have existed originally. In many cases the pollen itself may indicate that cultivation must have been in progress; a predominance of pollen of grasses and arable weeds, with the consistent occurrence of cereal grains, would be a direct pointer. If, however, there has only been one episode of cultivation, the effect on the pollen profiles will depend on the nature of the tool used. If it were a mould-board plough, the turning of the sod should be detectable. The pollen-rich surface can be seen at a depth of several centimetres, in the same way that an inverted turf shows up in a pollen sequence through a turf-mound. Hoes might be expected to show a similar effect, though the writer has never seen a pollen profile from a hoe-cultivated soil. Ard ploughs scratch the soil and as the furrows were often spaced widely, at about 40 to 50 cm apart, the disturbance of the profile in light soil may be localised so that there is a good chance that any one pollen profile is little influenced by the cultivation. In more cloddy soils, or in soils with ramifying roots, however, even an ard plough can produce heaving of the soil far beyond the furrow being cut. Further work on the pollen profiles of soils under primitive cultivation would be valuable.

(c) Occupation layers

Occupation layers can be so heterogeneous and so far removed from the natural equilibrium between a surface and the contemporary pollen rain that pollen analysis of them is not very informative. They are another layer which tends to the highly humic and is therefore regarded by many archaeologists as suitable for pollen analysis.

Occasionally interesting facts can emerge, however. For instance, in a number of instances (e.g. Thompson and Ashbee, 1957) high values of ivy (*Hedera*) pollen have been found. This is a plant which, being insect pollinated, does not produce much pollen and normally forms only about 5 per cent of the tree pollen count. In these occupation sites, values of several hundredfold greater than this have been found. It seems likely that the plant was either being stored as winter fodder – it is evergreen and flowers in late autumn – or that animals which had fed on it were being stalled on the sites. Some such explanation may account for other instances of abnormally high concentrations of particular pollen types on archaeological sites.

Occupation layers may be added to constructional materials which contain pollen. Perhaps the most obvious example is accumulated material from the destruction of adobe or mud-brick walls. Pollen can be present in mud-brick, though so far it has not been found in large quantity, but it may be a pollen suite which is alien to the immediate environs of the site itself. Thatch material, too, may contain pollen which was gathered with the material and brought to the occupation site.

(d) Earthworms

It has already been mentioned that where earthworms and pollen are both present in a soil any stratification, however crude, will be destroyed. This has been shown experimentally (Ray, 1959), and pollen profiles of soils containing earthworms show a characteristic pollen spectrum, both the absolute frequencies and the percentages being uniform throughout the depth of the profile. There may be a slight increase in quantity at the surface, presumably due to the time lag between pollen deposition and incorporation.

In calcareous soils, it may be that this mixing is not as damaging to interpretation as might be imagined. In such soils microbiological activity is great, so that pollen is rapidly destroyed. It is reasonable to conclude from this that all the pollen in such a soil is more or less coeval (Evans and Dimbleby, 1974), so that whether it is stratified or mixed up does not affect the interpretation. Examination of earthworm casts shows that pollen passes through the worm gut without destruction.

The Overton Down experimental earthwork has shown that where earthworms are present, pollen is progressively mixed from the old land surface into the soil beneath, in contrast to the Wareham earthwork on a podzol where there has been no movement in 10 years. At Overton Down it was also apparent that pollen could be moved *upward* into the mound; worms were moving freely into the chalk mound and casting in the interstices.

The archaeologist who is considering the applicability of pollen analysis to the various deposits on the site he is excavating may find it useful to consider three basic questions about each of the deposits:

(a)   how does the pollen get into the deposit?
(b)   how rapidly does it decay?
(c)   how is it transported within the deposit?

It is hoped that this review will help him to answer these questions, and in the light of the answers to decide whether it is worth while calling in a palynologist. It should be said, however, that these considerations are only guidelines. We are dealing with living biological systems and it is not always easy to forecast their effects. Pollen can be present when all the signs are that it should not be, and vice versa. There is a lot to be said for the trial sample in important contexts.

REFERENCES

Ashbee, P. and G.W. Dimbleby (1959) 'The excavation of a round barrow on Chick's Hill, East Stoke Parish, Dorset', *Proc. Dorset nat. Hist. Archaeol. Soc.* 80, 146-59.

Dimbleby, G.W. (1954) 'Pollen analysis as an aid to the dating of prehistoric monuments', *Proc. Prehist. Soc.* 20, 231-6.

Dimbleby, G.W. (1957) 'Pollen analysis of terrestrial soils', *New Phytol.* 56, 12-28.

Dimbleby, G:W. (1961a) 'Soil pollen analysis', *J. Soil Sci.* 12, 1-11.

Dimbleby, G.W. (1961b) 'Transported material in the soil profile', *J. Soil Sci.* 12, 12-22.

Dimbleby, G.W. (1962) 'The development of British heathlands and their soils', *Oxf. For. Mem.* No. 23.

Evans, J.G. and G.W. Dimbleby (1974) 'Pollen and land snail analysis of calcareous soils', *J. Archaeol. Sci.* 1, 117-33.

Evans, J.G. and S. Limbrey (1974) 'The experimental earthwork on Morden Bog, Wareham, Dorset, England: 1963 to 1972', *Proc. Prehist. Soc.* 40, 170-202.

Havinga, A.J. (1971) 'An experimental investigation into the decay of pollen and spores in various soil types', *in* V. Brooks *et al.* (eds.), *Sporopollenin,* 446-79.

Jewell, P.A. and G.W. Dimbleby (1966) 'The experimental earthwork on Overton Down, Wiltshire, England: the first four years', *Proc. Prehist. Soc.* 32, 313-42.

Martin, P.S. (1963) *The Last 10,000 Years.* Tucson.

Ray, A. (1950) 'The effect of earthworms in soil pollen distribution', *J. Oxf. Univ. Forest Soc.* 7, 16-21.

Thompson, M.W. and P. Ashbee (1957) 'Excavation of a barrow near the Hardy Monument, Black Down, Portesham, Dorset', *Proc. Prehist. Soc.* 23, 124-36.

Waterbolk, H.T. (1954) *De praehistorische mens en zijn milieu.* Groningen.

DISCUSSION

Mr. G. de G. Sieveking enquired about the examination of pollen from sediments in open calcareous caves where humus is not present. Professor G. Dimbleby thought that in certain climatic conditions, for example in semi-arid climates, pollen might be preserved although this would not normally be expected. In a cave environment, careful examination was required to show what the pollen remains meant since some might be intrusive and not part of the pollen rain of the region. Professor A. Straw asked if earthworms tended selectively to destroy pollen, but the speaker felt that this was not the case. Pollen recovered from the casts of earthworms showed no signs of decomposition. Dr. R. Webster enquired about the possibility of distinguishing old from new pollen by differential preservation. Professor G. Dimbleby thought that in a biologically active soil it might be possible, but that it would be difficult in other materials.

R. L. Jones

The activities of Mesolithic man: further palaeobotanical evidence from north-east Yorkshire

Only in comparatively recent times has the view emerged that Mesolithic man in Britain was not completely dominated by his environment. Pollen analytical studies of pre sub-Boreal sediments have revealed declines in arboreal pollen values accompanied by the presence of pollen of open habitat species (Dimbleby, 1962, 1963; Simmons, 1964, 1969a). Such trends are frequently encountered in sub-Boreal and later pollen diagrams when they are usually interpreted as indicators of Neolithic and subsequent anthropogenic activity (Godwin, 1956). Their occurrence on a smaller scale in pollen assemblages of pre-Neolithic age has led to suggestions that Mesolithic man influenced the vegetation cover to a greater extent than had formerly been considered possible (Dimbleby, 1962; Simmons, 1969b). The new evidence has been sufficiently convincing to provoke a re-appraisal of certain established concepts of early post-glacial vegetational change, notably by Smith (1970).

North-east Yorkshire contains the important and well-documented Mesolithic site of Star Carr in the Vale of Pickering (Clark, 1971). Here, a group of people whose culture was transitional in form between that of the Upper Palaeolithic and the Mesolithic existed at a lakeside habitat in early Flandrian (pre-Boreal and Boreal) time. Their occupation level has been dated by radiocarbon to 9488±209 B.P. and pollen analytical evidence (Walker and Godwin, 1971) indicates that these organised, early Maglemosean settlers had a negligible effect upon their environment.

At numerous localities over the North Yorkshire Moors, especially across their central watershed, Dimbleby (1962), Simmons (1969a) and Cundill (1971) have presented substantial proof of the activities of Mesolithic man in the modification of the vegetation cover. It is envisaged that this occurred during middle Flandrian (Atlantic) time, when occupance of the region is indicated by the presence of large numbers of microliths of late Mesolithic type at sites which are considered to have been hunting camps (Radley, 1969). Flints are sometimes found stratified within organo-mineral deposits containing charcoal and pollen sequences from such sediments demonstrate the occurrence of ruderal plants in what was generally a well developed, mixed oak forest. It has hence been postulated that man was using

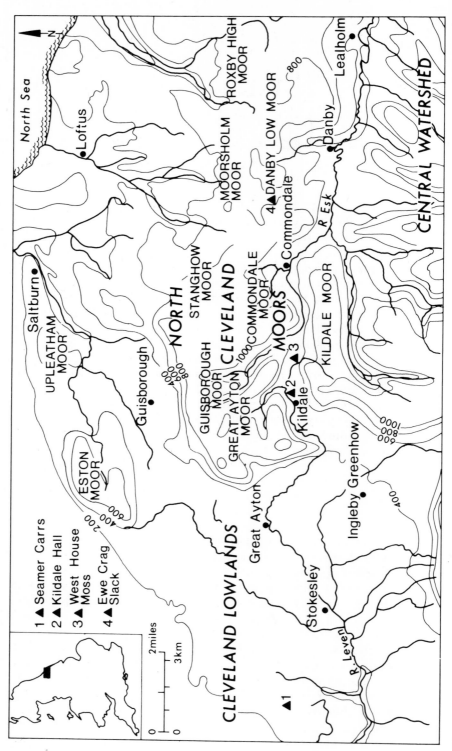

Figure 1. The Cleveland District. Topography and the location of sites.

fire, disturbing the vegetation and soils and encouraging the growth of weeds in cleared areas.

Palaeobotanical investigations in the Cleveland district of northeast Yorkshire (Jones, 1971), especially at sites upon the low moorland and in the dales which dissect it (Fig. 1), have provided support for the hypothesis of Atlantic forest disturbance and established a case for anthropogenic activity in early Flandrian times. Within the region, prolific flint sites, interpreted as late Mesolithic settlements, have recently been discovered (D.A. Spratt, personal communication) and *in situ* archaeological remains found in both organic and inorganic sediments have also aided in the elucidation of the early and middle Flandrian pollen record.

1. The evidence

Within Kildale (Fig. 1), Cameron (1878) recorded a peat deposit close to Kildale station from the lower part of which remains of reindeer and red deer were recovered. In 1969 near to Kildale Hall and a short distance from the first locality at 168 m O.D., an almost complete, well preserved, disarticulated skeleton of *Bos primigenius* was discovered embedded in marly peat (F.A. Aberg, personal communication). Subsequent stratigraphic investigations revealed the former existence of a small lake basin adjacent to the area of the bone finds (Fig. 2). Macro-fossil analysis of the peat encasing the bones showed it to be dominated by sedges, to contain silt and fragmentary mollusca, together with charcoal of *Betula* and Ericaceae.

A pollen diagram through this sediment indicated an early Flandrian (Zone IV/V) age (Fig. 3). The earliest pollen spectra are characterised by high values of *Empetrum,* Gramineae and Rosaceae

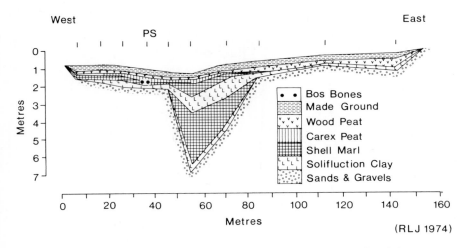

Figure 2. Kildale Hall. Stratigraphy showing the position of the *Bos* skeleton.

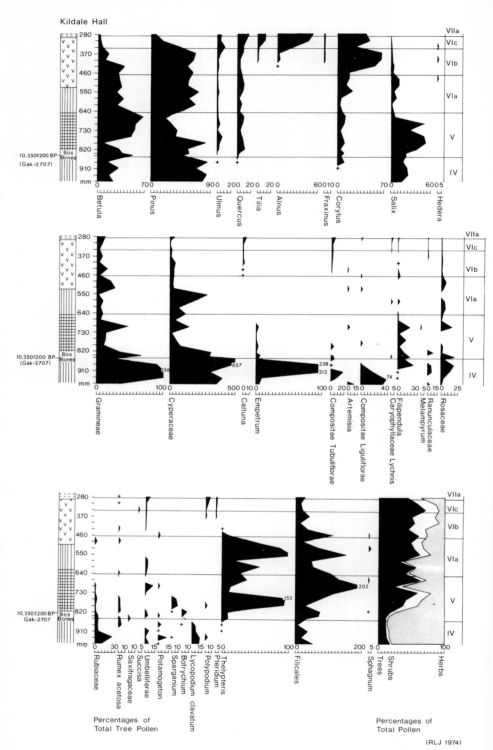

Figure 3. Kildale Hall. Pollen diagram from the early Flandrian sediments.

relative to those of *Betula*, *Pinus* and *Corylus*. This suggests that the landscape in the vicinity of the site (itself an ill-drained swamp marginal to a small lake) was clothed mainly by heath vegetation, with, at best, a scanty tree and shrub cover. A radiocarbon date of 10350±200 B.P. (Gak – 2707) from organic material in contact with the *Bos* bones confirms the early Flandrian time context; its slight excess age with regard to accepted chronology (Mitchell, Penny, Shotton and West, 1973) may be explained by error due to the incorporation of old carbon from the molluscan remains within the sediment. Later in the Boreal period (Zone VI) rising *Betula* pollen values are superceded by high totals of *Pinus* and *Corylus* indicating that tree cover increased, though open habitat species such as *Artemisia*, *Melampyrum* and Rosaceae continue to be represented in substantial amounts. It was not until the water body was succeeded first by swamp and later by fen-carr that woodland became well established around the mire.

At West House Moss, 178 m O.D. (Fig. 1), the palynological record from the early Flandrian sediments (Fig. 4) is not dissimilar to that of Kildale Hall. It shows that there was no major enclosure of the landscape by trees with a profuse shrub and herb vegetation persisting in the environs of a swampy hollow until Zone VI when local fen-carr and regional *Pinus-Corylus* forest took over, incorporating sporadic patches of *Betula* woodland. Minor changes in the composition of the forest are indicated within Zone VI when an expansion of *Melampyrum Succisa*, Chenopodiaceae and Ericales pollen and of *Pteridium* spores is followed by increased totals of Gramineae, Rosaceae and *Rumex*, both events being accompanied by declines in tree pollen values. The characteristic sharp rise in the curve for Alnus pollen which denotes the Boreal/Atlantic transition and which has been determined by radiocarbon to 6650±290 B.P. (Gak – 2706), post-dates these phenomena.

Ewe Crag Slack, 235 m O.D. (Fig. 1), also has sediments which extend back into earliest Flandrian time. Within them there are numerous silt layers (Fig. 5), the lowest of which attains 150 mm in thickness and contains fragments of charcoal. A pollen diagram through these basal levels shows a Boreal spectrum including declines in tree pollen values and increases in those of *Corylus* and certain herbs such as *Melampyrum*, *Rumex acetosa*, *Succisa*, *Artemisia*, Rosaceae and Ericales (Fig. 6). A further pollen diagram from a different location at the same site confirms these trends during what must have been a period of well developed coniferous forest cover (Fig. 7).

2. Discussion

No evidence has yet been forthcoming to suggest developments to the north of the North York Moors analogous with those at Star Carr to the south. There are, however, a number of large former lake sites with similar local and regional ecological histories (Jones, 1971). From one such locality, Seamer Carrs (Fig. 1) flint implements resembling those from Star Carr have been reported, but almost identical artefacts are encountered as components of later Mesolithic assemblages (D.A. Spratt, personal communication). The pollen record from the site provides no indication of vegetational changes which could be ascribed with confidence to human causes, though this was also the case at Star Carr at the Mesolithic levels.

Nevertheless, the accumulated palaeobotanical and archaeological data from the three sites described strongly hints at the possibility of human activity in the early Flandrian period. Within Kildale around 10000 B.P. at least one small lake surrounded by *Empetrum* heath and

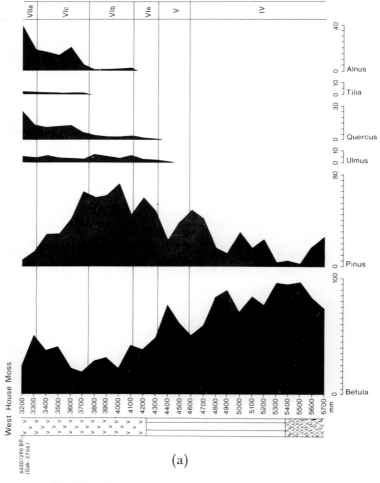

(a)

Figure 4 (a) to (c) West House Moss. Pollen diagram from the early Flandrian sediments.

scrub-woodland with *Betula* existed. Some interference with the vegetation is suggested by the pollen spectra at the levels of the incorporation of the *Bos* skeleton at this site, increased herbaceous cover developing as a result. Charcoal and silt within the sediment at this point indicate the use of fire and the incorporation of inorganic material into the mire stratigraphy, presumably from unstable, unvegetated adjacent slopes. The reasonably warm, dry Boreal climate and the presence of sandy heathland and open woodland would have encouraged the use and spread of fire, the consequences of which must have included the creation and maintenance of numerous rapidly changing, diverse, open plant communities. Such vegetational assemblages would have been ideal habitats for the co-

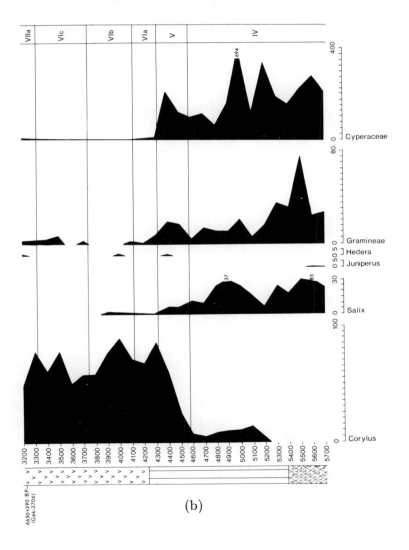

(b)

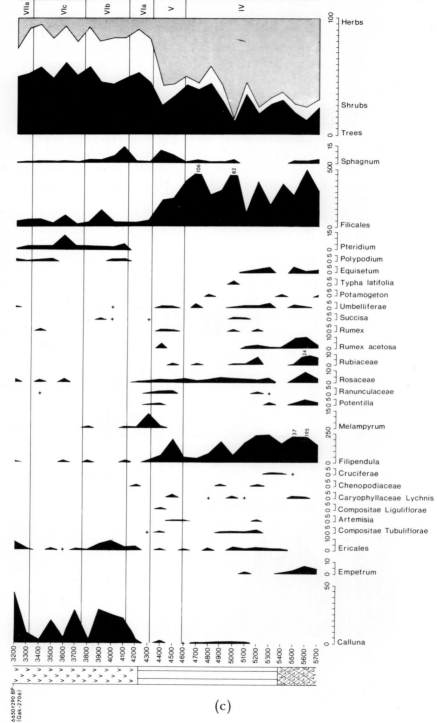

Percentages of Total Pollen

(RLJ 1974)

Percentages of Total Tree Pollen

(c)

West

East

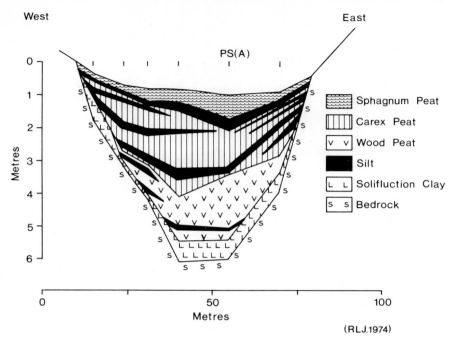

PS(A)

Metres

0
1
2
3
4
5
6

0   50   100
Metres

Sphagnum Peat

Carex Peat

Wood Peat

Silt

Solifluction Clay

Bedrock

(RLJ.1974)

Figure 5. Ewe Crag Slack. Stratigraphy showing the silt layers.

existence of ruminants and man. Wild cattle and red deer are the natural woodland successors to reindeer which prefers a tundra environment and Mesolithic people are known to have favoured dry, sandy terrain with watering places and a balance of open and forested communities where they could hunt and gather food while effecting easy passages over quite wide areas. The population of the Star Carr settlement is estimated to have ranged over at least 300 km² along the easiest routes in their quest for spring and summer sustenance (Piggott, 1965). It is possible therefore that small numbers may have reached Cleveland across the coastal lowlands which were considerably more extensive at this time (Agar, 1954).

West House Moss and Ewe Crag Slack do not provide such positive evidence for the presence of Mesolithic man in Boreal time, though the Kildale activity, only a few kilometres distant, may be reflected in slight, temporary changes in their pollen spectra during Zone V. Rising *Corylus* values are accompanied by expansions in *Melampyrum, Artemisia, Rumex acetosa*, Gramineae and Ericales pollen suggesting the advent of more open conditions. Zone VI shows similar transient trends at both localities, the most important feature within its time-span being the silt layer containing charcoal at Ewe Crag Slack. The corresponding pollen records denote falling tree pollen totals with increasing amounts of *Corylus* and certain herbaceous species. There can be little doubt that the coniferous forest cover was disturbed with

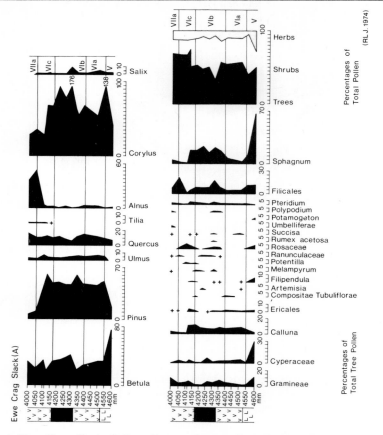

Figure 6. Ewe Crag Slack. Pollen diagram from sediments including the basal silt
   layer.

fire used as an adjuvant. The small cleared areas which were created
as a result yielded sediment deposited as colluvium within the nearby
mire, leaving patches where a mosaic of plant communities including
ruderals, grassy sward and *Corylus* thicket soon developed in
regeneration complexes.

Conclusions

In areas where Mesolithic man is thought to have influenced the
vegetation cover, woodland clearance and the creation of patches of
unstable, quickly changing plant communities has been postulated
(Dimbleby, 1962; Simmons, 1969b). It has also been suggested that
major changes in the frequency of pollen of certain species in pre-
Neolithic times may be ascribed to human rather than climatic or
edaphic causes (Smith, 1970). Alternatively, records of pollen from
open habitat species in diagrams showing mainly a closed forest cover

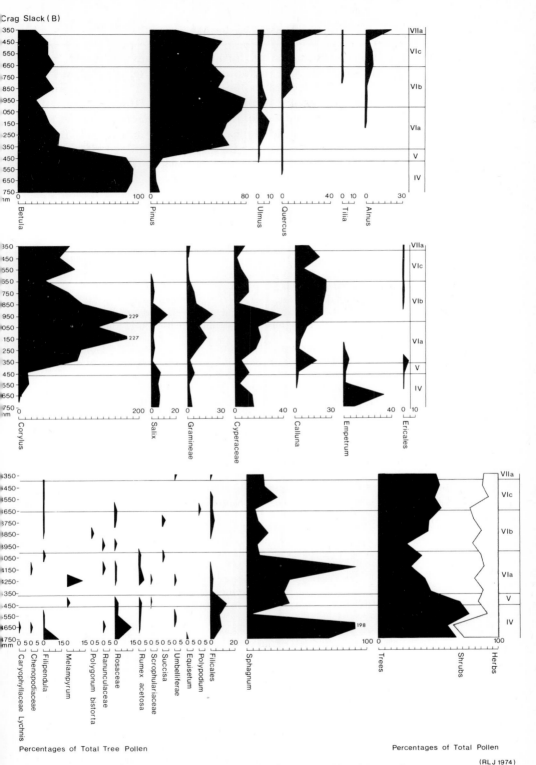

Figure 7. Ewe Crag Slack. Pollen diagram from the early Flandrian sediments.

(RLJ 1974)

could simply be remnants of former montane vegetation which survived until the trees came to dominate the landscape (Tallis, 1964).

The trends of the Boreal pollen curves in Cleveland could be regarded with respect to any of these hypotheses. However, the nature and consistency of the palaeobotanical and archaeological evidence, accompanied by a well founded case for later Mesolithic activity both within the region and elsewhere based upon similar criteria, lends towards an interpretation amounting to positive if minor effects by man upon the early Flandrian landscape.

REFERENCES

Agar, R. (1954) 'Glacial and post-glacial geology of Middlesborough and the Tees estuary', *Proc. Yorks, Geol. Soc.* 29, 237-53.

Cameron, A.G. (1878) 'Notes on some peat deposits at Kildale and West Hartlepool', *Geol. Mag.* 5, 351-2.

Clark, J.G.D. (1971) (ed.) *Excavations at Star Carr*. Cambridge.

Cundill, P.R. (1971) 'Ecological history and the development of peat on the central watershed of the North Yorkshire Moors', unpublished Ph.D. thesis, University of Durham.

Dimbleby, G.W., (1962) *The Development of British Heathlands and Their Soils*. Oxford.

Dimbleby, G.W. (1963) 'Pollen analysis of a Mesolithic site at Addington, Kent', *Grana Palynol.* 4, 140-8.

Godwin, H. (1956) *The History of the British Flora*. Cambridge.

Jones, R.L. (1971) 'A contribution to the late Quaternary ecological history of Cleveland, north-east Yorkshire', unpublished Ph.D. thesis, University of Durham.

Mitchell, G.F., L.F. Penny, F.W. Shotton and R.G. West (1973) 'A correlation of Quaternary deposits in the British Isles', *Geol. Soc. Lond.* Special Report No. 4.

Piggott, S. (1965) *Ancient Europe*. Edinburgh.

Radley, J. (1969) 'The Mesolithic period in north-east Yorkshire', *Yorks. archaeol. J.* 167, 314-27.

Simmons, I.G. (1964) 'Pollen diagrams from Dartmoor', *New Phytol.* 63, 165-80.

Simmons, I.G. (1969a) 'Pollen diagrams from the North York Moors', *New Phytol.* 68, 807-27.

Simmons, I.G. (1969b) 'Evidence for vegetation changes associated with Mesolithic man in Britain', *in* P.J. Ucko and G.W. Dimbleby (eds.), *The Domestication and Exploitation of Plants and Animals*, London, 111-19.

Smith, A.G. (1970) 'The influence of Mesolithic and Neolithic man on British vegetation: a discussion', *in* D. Walker and R.G. West (ed.), *Studies in the Vegetational History of the British Isles*, Cambridge, 81-96.

Tallis, J.H. (1964) 'The pre-peat vegetation of the southern Pennines', *New Phytol.* 63, 363-73.

Walker, D. and H. Godwin (1971) 'Lake stratigraphy, pollen analysis and vegetational history', *in* J.G.D. Clark (ed.), *Excavations at Star Carr*, Cambridge, 25-68.

DISCUSSION

Mr. J. Bintliff enquired whether the use of water bodies caused confusion in distinguishing natural clearances induced by fluctuations in the size of the water body and associated areas from human interference. The speaker thought that Mesolithic man was likely to use these water bodies as focal points for his activities. Professor G.W. Dimbleby asked the speaker if he regarded the Boreal hazel maximum as possibly an anthropogenic effect and the speaker replied in the affirmative.

P.C. Buckland

The use of insect remains in the interpretation of archaeological environments

The Arthropoda represent the most diverse and successful phylum of the animal kingdom. Within it, the Class Insecta has, since the early Devonian over 390 million years ago, colonised every available habitat from the intertidal zone to the mountain peaks; nearly a million species have been described, outnumbering all other animals. Throughout most of their geological record, however, fossils are rare, if locally common, and it is only in the poorly consolidated peats and organic muds of the Pleistocene that insects become the most frequent animal remains. Although there have been considerable numbers of papers on fossil insects, it is only in the last fifteen years, since the study of a mid-Devensian fauna from Upton Warren, Worcestershire (Coope, Shotton and Strachan, 1961), that the importance of insect faunal assemblages for the reconstruction of palaeoenvironments has become fully realised. The research of Coope and his co-workers has centred on the interpretation of the sequence of environmental and climatic changes during the Pleistocene (Coope, Morgan and Osborne, 1971), but Osborne (1969, 1971) has applied the methods to a number of archaeological contexts. As early as 1911 a few identifications of insect remains were made from Roman Caerwent (Lyell, 1911).

The chitinous cuticle of many insects, particularly the Coleoptera (beetles), is quite resistant to decay. The heads, thoraces and elytra of beetles eaten by owls or foxes are still clearly recognisable in the animals' droppings. Whilst fragments will not survive in damp, aerobic conditions, good insect assemblages can be recovered from most anaerobic, particularly permanently waterlogged, deposits. Burning may carbonise any included insects, as in burnt grain deposits, and the dry arid environment of Egypt and similar regions preserves the insect evidence as well as the seed (Zacher, 1937). Occasionally superficial calcification can preserve recognisable remains in an otherwise unfavourable environment, as the example figured from a seventeenth-century drain in York (Fig. 1).

The waterlogging of deposits is, in northern Europe, the most frequent means of preservation. On some sites this is restricted to the lower layers in wells, but in districts close to sea level, prehistoric sites occur which are totally waterlogged. On the east coast, aggradation

Figure 1. Calcified puparia in a seventeenth-century drain, Bishophill, York.

up from a Late Glacial low sea level, probably accompanied by changes in the tidal regime of the North Sea, has resulted in the widespread preservation of organic deposits over large areas, including some major urban centres. At London Bridge the present river level is estimated to be up to 7.0 m above that of the early Roman period and in the Humber Basin there are up to 5.0 m of saturated deposits from the Late Bronze Age to the post-Medieval period, when artificial drainage halted the process. Despite the fact that insect fragments are often clearly visible during excavation, they have been almost totally neglected by a profession which has always tended to be artifactually rather than environmentally orientated.

1. Methods

As in other sedimentological and biological disciplines, it is axiomatic that a meticulous sampling procedure should be adopted. In peat and silt successions which are not markedly differentiated, a sampling unit of 50 mm is usually the minimum feasible; at lesser thicknesses minor sedimentary structures and the reworking effects of any bottom fauna become dominant. In natural deposits disconformities have to be isolated, as pits, ditches and postholes have to be in archaeological strata; the careful recording of sections with as much regard to sedimentology as to archaeological detail must be stressed. In a sequence of rapidly changing environments channelling can result in the removal of a significant faunal unit. The isolation of intrusive features is often particularly difficult on waterlogged sites. Contamination can create problems. The thorough cleaning of tools between samples is essential and samples can only be taken from well sealed, newly exposed horizons. Modern contamination can introduce species which are now becoming cosmopolitan through trade, like *Aridius bifasciatus* from Australasia, into older horizons and unclean tools precipitate arctic individuals into temperate assemblages. It should be apparent why most research workers in this field prefer to take their own samples.

It is not possible to recommend a standard sample size; this varies with the character of the deposit. When sampling has to be done in the absence of the person who is to carry out the research, as much material as possible should be retained in sealed polythene bags. A 10 kg sample of peat from between the timbers of a Late Bronze Age trackway on Thorne Moor, Yorkshire, produced several thousand individuals, whereas a 20 kg block of coarse fluviatile silt of the same age from the former River Don nearby yielded less than a hundred individuals. Peats, provided that they have not been dried out, are usually fairly rich in insect remains, although often difficult to disaggregate; fen peats tend to be richer than acid peats, more as a result of the greater diversity of habitats offered than as a factor of preservation. Organically rich clays and silts vary considerably in quantity of insect remains, depending on the rate and mode of deposition. Wells, unless rapidly backfilled, act as large pit-fall traps and frequently produce an embarrassment of remains; cess pits and deep rubbish pits act in a similar way. Well faunas can be of particular value in the interpretation of both regional and immediate conditions. Osborne (1969) was able to show that the Bronze Age Wilsford Shaft, regarded as ritual by the excavator (Ashbee, 1966), contained an insect fauna commensurate with its use primarily as a source of water for cattle in a landscape of open pasture. Although the archaeological propensity for a token kilogram is to be discouraged, it must be remembered that samples may require many hours of research in the laboratory.

Insect remains can be picked from deposits by careful splitting along bedding planes. This method was employed by previous workers in this field and has the advantage that complete individuals, rather than disarticulated fragments, are recovered. It is, however, tedious and liable to bias the final result towards the larger and more brightly coloured species. Various techniques have been tried and it is apparent that the cheapest efficient method is that outlined by Coope and Osborne (1968).

About 5 kg of sample is placed in a bowl and disaggregated. In some cases this can be accomplished by adding warm water and gently working the material with the fingers, splitting along the bedding. Too heavy handling must be avoided since this damages the insect remains. Silts and clays frequently need steeping overnight in hot water and harder deposits may require the addition of a deflocculant. Gentle boiling in dilute caustic soda can be used for fairly dry sediments, but this must be carried out with care. The sample is then washed over a 300 micron sieve; this is sufficiently fine to retain all recognisable insect fragments as well as plant debris. Any large objects, pottery, stones and wood can be picked out by hand; wood should be closely examined for evidence of insect borings. The residue retained on the sieve is drained of excess water and returned to the bowl. Paraffin (kerosene) is added and gently worked into the material. Surplus paraffin is drained off and warm water added, preferably such that large numbers of air bubbles are included. The paraffin, which has adsorbed onto the insect fragments, causes these to float to the surface. Most seeds and a proportion of any other plant material also rise to the surface but, in most samples, a clear separation of an insect rich flotant from the inorganic sediment and plant debris is obtained. After the sample has stood sufficiently long for this separation to take place, the flotant is tipped off into the sieve and washed, first with hot water and detergent and then with alcohol. The material is stored in alcohol until it can be sorted under a binocular microscope. After the float, part of the material which has sunk must be sorted to check the efficacy of the floats. Particularly with fibrous peats, it is sometimes not possible to obtain a clear

Figure 2. Insect fragments from the filling of a Roman sewer, Church Street, York

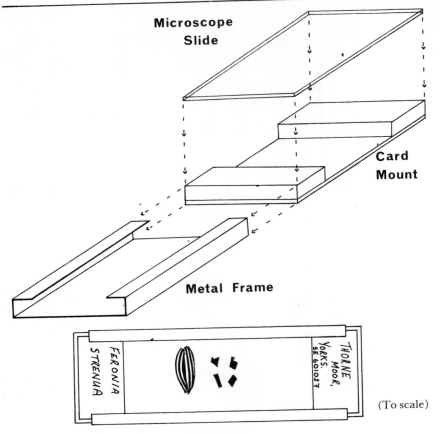

Figure 3. Slide frames for mounting insect remains.

separation and the whole sample has to be sorted. Care must be taken
to recover all fragments since the propensity for the heavy elytra of the
larger beetles to sink can result in a significant bias in the fauna
recovered. The problems of sample bias are a major deterrent to
attempts to adapt any of the seed machines, retailed or 'heath-
robinsoned', so beloved of many palaeobotanists.

After sorting, the insect fragments (Fig. 2) are mounted with a
water-soluble glue on card mounts, which are covered by microscope
slides and sealed by thin metal frames (Fig. 3). Some Coleoptera,
however, cannot be affixed with this medium since they tend to curl
up. This applies particularly to the dung beetles and these, with insect
genitalia, wings and similar fragile parts, have to be mounted in a
more viscous medium which will maintain the shape of the object.
D.M.H.F. (Dimenthylhydrantoin formaldehyde resin) has proved the
most effective, although Euparal is suitable.

There are considerable difficulties in identification and it is seldom
that the characters utilised in the keys for complete individuals survive

as fossil fragments, although the standard reference works, Joy (1932) and the Handbooks for the Identification of British Insects, published by the Royal Entomological Society (1950, *et seq.*), are indispensable. Identification has to be achieved by a process of direct comparison with as complete a collection of the British insect fauna as possible. It has always to be remembered that species which were common before the various public health acts drove certain insects from our homes may now be quite rare and not figure in a modern collection. On a Late Bronze Age site like Thorne Moor nearly 20 per cent of the forest insects can consist of species with a very limited distribution at the present day (Buckland, 1973) and others may no longer be native to this country (Buckland and Kenward, 1973). At least since the Roman period many species have been introduced by way of trade. Where any doubt exists over the determination of an individual a major national collection has to be consulted. Although the majority of fragments recovered tend to be Coleoptera, other groups may be common, including the Hemiptera-Homoptera, the bugs. Wings of some groups of Diptera (two-winged flies), Trichoptera (caddis flies) and some Hymenoptera (bees, ichneumons, etc.) can be identified using the relevant keys, although there are many groups in which wing venation is not diagnostic. Dipterous pupae can be determined but certain groups have yet to be described.

After identification, individuals are listed and counted and an attempt made to reconstruct the palaeoenvironment from the mosaic of modern ecological data, subdivision into specific or overlapping communities being often necessary to interpret the fauna. There are many species, however, for which the available modern habitat information remains inadequate.

2. Interpretation

Rather than attempt to discuss the theoretical implications and potential of insect faunas from archaeological sites, the preliminary results from a number of sites with differing archaeological environments will be discussed. Although it can be demonstrated from the fossil evidence that there has been no major morphological evolution in the insects during the last seventy thousand years, the possibility that physiological changes have occurred, allowing certain species to change their biotope, has to be entertained. Coope (1970) has argued that this has not occurred to a significant extent. Where it has been possible to reconstruct an environmental model from other evidence, the fossil insect fauna has been found to fit within it. During the past five thousand years, however, man has created many wholly artificial biotopes, which have been colonised by insects, and it might be expected that physiological changes have taken place to adapt to these, particularly in the strongly synanthropic species. A careful comparison of natural and artificial habitats suggests that, whilst the two may seem radically different in terms of the overall environment, the niches, conditioned by the microclimate, occupied by the insects, are very similar, if not identical. By way of example, many species associated with man, the fungivorous species and others found in stored products residues and houses, are also found in the similar habitats of birds' nests. Some species, like the grain weevil, *Sitophilus granarius* which are obligate synanthropes in the extreme parts of their range, have not been studied in sufficient detail under natural conditions and the possibility of slight physiological modifications to artificial habitats has to be considered. This, however, does not affect the general archaeological interpretations.

The sites selected to illustrate the use of insect remains in the interpretation of archaeological environments lie in the Vale of York (Fig. 4) and range from the Upper Palaeolithic, where there is no evident human interference in the environment, through the Late Bronze Age, when forest clearance was beginning to have radical effects on native insect fauna, to the artificial environment of the sewer beneath the Roman fortress at York and of Medieval sites in that city.

3.   The Cover Sand of north Lincolnshire

The Cover Sands are a Late Devensian aeolian formation, whose outcrop extends from Northallerton southwards along the west-facing scarps of the Vale of York and Trent Basin as far south as Torksey. Over much of north Lincolnshire the Sands, which vary in thickness from a few grains in the topsoil to over 5 m in the river valleys, are underlain by a thin peat bed. At Messingham, 5 km south of Scunthorpe, a flint end scraper and the astragalus of a large ruminant, probably *Megaloceras giganteus*, the extinct Irish 'elk', were recovered from this peat. The scraper cannot be regarded as diagnostic of any particular industry, but the C14 date of 10280±120 B.P. (Birm. 349) on the peat suggests contemporaneity with the Creswellian. Since Armstrong (1956) mentions a number of probable open sites of this industry along the Lincoln Edge, an examination of the peat deposit would allow a reconstruction of the palaeoenvironment for at least one interval in time, whilst recognising that the Creswellian probably covered a considerable period during a phase of rapidly changing conditions. Although the Late Glacial has been examined in some detail, at Glanllynnau in North Wales (Coope and Brophy, 1972) and at Church Stretton, Shropshire (Osborne, 1972), there had not previously been any direct association with archaeological material.

Samples were recovered from Messingham and from Flixborough, 4 km north of Scunthorpe. The compact, fibrous peat consisted virtually entirely of moss remains and only a single leaf of *Salix*, probably *herbacea*, the dwarf willow, was recognised. The insect faunas, wholly Coleoptera, were fairly rich in individuals but low in diversity (Table 1). The Staphylinids, *Arpendium brachypterum* and *Olophrum fuscum*, were

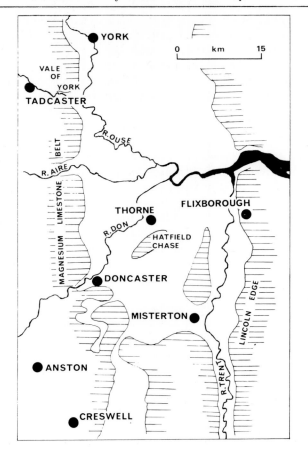

Figure 4. Map showing sites discussed in text.

present in large numbers and most other species were represented by one or two individuals. *O. fuscum* is rather eurytopic, if northern, in its distribution, whilst *A. brachypterum*, which occurs southwards across the North German Plain into the Harz and Saxony, is restricted in Britain to the north; in fossil assemblages it is characteristically dominant in deposits laid down under extremely cold conditions (Coope and Brophy, 1972). Three species occur which are no longer found in Britain and an idea of the climatic regime can be gained by examining their Scandinavian distribution, since the information on this fauna is fairly complete (Hansen *et al.*, 1960). The ground beetle *Diachila arctica* is a strongly arctic stenotherm, restricted to northern Lapland, Finnmark and the Kola Peninsula to the shores of the White Sea, all within the Arctic Circle (Fig. 5) and eastwards in Siberia and North America; it is found in wet moss from the edge of the coniferous forest northwards onto the tundra (Lindroth, 1945). The Hydrophilid 'water beetle' *Helophorus sibiricus* occurs throughout Siberia and westwards to the mountains of Norway as far south as Hedmark (Fig. 5). There is some variation in habitat shown over this wide geographical range; it is more common by the sides of rivers in Scandinavia than in Siberia where it is frequently found by grassy pools (Angus, 1973); it is this latter habitat which accords best with the Late Glacial evidence. The uncertainty over the identification of the non-British *Aphodius* species means that little weight can be put on the distribution of *A. piceus* in interpreting the fauna. It does, however, tend to be montane and northern continental.

The fauna, taken as a whole, gives an impression of a harsh, treeless tundra landscape with the ground covered by a carpet of moss with frequent pools during the summer months. The only higher plants suggested by the beetles are willow, already identified from macro-plant remains, which is the host for species of *Phyllodecta,* and Cyperaceae, probably cotton grass, *Eriophorum* spp., for the weevil *Limnobaris pilistriata.* Pollen is usually poorly preserved in these deposits but this adds wormwood, *Artemisia* spp., and meadow rue, *Thalictrum* spp. to the list. From the north Lincolnshire sites there is no evidence for the presence of birch at this stage, late in pollen zone III, but at Tadcaster, on the other side of the Vale of York, both tree and dwarf birch (*Betula nana*) are represented in the pollen diagram (Bartley, 1962). It is possible that these more developed vegetation communities were restricted to sheltered localities.

Although it is hazardous to argue from a limited fauna, it is of some significance that the whole assemblage can be found today in Scandinavia only in the provinces of Torne Lappmark in the very

Table 1. Beetles from the Cover Sands and their common habitats. Data from Angus (1973), Joy (1932), Lindroth (1945) and Skidmore (pers. comm.)

Taxa	Habitat
Diachila arctica Gyll.	in wet moss.
Feronia sp.	
Agonum fuliginosum Panz.	in wet and marshy places.
Hydroporus sp.	in pools, open water.
Helophorus sibiricus Motsh.	by grassy pools, rivers, etc.
Olophrum fuscum Grav.	in moss.
Arpedium brachypterum (Grav.)	in moss.
Acidota crenata (F.)	in moss, etc.
Stenus spp.	in damp places, moss, etc.
Quedius/Philonthus sp.	
Cymnusa brevicollis (Payk.)	in wet moss, usually *Sphagnum*.
Aleocharinae indet.	
Helodidae indet.	in wet moss.
Byrrhus sp.	at roots of moss.
Aphodius prob. *piceus* Gyll.	in herbivore dung.
Phyllodecta sp.	on willows (Salicaceae).
Limnobaris pilistriata (Steph.)	usually on cotton grass (*Eriophorum* spp.)

* indicates a non-British species.

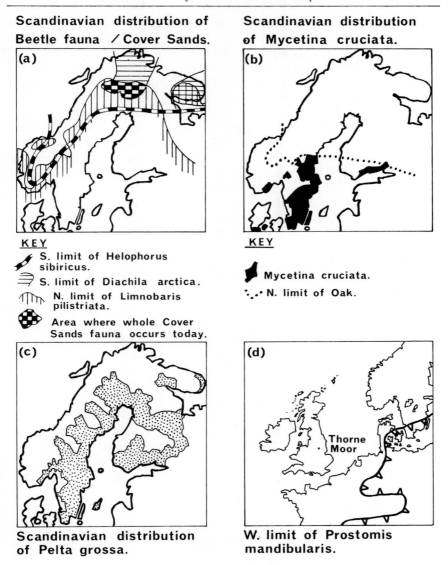

Figure 5. Distribution map of insect species discussed in text.

north of Sweden and Lapponia Kemesis in Finland. This region marks the northern limit of *L. pilistriata*, a limit which appears to relate to climate rather than the distribution of its host plants, and the southern limit of *D. artica*. Climatically both *D. arctica* and *H. sibiricus* imply rather continental conditions with a high degree of contrast between summer and winter temperatures; such a regime could in part be engendered by a low sea level during this period, for which there is some evidence in the Humber Basin (Gaunt *et al.*, 1971). At Karasuando (Lat. 68° 27′) on the border between Sweden and

Finland in this part of Lapland, the average January temperature is
−13°C and the July average is +15°C, only a degree or so below
summer average for England, but representing a winter depression of
the order of 16°C. Under such conditions the ground is permanently
frozen in winter: in north Norrland, Sweden, the daily temperature
remains below freezing for an average of 150 days each year (O'Dell,
1963). Some indication of conditions is also provided by the
geomorphology of the Cover Sands. At all localities examined in the
Scunthorpe region and on the Isle of Axholme, both Sands and
underlying peat bed were found to conform to the broad contours of
the land, sweeping down dip and up scarp slopes with dips sometimes
in excess of 5°. The peat also overlies substrates of varying porosity
with no change and for this to occur ground water must have been
permanently frozen with only superficial thawing in summer, when
any plant growth and insect activity took place in the wet moss and
small pools on the superficially thawed surface. From the existence of
the peat bed and the C14 date, 10280±120 B.P., it is apparent these
conditions only developed towards the end of zone III and evidence for
the phase preceding this tundra environment is at present lacking in
the area. The upper part of the peat has sandy intercalations and
these mark a change to a regime dominated by westerly winds
blowing across the Vale of York, a direction attested by the heavy
mineral suite of the Sands (Loughlin, pers. comm.). These strong,
probably fohn-like winds deposited several metres of sand against the
west facing scarps, in thin laminae, trapped on the perennially wet
surface. Thin organic mud lenses occur within the Sands and these
have produced no faunal evidence, only poorly preserved moss
fragments, perhaps representing very transient and local phases of
non-aeolian deposition. The whole phase of Cover Sand accumulation
was of short duration; a C14 date on a gyttja near the top of the Sands
at Sutton on the Forest, near York, was 9950±180 B.P. (N. 820)
(Matthews, 1970) and a rapid change in conditions with rising
temperature must have marked the end of zone III.

Although the direct comparison of conditions in this area during
the Late Glacial with parts of Lapland gives a good impression of the
climatic regime, it is probable that a better analogy could be obtained
further east in parts of Siberia, where the annual variation in
temperature could be matched in a district at a similar latitude to the
Vale of York; the necessary data, however, are not readily available.

An environmental model can thus be created in considerable detail
in which to fit, in this instance, a solitary artefact. The only other
securely dated Upper Palaeolithic site in the East Midlands is Anston
Cave in south Yorkshire. Here a Creswellian industry of a few flakes
and tools has three dates: 9850±115 B.P., 9940±115 B.P. and
9750±140 B.P. (Mellars, 1969), dates which compare well with
Messingham. Reindeer was identified from this site and a small,
mobile human population following these herds from their summer

feeding grounds on the tundra southwards to the edge of the taiga can be postulated. It is possible that Irish 'elk' showed similar movements but the fossil data is insufficient. No other species can at present be added to the zone III mammal fauna in the area, since Armstrong's Creswell material appears to be thoroughly mixed (Kitching, 1963); at Flixton in the Vale of Pickering horse bones were associated with a shouldered point and the C14 date of 10412±210 (Q.66) overlaps with those quoted above, although the pollen suggests a zone II date (Clark, 1954). In the west, at High Furlong in Lancashire, an elk skeleton, with uniserial bone points, has been recovered from a deposit of zone II (C14 date of 11665±140 B.P. St. 3836, 12200±160 B.P. St. 3832) (Hallam *et al.*, 1973). In the absence of contemporary archaeological sites with good preservation to the south, it is difficult to suggest how far south the northern limit of birch forest in England was. Osborne (Peake and Osborne, 1971) examined a fauna from a locality at Croydon, south of London, with a C14 date of 10130±120 B.P. (Birm. 101), which seemed to have a greater vegetation cover, but this could have been a sheltered locality. Assuming that reindeer in the Late Glacial did migrate, as many modern reindeer and caribou do, then an annual movement in excess of a hundred miles would be quite feasible, implying a human mobility which perhaps accounts for the paucity of Creswellian sites.

4. Thorne Moor

Between the end of the Palaeolithic and the Late Bronze Age, the balance between man and environment changed radically and his effects on the native fauna and flora became progressively more evident. The problems posed by the prolific insect faunas from the Late Bronze Age site on Thorne Moor are considerably different from the relatively simple essay in environmental reconstruction presented by the Cover Sands. Thorne Moor lies roughly 15 km north-east of Doncaster. It is a much modified remnant of a once extensive series of raised bogs along the northern edge of Hatfield Chase, a royal deer park until the early seventeenth century. In 1972 new dykes were cut in an attempt to drain the Moor and in the upcast from one of these a number of large, superficially charred trees were noted, a feature which had been previously observed on the Chase in the late seventeenth century (de la Pryme, 1701). Examination showed that over a length of about 50 m all the large timbers at the base of the peat were burnt; a further area showing similar burning was also located. A number of apparently split parallel timbers in one ditch section proved, on excavation, to be part of a rough timber trackway at the same horizon as the burnt trees. The features were interpreted as a phase of 'landnam', small temporary clearances in the mixed oak forest, fossilised by peat growth; the associated settlement was not located and there were no datable artefacts. C14 determinations on the charcoal from the surface of a burnt tree and on the bark of a small pine incorporated in the trackway gave ages of 3080±90 B.P. (Birm. 336) and 2980±110 B.P. (Birm. 358) respectively, giving a Late Bronze Age date for these activities, which is in agreement with the earliest clearances suggested on palynological grounds (Turner, 1965). A further date of 3260±100 B.P. (Birm. 335) was obtained on a small birch, rooted on the underlying silts.

Samples were collected from a *Sphagnum* rich horizon with many 'cranberry' stems over the trackway, from between the timbers of the trackway, and from the side of a stump in position of growth on the forest soil beneath the peat. The pollen diagram (Fig. 6) includes cereal pollen at the clearance horizon, and it is probable that the track was constructed in an attempt to continue cultivation on the slightly higher areas of the moor in the face of increasingly wet conditions. This activity was rapidly curtailed by flooding, which occurred over a sufficiently short period of time to preserve insects that had lived on the leaves of the trees. Faunas from both the trackway and the tree stump were dominated by species from two distinct communities: many species were associated with small pools and wet decaying vegetation, of fen rather than acid bog aspect; the other assemblage, which overlapped to a certain extent, was characteristic of moribund timber. The widespread flooding seems to have been the result of cumulative changes in the Humber Basin and increased runoff from the uplands, created by forest clearance (Buckland, 1973). The flooding and waterlogging, killing the trees and leaving them standing, provided a massive expansion in available habitat for the insects associated with timber. This fauna of only slightly disturbed natural forest, something which no longer exists in Britain today, is markedly different from that of any modern British woodland. Five species are no longer found in Britain (Buckland and Kenward, 1973) and nearly 20 per cent of the remainder now have a restricted distribution. *Prostomis mandibularis* was the most abundant beetle associated with the timbers of the trackway; it is now extinct in Britain. On the continent it occurs in the extreme south of Sweden, on Zealand, north-east and southern Germany, southwards and eastwards to the U.S.S.R. and North America. Both larvae and adults are found in very rotten wood, often oak, locally in some numbers, but, on the whole, the species is rare, all the north German records being before 1910 (Horion, 1951). It is regarded as an insect of old, undisturbed forests (Horion, 1960). A single specimen of the Endomychid *Mycetina cruciata* was found between the timbers of the trackway. In Sweden its range extends to the northern limit of the oak zone, but it is absent from the west coast of Norway and from Denmark. It is associated with bracket fungus, and is locally common in Germany beneath the bark of both deciduous and coniferous trees. In splitting timber from various localities at the base of the peat on the Moor, a number of complete individuals of the

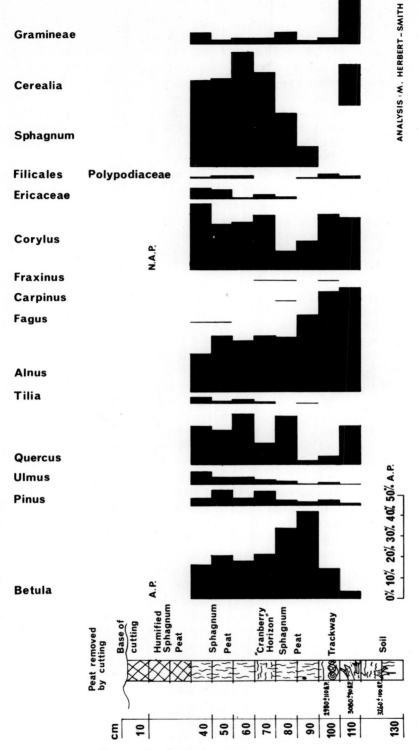

Figure 6. Thorne Moor: pollen diagram.

Eucnemid *Isorhipis* (Tharops) *melasoides* were recovered and its characteristic galleries and larval head capsules were frequently encountered. The species, absent from Scandinavia, is usually found in beech and its present distribution centres on areas of old beech forest in France and Central Europe, avoiding the western seaboard and the Mediterranean (Horion, 1953). *Rysodes sulcatus* is now a rather rare insect, occurring in south east Sweden, southern Germany, Austria and Czechoslovakia, although most records are old. Also known from Anatolia and the Caucasus, it is usually found in rotten wood, where it appears to be a fungal feeder. Whilst these species now have a predominantly southern distribution, *Pelta* (*Zimioma*) *grossa*, a complete adult of which occurred in its pupal cell beneath the bark of a birch tree, is known from much of Sweden and Finland as far north as the southern edge of the tundra (Fig. 5) and eastwards to the U.S.S.R. It is usually found in birch or pine, frequently in old woodland which has been ravaged by fire (Horion, 1960). It is interesting to note that 'slash and burn' clearance for agriculture was practised until recently in the northern and montane forests of Scandinavia, providing abundant habitats for any species which was in any way attracted to burnt ground. One other species from Thorne, still found in Britain, shows a parellel northward trend. The bark beetle *Scolytus ratzburgi* occurs in birch and is now restricted to Scotland, principally to the east. Like *P. grossa*, in Scandinavia, it is recorded as far north as Finnmark but avoids the west coast of Norway. Osborne (1972) has found a further species of nothern and montane forests: *Agabus wasastjernae* from deposits of pollen zone VII(b) age at Church Stretton in Shropshire. It is not tied to a particular species of tree, but seems to require the acid pools and litter of coniferous forest. Most of the Thorne Moor fauna, however, tends to show a south and eastward retraction. *Dryophthorus corticalis* is today only recorded in Britain from Windsor Forest (Donisthorpe, 1938), usually in thoroughly rotted oak, although it is recorded on the continent from other woods (Kopponen and Nuorteva, 1973), frequently with ants (*Lasius* spp.). It must have been fairly widespread in the Bronze Age since it has been reported from Church Stretton (Osborne, 1972), Alcester and Shustoke in Warwickshire (Osborne, 1965), as well as Thorne. The small Histerid beetle *Microlomalus parallelepipedus* is regarded as one of our rarest beetles (Allen, 1971) and, besides old records in the New Forest, there is only one recent (1952) find, from beneath bark near Canterbury. Many other species show a similar retraction, including *Trinodes hirtus, Colydium elongatum, Hypulus quercinus, Gastrallus immarginatus* and *Leptura scutellata*. A relict population of the large Alleculid *Prionychus fairmairei*, which is western in its continental distribution, occurring in France but not Germany, survives in Sherwood Forest, a few kilometres south-west of the Thorne Moor site, and the Colydiid *Teredus cylindricus* also reaches its northern limit there. These species, with those no longer found in

Britain, are restricted to old forest area; the German term *Urwaldtieren* – old forest animals – (Palm, 1960) is most appropriate.

A superficial examination of both British and continental distributions of these species might suggest a direct correlation with a deteriorating climate, perhaps the change at the sub-Boreal sub-Atlantic transition (pollen zones VII(b) and VIII) to colder conditions. A simple (perhaps worldwide) drop in temperature, however, does not seem to be the answer. As has been outlined some species show a withdrawal northwards whilst the majority seem to have gone south. A transition from continental to more oceanic atmospheric circulation has been suggested on both botanical grounds and the evidence for glacial advance in the sub-Atlantic (Lamb *et al.*, 1966). One species in the Thorne assemblage seems to contradict this; *Platypus cylindrus* still occurs in fairly wet, oceanic areas in the west, around Merioneth (Skidmore and Johnson, 1969), but is absent from the more continental east and in Scandinavia is restricted to the very southern tip of Sweden; it is apparent that other factors are operating. Kenward (pers. comm.) has suggested that the warmer winters experienced on the western seaboard could result in winter activity with consequent consumption of food reserves. This is unlikely for wood-boring species, which, at least whilst in the wood, are largely insulated against such problems. Early emergence from the pupal cell to be killed by a subsequent cold spell may occur and Skidmore (pers. comm.) considers that a temperature threshold reached more readily in a continental regime with more markedly contrasting seasons, may be involved. Schimitschek (1948) found that the temperature of 20°C was necessary for swarming of the bark beetle *Ips typographus* to occur. In the absence of detailed physiological studies on the insects involved, it is difficult to form concrete ideas about any climatic effects at this juncture. If only the macroclimate is to be considered, it remains possible that certain species survived until the 'Little Ice Age', historically attested (Ladurie, 1972) in the early post-medieval period.

It remains to discuss some of the factors other than climate which must be involved. That anthropogenic factors could have had major effects on the forest insect fauna was considered by Osborne (1965) and it is apparent from the Thorne and subsequent evidence that these effects have perhaps been surprisingly drastic in this country. The list of localities where the rare insects remain in England is significant. The majority occur in either or all of Windsor Forest, the New Forest and Sherwood Forest, areas which at once betray the hand of the Norman aristocracy; and, although it should be remembered that the word forest does not necessarily refer to trees, their action in creating large areas preserved from pastoral and arable exploitation provided the requisite habitats for insects which might otherwise have disappeared from Britain. The historical geographers' concept of early medieval England – 'If one walked half a mile, a mile at the most, out from the village, one came to the edge of the wild, to . . . the massed tree trunks of the primaeval woods still waiting the axe' (Hoskins, 1955) – is undergoing considerable modification, although it is still often repeated (cf. Bennett, 1969; and the reprint of Darby, 1936, 1969). Aerial photography in the north-east and east Midlands (cf. Riley, 1973) is showing that virtually all the 'primaeval forest' had been cleared by the end of the Roman period. The areas that remained and areas of natural regeneration, although probably always considerable, were subject to grazing pressures and various demands for timber by the population. This secondary woodland was that which was 'economically necessary to the continuance of the community' (Pickering, pers. comm.), a precariously balanced ecosystem which was finally upset by an industrialised society.

Clearance, even if selective or by coppicing, modified the microclimate within closed forest, which was cushioned against both diurnal and annual variations (Geiger, 1965). Many species which are dependent upon mature forest have a low dispersal potential, being either flightless or very poor fliers, and as clearance proceeded populations became smaller and more isolated, more vulnerable to habitat disturbance by any anthropogenic factors or slight fluctuations in climatic parameters. For some, survival was only possible in those areas which were set aside as hunting reserves for the nobility in the eleventh century, protected from the destructive pannage of semi-feral pigs, effective clearers of the forest floor, and from coppicing, charcoal burning and other activities in timber collection. It is perhaps due to that much maligned conservationist, William the Conqueror, that British insect fauna remains as diverse as it is, a diversity which is being curtailed by urban expansion and the over-tidy, conifer-oriented practices of the Forestry Commission.

Archaeologists who are prepared to accept that their findings can be fitted into a palaeoenvironmental setting frequently resort to a token pollen diagram and relate their local results to the more regional picture presented by the pollen rain. The many pitfalls which may be consequent upon this can be examined in the light of the Thorne Moor insect faunas, in comparison with the pollen diagram (Fig. 6). It should be emphasised that similar traps are available for those who use the insect data in isolation from the botanical information; integrated studies in palaeoecology are time-consuming but more effective. Considerable disparities between pollen diagrams and insect data have been noted previously, particularly in periods of rapid climatic change. In the Late Glacial non-phytophagous beetles reacted with alacrity to the improved thermal regime during pollen zone I and a temperate faunal assemblage occurs in the absence of most of the temperate plants, which were immigrating much more slowly (Coope and Brophy, 1972). Only during pollen zone II, when the temperatures, as implied by the insect fauna, were again declining, did birch forest establish itself and the faunal and floral elements become in phase. At Thorne phytophagous species occur which require trees not represented in the diagram and faunal changes record the evolution of the local environment from mixed oak forest through fen to acid raised bog. These changes are apparent in an 'absolute' diagram, but are more difficult to observe in the percentage arboreal diagram normally employed. In the diagram single grain occurrences of non-arboreal pollen have been omitted and there is slight rearrangement to bring out the salient features of clearance and expansion of bog. The fauna, however, cannot be compared directly in the same quantitive way, since the samples from the side of an oak stump will clearly largely reflect the fauna of that tree and the trackway fauna represents a human selection of available timber. Many species, moreover, may be found in more than one type of tree. Both at the trackway horizon and below, beetles associated with oak dominated the assemblage, with the bark beetle *Dryocoetes villosus* occurring in large numbers; *Rhynchaenus quercus*, a weevil which feeds on the foliage of oak, is also present. The clearance and regeneration, so evident in the pollen diagram, with high peaks in cereals and grasses, followed by an expansion of birch and the appearance of ash (*Fraxinus*), a light-demanding tree, cannot be traced by changes in the insect fauna – the only species loosely associated with burnt ground, *Pelta grossa*, appears at a higher level. The development of acid bog with some drier areas of Ericetum, shown in the diagram by the expansion of *Sphagnum* and the Ericales, with a reappearance of Filicales, is very well defined by the insect faunas. From several hundred individuals associated with timber in its various forms at the trackway horizon, this component declines to a Scolytid, which is probably adventitious. Over the whole bog, an area of several square kilometres, few trees were observed at this level. The water beetles change to a more acid group of species, the Donaciine *Plateumaris sericea* is replaced by *P. discolor*, more characteristic of *Sphagnum* bog; heath and bog ground beetles, like *Bradycellus ruficollis*, replace woodland and fen species, *Agonum obscurum* and *A. livens*. These changes tend to be masked in a percentage pollen diagram. The overall drop in the amount of tree pollen is not evident and this pollen must now be derived from further afield, perhaps from as far away as the Magnesian Limestone dip-slope, 15 km to the west (Turner, 1965), where lime (*Tilia*), which reappears at this level, today forms part of the semi-natural woodland. In the Late Neolithic, at both Thorne Waterside (4230±120 B.P. (Birm. 359)) and Misterton Carr (4330±100 B.P. (Birm. 328)). Lime, as pollen evidence from Hatfield Moors suggests (Smith, 1958), was an important constituent of the forest cover. The small bark beetle *Ernoporus caucasicus*, now known from only one British locality is common in samples from both sites but absent from the Thorne Moor material. Conversely beech (*Fagus*), which does not appear in the pollen diagram until above the trackway horizon, seems to be indicated earlier by *Isorhipis melasoides*, *Eucnemis capucinus* and *Colydium elongatum*, which are usually found in this tree.

Although none are exclusively beech feeders, they suggest, along with other species, that this tree was part of the forest. Plants which are insect pollinated are also poorly represented in the pollen rain. The willows (Salicaceae) do not occur in the diagram, but are abundantly indicated by such insects as the saw-fly *Xiphydria prolongata* and the Chrysomelid beetles of the genus *Phyllodecta*.

A more detailed, if more localised environmental picture can be constructed from the insect evidence, in contrast with the fluctuating regional picture presented by the pollen; both lines of approach should be integrated with the evidence of plant macrofossils, since major elements in the flora may not be identified in any other way.

5.   *York: the Roman sewer*

The sewerage disposal tunnels beneath the Roman fortress of Eburacum offer an example of a wholly artificial palaeoenvironment. The autochthonous fauna was fairly limited and most of the interest centres on the species washed into the system during its final phase of use. Apparently built during the second century, the system consists, in the excavated part, of a main channel, in part realigned at a later date, with a number of side channels of varying character. Built of large Millstone Grit blocks (Whitwell *et al.*, 1974), the sewer was not completely sealed against the ground fauna and there had been some re-sorting of the sediment, which varied from a very fine gyttja-like deposit to a coarse gravel with large pieces of limestone and mortar. The soil fauna, however, was limited and careful collecting shortly after discovery produced only three species of beetle, the ground beetle *Trechoblemus micros*, an Aleocharine and a Ptiliid. Further collecting after the system had stood open some time added a species of *Cercyon* and the rove beetle *Quedius mesomelinus*. Since examples of all these were found in the samples of sediment, some reservations had to be made about the interpretation of the fauna, whether the fauna was contemporary with the last period of usage of the system or whether it represented subsequent and modern intrusions, particularly since the roof lintels had been hacked through in a number of places in the nineteenth century. Fortunately the greater part of the contemporary sediment, containing much Roman pottery, and up to 75 cm thick, was well sealed by a humate-rich mud on the surface, the result of limited worm action. Most fragments, moreover, were of species belonging to wholly different biotopes washed into the system.

Since the structures above the sewer could not be excavated, a study of the insect fauna and other biological materials promised some clues as to the functions of the buildings during the final period. This part of the system seems to have been allowed to silt up in the late fourth or early fifth century, although this was a gradual process and there is some suggestion that some parts had gone out of use earlier, perhaps by the late third century. The primary problem to be answered was whether the sewers had been constructed principally as a freshwater overflow from the bath-house, which was known to be nearby (R.C.H.M., 1962), and as a storm drain, or whether they had also carried foul water with human sewerage. No insect is specifically associated with human excreta, but in the course of working through the 20 kg samples, many spicules of sponges were encountered. These must have been imported into York from the Mediterranean. There are classical references to the use of sponges as toilet paper; Martial, commenting on the fate of a consumed dish of mushrooms, has, '... *nihil, quod sciat infelix damnatae spongea virgae*' (Epigrams 12, 48, 7). The putrid nature of the sediment in the sewer is indicated by part of the insect fauna. In one side channel, partly blocked by an excess of silt in the main channel, a small, muddy 'backswap' deposit had formed and this contained many thousand pupal fragments of Psychodid flies. These are sufficiently numerous on modern sewerage farms to be termed 'sewerage flies' and breed in liquid organic material. Their maggots, along with those of a few Phorid and Borborid flies, were subject to the predation of two species of large rove beetle, *Coprophilus striatulus* and *Quedius mesomelinus*, and the Rhizophagic, *Rhizophagus parallelocollis*. All are found in foetid, rotting organic material and the latter has previously been recorded from Thorne Moor and from a mediaeval burial in London (Stafford, 1971). Both *Q. mesomelinus* and *R. parallelocollis* were found in large numbers in the tomb of Archbishop Greenfield (ob. 1316) in the Minister.

The remaining part of the insect fauna represents material swept into the system and it proved possible by detailed sampling to localise entry points for the various components in not only the insect fauna, but the whole vertebrate and invertebrate assemblage, as well as the plants. Freshwater snails occurred in two channels entering the main channel from the side where the legionary bath-house was known to lie; a few diatoms also were recovered from one of these. These outfalls were interpreted as the waste water channels from the baths, which, as is known from similar suites on other sites, probably also served to flush out the toilets. Only one water beetle was found, somewhat unusual in view of the amount of water concerned. Any standing body of water, or even a shiny, dark motor vehicle will attract them. The explanation for this near absence must be that both baths and water supply, probably an aqueduct, were completely roofed over. The sources of the fortress water supply is not known but transport over some kilometres in concrete-cased pipes, as at Lincoln, can be postulated.

Most channels were short, ending with a small circular shaft down through the roof. Initially these were thought of as outlets from toilets, but the included insect faunas suggested that they acted as ordinary drains. Beetles associated with stored products were common and these occurred in samples from beneath the chutes. Two of these, the grain weevil, *Sitophilus granarius*, and the saw-toothed grain beetle,

Oryzaephilus surinamensis, are our commonest pests of grain stores at the present day and have previously been recorded from a Roman pit at Alcester, Warwickshire (Osborne, 1971), from burnt grain at Droitwich, Worcestershire (Osborne, 1974), and from the so-called destruction deposit at Malton, Yorkshire (Buckland, in prep.). *O. surinamensis* was also found in the well on the Roman villa site at Barnsley Park Gloucestershire (Coope and Osborne, 1969). Although the latter is occasionally recorded from fungus in the south of England (Joy, 1932) and both species are capable of overwintering in unheated stores (Solomon and Adamson, 1955), there can be little doubt that these were introduced to this country by man. The earliest occurrence so far recorded in Britain is from the mid to late second century pit at Alcester (*op. cit.*). Insects associated with stored products occurred in the tomb of Tutenkhamen in Egypt (Zacher, 1937), dating back to the mid fourteenth century B.C., and at Herculaneum in A.D. 79 (Dal Monte, 1956). It is possible that their introduction to Britain may date back to the initial seed corn of the Neolithic colonists, although this group of pests must have been replenished many times by trade, which has now made these species thoroughly cosmopolitan. *Laemophloeus* sp., another pest of stored products, also occurred in the sewer; this had previously been recorded from Roman deposits at Fishbourne (Osborne, 1971b) and Droitwich (*op. cit.*), but, since some species have also been found under bark (Donisthorpe, 1938), these may represent a movement into an artificial biotope by one of the native fauna. The presence of these beetles in some numbers, both in the fourth century and earlier parts of the fill, shows that grain was being stored in one of the buildings drained by the sewer; from what is known of the plan of the fortress, it does not appear that the main granaries were nearby.

From two side channels in the system·large numbers of the ant *•Ponera punctatissima* were recovered. This is a doubtfully native species, accepted by Donisthorpe (1927) and Collingwood (1964) on the strength of two records away from human habitation in Kent and one from South Wales. An additional outdoor record now exists for Hatfield Lings, near Doncaster (Burns, pers. comm.). In Europe the species tends to be southern in its distribution (Schmiedeknecht, 1930) and in the north it is almost wholly synanthropic, having become widely distributed by trade. The York ants' nests may be therefore a result of a local native population moving into the building or, perhaps more probably, have been introduced from continental Europe by Roman trading activities. It is improbable that the ants were nesting in the system, indeed their distribution in samples precludes it, but the bath-house would have provided an ideal habitat. In Kent, a nest has been recorded in decaying timber down a coal mine at Betteshanger, 'with condensation water much in evidence . . . the temperature taken was 25.4°C' (Yarrow, 1967) – very similar conditions to those in an artificially heated Roman bath-house. The woodworm, *Anobium punctatum*, in some samples shows that some timber was present.

Some idea of the dates of introduction of many species to this country has yet to be gained. Whilst some, like *Aridius bifasciatus*, first recorded in 1951 (Allen, 1951) are well documented, others, frequently regarded as recent introductions, may turn out to be native or ancient immigrants. The small, eyeless, flightless Colydiid *Aglenus brunneus* was found in the Roman sewer and on the Anglo-Danish site beneath Lloyds bank; it had been thought of as a fairly recent introduction from North America (Peyerimhoff, 1945). Occurring in birds' nests in southern Europe, it is completely synanthropic in the north (Horion, 1961) and is found feeding on fungi in damp grain bin residues, tanning pits, chicken coops, stables, etc., (Hinton, 1945). This species is able to survive in the artificial habitats created by man well to the north of its natural range, having been introduced to both Greenland and Iceland (Larsson, 1959). Whilst the native European status of this species has been proven from the archaeological record, others still create problems. *Aridius nodifer*, as the rest of the genus, is regarded as an import from Australasia. A specimen was, however, described from the Cromerian clays of the Durham coast (Trechmann, 1920) by Lesne (Bell, 1922), who, like most of his contemporaries, had a propensity for ascribing new species, even genera, to fossil material (Coope, 1968). If this insect did retreat southwards and eastwards during the course of the Pleistocene, it provides an interesting parallel to the withdrawal of many plants in the same direction since the Tertiary (Pennington, 1969). A retraction to eastern Siberia has been demonstrated by Angus (1973) for two species of *Helophorus* from deposits of the Last Glaciation in Starunia, perhaps a temporary retreat until the next glaciation. In view of the extreme mobility of even flightless species during the Pleistocene, it is difficult to envisage a series of geographical accidents within the sequence of admittedly catastrophic temperature changes, which would leave a species with an Australasian distribution, to be re-imported by man to a wide range of suitable habitats in England during the early years of the nineteenth century. On a lesser scale perhaps the restriction of *Oxytelus gibbulus* to a small area in the Caucasus since the Last Glaciation might be thought of as similar (Coope, 1970). In the absence of Lesne's original specimen, it would be unwise to place too much weight on this discussion, but a single elytron of *A. nodifer* from the Late Bronze Age site on Thorne Moor can only be interpreted when more fossil evidence is available. This specimen, at present regarded as a contaminant, serves to stress the need for care in taking samples, storage and washing out.

6. *York: Anglo-Danish and medieval sites*

A preliminary report has already been published on the environmental interpretation of the Anglo-Danish site beneath Lloyds

Bank (Buckland *et al.*, 1974). Rather than reiterate the conclusions, it is perhaps of more value to examine some of the problems associated with the interpretation of complex sites. In assemblages where all taxa reflect in some way the effects of man, the primary problem is to separate off the autochthonous, that part of the fauna which was native to the site, from the allochthonous communities, the adventitious species and those brought in by man with some commodity. This latter point has already been illustrated by the finds from the Roman sewer, and Osborne's (1971) record of a Mediterranean longhorn beetle, *Hesperophanes fasciculatus*, from the Alcester pit expands this group to imported timber. In the simplest of cases, all the taxa present may form a biocoenosis, a life assemblage. This can be seen from the limited fauna recovered from the lead-lined stone coffin of Archbishop Greenfield from within the north transept of the Minster. Here, the whole community survived: coffin flies, which had fed on the decaying corpse, with their predators, *Quedius mesomelinus* and *Rhizophagus parallelocollis*, and a few fungal feeders. The closely sealed nature of this deposit, however, suggests that the initial infestation, probably the laying of eggs, took place before he was buried, probably when he was lying in state. On most sites such a complete and closed system does not survive and one is faced with interlocking and overlapping fragments from several different biotopes. On Anglo-Danish and medieval sites sampled in York, the overwhelming impression is one of squalor. Rush and reed flooring materials were rarely changed and well developed communities of insects associated with compost and organic rubbish lived in, on and around the floors. In part, however, this impression may be fostered by sample bias since *in situ* preservation is only obtained on sites near the water table and this, of itself, may dictate the character of the occupation. In this light, a comparison of the faunal results from the Lloyds Bank site in York with those from a group of pits in Saxon Southampton provides some interesting contrasts. On the latter site, pits, apparently close to a structure, appeared during excavation to contain dung. Examination of samples (Holdsworth and Buckland, in prep.) supported this interpretation and, by its texture, seed evidence, a considerable assemblage of dung and straw litter insects, with the Histerid beetle *Onthophilus striatus*, further suggested that the dung was from horses. The shape of the deposit in the pits implied that large single bucketfulls of dung and straw from mucking out a stable were being brought out and systematically buried. How frequently this was carried out cannot be gauged, but it contrasts markedly with the York tannery and leatherworkers' tenement, where all the filth from the processing was allowed to accumulate on the floor. It is probable, however, that the Southampton household with its horses, belonged to a higher echelon of society and a direct comparison, if informative from the point of view of faunal interpretation, is not justified. Burial of dung seems to further imply a lack of agricultural activity in this part of the Saxon town.

Considerable overlap occurs between the insect communities from each archaeological site. At Lloyds Bank the autochthonous rubbish community was in part inseparable from species brought in with the flood debris to spread on the floor. It was possible to show on this site that what appeared on the archaeological evidence to be a leather workers' shop was also a tannery; hair scraped from the skins survived in quantity, as well as plants and wood ash used in the processing. Some of the insects were associated with the waste products from leather preparation and tanning. The many thousands of puparia of the house fly, *Musca domestica*, and the biting stable fly, *Stomoxys calcitrans*, derive from maggots nurtured on the flesh and fat scraped from the skins and thrown into a corner. *Aglenus brunneus* was found in some numbers, a pest in tanneries, and the Scarabaeid *Trox scaber* probably lived on the animal waste, since it has been taken in traps baited with cat meat (Kenward, pers. comm.). Both, however, occur in other habitats and the presence of feathers and eggshells, probably from chickens, might be taken to suggest that they formed part of the allochthonous fauna. *A. brunneus* has been taken in the litter in chicken coups and *T. scaber* caught in large numbers with feather-baited traps (Spector, 1943). It is fortunate in this case that this need not modify the model constructed on other evidence for there is good ethnographic evidence for the use of chicken dung in stripping skins. Other introduced communities were more sharply demarcated; in one corner of a room on this site were a few sprigs of heather and a considerable heathland insect community, suggesting that this plant was brought in in some quantity, perhaps to serve as bedding as it did until recently in parts of Ireland (Evans, 1957), although the many biting ants would be poor bedfellows. On all sites beetles associated with the timberwork were common, the ubiquitous woodworm being supplemented by deathwatch, *Xestobium rutovillosum*, the powder post beetle, *Lyctus brunneus*, in major structural timbers and two longhorns, *Saperda populnea* and *Gracilia minuta*, in the wattling. The latter, often found in basketwork, is now wholly synanthropic, regarded as an introduction in southern Scandinavia (Hansen *et al.*, 1960). It perhaps lived outdoors in willows until the end of the 'little climatic optimum' in the fourteenth century (Ladurie, 1972), but has only maintained a tenuous existence in England indoors since. There is a little other insect evidence for this climatic change, although it remains difficult to isolate it from anthropogenic factors and insufficient material has yet been studied to form any conclusions.

Acknowledgments

Aspects of the use of insects in the interpretation of archaeological sites have been outlined. Such studies cannot be attempted in the absence of co-operation from archaeologists and other scientists.

Amongst the former, the writer is particularly indebted for their interest to P.V. Addyman and J.B. Whitwell of the York Archaeological Trust, D. Philips of the Minster Archaeological Group, M.J. Dolby of Doncaster Museum, and P. Holdsworth of the Southampton Archaeological Research Unit. Much assistance has been provided by P.J. Osborne, who has pioneered these methods of research in archaeology, and by P. Skidmore, of Doncaster Museum, who also identified the Diptera. The help of Prof. F.W. Shotton, R.B. Angus, G.R. Coope, J.R.A. Greig, H.K. Kenward and C. O'Toole is gratefully acknowledged. Classical references to sponges were kindly provided by J.P. Wild and M. Herbert-Smith provided the Thorne pollen analysis. D. Jefferies of the Trust provided the photographs and D. Bentley of Doncaster Museum prepared the figures. The author is grateful also to the Doncaster Corporation Amenities Committee and Mr. J. Barwick, The Museum's Curator, for allowing these perhaps less than traditional methods of archaeological research to be pursued. Last, but not least, the typing ability of his wife is much appreciated.

REFERENCES

Allen, A.A. (1951) '*Lathridius bifasciatus* Reitter (Col. Lathridiidae), An Australian beetle found wild in Britain', *Entomologist's mon. Mag.* 87, 114-15.

Allen, A.A. (1971) '*Microlomalus parallelepipedus* Hbst. (Col. Histeridae) in Kent', *Entomologist's mon. Mag.* 107, 80.

Angus, R.B. (1973) 'Pleistocene *Helophorus* (Coleoptera, Hydrophilidae) from Borislav and Starunia in the Western Ukraine, with a reinterpretation of M. Lomnicki's species, description of a new Siberian species and comparison with British Weichselian faunas', *Phil. Trans. R. Soc.* B, 265, 299-326.

Armstrong, A.L. (1956) 'Palaeolithic, Neolithic and Bronze Ages', *in* D.L. Linton (ed.), *Sheffield and its Region*, Sheffield, 90-110.

Ashbee, P. (1966) 'The dating of the Wilsford Shaft', *Antiquity* 40, 227-8.

Bartley, D.D. (1962) 'The stratigraphy and pollen analysis of lake deposits near Tadcaster, Yorkshire', *New Phytol.* 61, 277-87.

Bell, A. (1922) 'On the Pleistocene and Later Tertiary British insects', *Ann. Rep. Yorksh. Phil. Soc.*, 41-51.

Bennett, H.S. (1969) *Life on the English Manor*, Cambridge.

Buckland, P.C. (1973) 'Archaeology and environment in the Vale of York', *S. Yorksh. Studies in Archaeology and Natural History* 1, 6-16.

Buckland, P.C., J.R.A. Greig and H.K. Kenward (1974) 'An Anglo-Danish site in York', *Antiquity* 48, 25-33.

Buckland, P.C. and H.K. Kenward (1973) 'Thorne Moor: a palaeoecological study of a Bronze Age site', *Nature, Lond.* 241, 405-6.

Clark, J.G.D. (1954) *Star Carr*. Cambridge.

Collingwood, C.A. (1964) 'The identification and distribution of British ants', *Trans. Soc. Brit. Ent.* 16, 93-121.

Coope, G.R. (1968) 'Coleoptera from the 'Arctic Bed' at Barnwell station, Cambridge', *Geol. Mag.* 105, 482-6.

Coope, G.R. (1970) 'Interpretations of Quarternary insect fossils', *Ann. Rev. Ent.* 15, 97-120.

Coope, G.R. and J.A. Brophy (1972) 'Later Glacial environmental changes indicated by a coleopteran succession from north Wales', *Boreas*, 1, 98-142.

Coope, G.R., A. Morgan and P.J. Osborne (1971) 'Fossil coleoptera as indicators of climatic fluctuations during the Last Glaciation in Britain', *Palaeogeography, Palaeoclimatol., Palaeoecol.* 10, 87-101.

Coope, G.R. and P.J. Osborne, (1968) 'Report on the coleopterous fauna of the Roman Well at Barnsley Park, Gloucestershire', *Trans. Bris. and Gloucs. Arch. Soc.* 86, 84-7.

Coope, G.R., F.W. Shotton, and I. Strachan (1961) 'A late Pleistocene flora and fauna from Upton Warren, Warwickshire', *Phil. Trans. R. Soc.* B, 244, 379-420.

Dal Monte, G. (1956) 'La presenza di insetti dei granai in frumento negli scari di Ercolaneo', *Redia* 41, 23-28.

Darby, H.C. (1936) ed. *Historical Geography of England before 1800.* Cambridge.

Dinsthorpe, H.St.J.K. (1927) *British Ants.* London.

Dinsthorpe, H.St.J.K. (1939) *A Preliminary List of the Coleoptera of Windsor Forest.* London.

Evans, E.E. (1957) *Irish Folk Ways.* London.

Gaunt, G.D., R.A. Jarvis and B. Matthews (1971) 'The late Weichselian sequence in the Vale of York', *Proc. Yorks. Geol. Soc.* 38, 281-4.

Geiger, R. (1965) *The Climate near the Ground.* Harvard.

Hallam, J.S., J.N. Edwards, B. Barnes, and A.J. Stuart (1973) 'A Late Glacial elk with associated barbed points from High Furlong, Lancashire', *Proc. prehist. Soc.* 39, 100-27.

Hansen, V., E. Klefbeck, P. Sjoberg, G. Stenius and A. Strand, revised by C.H. Lindroth (1960) *Catalogus Coleoptorum Fenno-Scandiae et Daniae.* Lund.

Hinton, H.E. (1945) *A Monograph of the Beetles Associated with Stored Products*, I. London.

Horion, A. (1951) *Verzeichnis der Käfer Mitteleuropas.* Stuttgart.

Horion, A. (1953) *Faunistik der Mitteleuropäischen Käfer*, III, *Malacodermata.* München.

Horion. A. (1960) *op. cit.* VII, *Clavicornia* I, Uberlingen-Bodensee.

Horion, A. (1961) *op. cit.* VIII, *Clavicornia* II, Uberlingen-Bodensee.

Hoskins, W.G. (1955) *The Making of the English Landscape.* London.

Joy, N.H. (1932) *A Practical Handbook of British Beetles.* Edinburgh.

Kitching, J.W. (1963) *Bone, Tooth and Horn Tools of Palaeolithic Man.* Manchester.

Koppenen, M. and M. Nuorteva (1973) 'Uber subfossile Waldinsekten aus dem Moor Piilonsuo in Südfinnland', *Acta ent. fenn.* 29, 1-84.

Ladurie, E. Le Roy (1972) *Times of Feast, Times of Famine.* London.

Lamb, H.H., R.P.W. Lewis and A. Woodroffe (1966) 'Atmospheric circulation and the main climatic variables between 8000 and 0 B.C.: meteorological evidence', *Proc. Int. Symp. on World Climate*, 174-217.

Larsson, S.G. and M. Gigja (1959) 'Coleoptera', *Zoology of Iceland* III, 46.

Lindroth, C.H. (1945) 'Die Fennoskandischen Carabidae, Eine Tiergeographisce Studie', *I.K. Vetensk.O. VitterSamh. Handl.* 6, Ser B, 4.

Lyell, A.H. (1911) 'Appendix on the insect remains. Ashby, T., A.E. Hudd and F. King. Excavations at Caerwent, Monmouthshire on the Site of the Romano-British City of Venta Silurum, in the years 1909 and 1910', *Archaeologia* 62, 445-7.

Matthews, B. (1970) 'Age and origin of aeolian sand in the Vale of York', *Nature, Lond.* 227, 1234-6.

Mellars, P.A. (1969) 'Radiocarbon dates for a new Creswellian site', *Antiquity*, 43, 308-10.

O'Dell, A. (1963) *The Scandinavian World.* London.

Osborne, P.J. (1965) 'The effect of forest clearance on the distribution of the British insect fauna', *Proc. XII Int. Congr. Ent., London 1964*, 456-7.

Osborne, P.J. (1969) 'An insect fauna of Late Bronze Age date from Wilsford, Wiltshire.' *J. Anim. Ecol.* 38, 555-66.

Osborne, P.J. (1971a) 'An insect fauna from the Roman site at Alcester Warwickshire', *Britannia.* II, 156-65.

Osborne, P.J. (1971b) 'Insect fauna from the Roman harbour,' *in* B. Cunliffe, Excavations at Fishbourne 1961-1969', *Research Rep. Soc. of Ants.* 27, 393-6.

Osborne, P.J. (1972) 'Insect faunas of Late Devensian and Flandrian age from Church Stretton, Shropshire', *Phil. Trans. R. Soc.* B, 263, 327-67.

Osborne, P.J. (1974) 'Insects associated with burnt grain from a Roman site at Droitwich, Worcestershire', *J. stored Prod. Res.* (in press).

Palm, T. (1960) 'Holz- und Rinden-Käfer der süd- und mitterschwedischen Laubbäume', *Opûsc. Ent.* Suppl. 16, Lund.

Peake, D.S. and P.J. Osborne (1971) 'The Wandle gravels in the vicinity of Croydon', *Proc. Croydon Nat. Hist. Sci. Soc.* 14, 145-76.

Penington, W. (1969) *The History of the British Vegetation.* London.

Peyerimhoff, P. de (1945) 'Les genres de Coléoptères importés ou acclimatés dans la faune Euro-mediterranéenne', *Revue fr. Ent.* 12, 5-11.

Pryme, A. de la (1701) 'Concerning trees found underground on Hatfield Chase', *Phil. Trans. R. Soc.* 22, 980-92.

R.C.H.M. (1962) *An Inventory of the Historical Monuments in the City of York,* ,I, Eburacum, *Roman York.*

Riley, D.N. (1973) 'Aerial reconnaissance of the West Riding Magnesian Limestone country', *Yorks. archaeol. J.* 45, 210-13.

Schimitscheck, E. (1948) 'Bioklimatische Beoabachtungen und Studien bei Borkenkäferauftreten', *Wetter u. Leben,* 1, 97-104.

Schmiedeknecht, O. (1930) *Die Hymenopteren Nord- und Mitterleuropas.* Jena.

Skidmore, P. and C. Johnson (1969) 'A preliminary list of the Coleoptera of Merioneth, Northern Wales'. *Entomologist's Gaz.* 20, 139-225.

Smith, A.G. (1958) 'Post-glacial deposits in south Yorkshire and north Lincolnshire', *New Phytol.* 57, 19-49.

Solomon, M.E. and B.E. Adamson (1955) 'The powers of survival of storage and domestic pests under winter conditions in Britain', *Bull. ent. Res.* 46, 311-55.

Spector, W. (1943) 'Collecting beetles (Trox) with feather bait traps, (Coleoptera; Scarabaeidae)', *Ent. News,* 54, 224-9.

Trechman, C.T. (1920) 'On a deposit of interglacial loess and some transported pre-glacial freshwater clays on the Durham coast', *Q. Jl. geol. Soc. Lond.* 75, 173-203.

Turner, J. (1962) 'The Tilia decline; an anthropogenic interpretation', *New Phytol.* 61, 328-41.

Whitwell, J.B. *et al.* (1974) *The Roman Sewer at York.*

Yarrow, I.H.H. (1967) '*Ponera punctatissima* Roger (Hym. Formicidae) down a coal mine in Kent', *Entromologist's mon. Mag.* 103, 39.

Zacher, F. (1937) 'Vorratsschädline und Vorratschutz, ihre Beduetung für Volkersernährung und Weltwirtschaft', *Z. hyg. Zool.* 29, 1-11.

J.G. Evans
Subfossil land-snail faunas
from rock-rubble habitats

The analysis of sub-fossil land-snail assemblages from archaeological deposits enables the reconstruction of ancient environments to be made (Evans, 1972). The assemblages are generally contemporary with the deposition of the sediments in which they occur, exceptions, such as the subterranean species *Cecilioides acicula* (Müller), being easy to detect. However, under certain circumstances, whole assemblages may be younger, perhaps by hundreds or thousands of years, than the associated sediments, and it is the purpose of this short paper to draw attention to this phenomenon. It applies mainly to limestone-derived (as opposed to chalk-derived) sediments, and in particular to coarse rock-rubble, typical situations being the primary fill of ditches, collapsed wall debris, thermoclastic scree in caves and at the foot of cliffs, and stone-built cairns (Evans and Jones, 1973).

The problem first arose at Cathole, a cave in the Gower Peninsula, and at Cadbury-Camelot, an Iron Age hillfort in Somerset. At Cathole the mammalian fauna and Upper Palaeolithic artefacts in a thermoclastic scree suggested a Late Glacial origin in a period of cold climate, as indeed did the deposits themselves. The molluscan assemblage, however, indicated temperate climatic conditions and a woodland environment, being virtually devoid of all the usual cold-climate and open-country species found in Late Glacial deposits. At Cadbury-Camelot, layers of rock-rubble in the ditch fill, which were laid down under conditions of surface instability and, one assumes, sparse vegetational cover, contained land-snail faunas in which open-country species amounted to less than 10 per cent. Many other sites on limestone were found to conform to a similar pattern.

These rock-rubble faunas are characterised by a predominance of shade-loving species, but they have peculiarities which set them aside from normal assemblages from leaf-litter habitats on a woodland floor. Three groups predominate – *Discus rotundatus* (Müller), *Oxychilus*, generally *O. cellarius* (Müller), and *Vitrea contracta* (Westerlund). Two other groups – *Carychium tridentatum* (Risso) and *Retinella* (*R. pura* (Alder) and *R. nitidula* (Draparnaud)) – are rare or absent altogether, although generally abundant in normal woodland faunas. A further group of less fastidious species, such as *Punctum pygmaeum* (Draparnaud), *Vitrina pellucida* (Müller) and *Retinella radiatula* (Alder), are also infrequent or rare. Nor are species which live on exposed rock surfaces such as cliffs and walls (rupestral species) at all well represented. In other words concern is with a low number of species, which in some cases occur in extreme profusion. Faunas of this type are indicative of specialised habitats.

Two of the three predominant groups – *Oxychilus* and *Vitrea* – are facultative carnivores; in other words they are able to live entirely on a diet of animal matter such as insects and worms. Moreover, all three groups are consistently recorded as the predominant forms in modern cave faunas, and there is even one instance of predation on live Lepidoptera. *Oxychilus cellarius* owes its name to its habit of living in underground places, and both its carnivorous and troglophile tendencies are enshrined in a number of instances where it has been found among human skeletons in Neolithic burial chambers, attracted no doubt by the odour of rotting flesh. *Retinella nitidula* and *R. pura*, on the other hand, although very similar in many respects in their habitat requirements to *Oxychilus*, lack the specific enzymes (e.g. chitinase) necessary for dealing with animal tissues, and are thus confined to above ground habitats and a vegetarian diet.

It is suggested, therefore, that the curious composition of subfossil rock-rubble assemblages can be explained as being essentially caused by the environmental conditions of the underground habitat. But it is difficult to explain why the phenomenon arises only in limestone-derived sediments, and not in those derived from chalk. Limestone is much harder than chalk and weathers more slowly. Deposits made up largely of coarse angular fragments may maintain an open lattice structure almost indefinitely thus providing the conditions of high humidity, shade, and low temperature variation so much required by woodland species. Chalk, on the other hand, has a much greater tendency to crumble so that larger lumps are quickly broken down to smaller fragments and the interstices filled in with fine debris creating a much more compact sediment.

The archaeological implications of this are manifest. Subfossil molluscan assemblages from uncompacted limestone rubble will be not necessarily contemporary with, but generally later than, the deposits. Only when the sediments are securely sealed, for example by a thick turf-line or stalagmite layer, or are themslves consolidated to form a breccia, will snails cease to penetrate their depths. Pleistocene records of Mollusca from cave sites must be treated with extreme caution, and many anomalies in the earlier literature are almost certainly due to a lack of appreciation of this phenomenon. Furthermore, the woodland facies of rock-rubble assemblages is a function of the micro-environment of the habitat and must not be taken necessarily to indicate a woodland environment at any stage in the history of the site.

The writer does not want to discourage work on faunas from limestone sites. On the contrary, far too little work has been done either in Britain or elsewhere, and it is certain that assemblages from cave sediments, in particular, would yield interesting results. But the work must be done bearing in mind that, against the general environment of an area, there is the special environment created by the sediments themselves.

REFERENCES

Evans, J.G. (1972) *Land Snails and Archaeology.* London.

Evans, J.G. and H. Jones (1973) 'Subfossil and modern land-snail faunas from rock-rubble habitats', *J. Conchol.* 28, 103-30.

INDEX OF AUTHORS

GENERAL INDEX

aerial photography, 63
Agiofarango Gorge, 269-73
Anglesey, 119-20
Apulia, 129-32
arc spectrography, *see* trace element analysis
Argolid peninsula, 269-73
Argos plain, 269-73
artefacts, *see* provenancing

Bandkeramik culture, 51
Barley, pits from, 17
Barrows, pollen analyses from, 347-54
Bedd Branwen, 120
bricks, 230-52, 262, 278
Bryn y Gefeiliau, 126
Bryn yr Hen Bobl, 120
bulk density measurement, 69-70
burial; in pits, 10; in urns, 119-20
buried soils, *see* soils
Burnham, 329

Cae Mickney, 120, 125
Caerwent, 369
carbon, 70, 106, 113
carbonates; mobilisation, 69; analysis, 70;
    removal, 72; solution/precipitation, 143,
    148-52
carse deposits, 179, 189
Caversham, 329
caves; sedimentation, 30, 137-54, 284-314;
    bones from, 30; vault studies in, 40-4;
    atmosphere, 43; pollen from, 315
cementation of sediments, 15, 70
ceramics, *see* bricks, pottery etc.
charcoal; recovery from pits, 10; analysis of,
    71, 280, 361
chronology; 217-18, 234-6, 310-14, 355, 380-1;
    of Holocene deposits, 180-6; of sea level
    change, 159; *see also* dating, radiocarbon
    dating
Clacton, 329-30
clay, 81-2, 259, 269, 270; in ceramics, 116;
    estuarine, 177; in loam, 197, 199, 259;
    particle shapes, 84; particle size analysis,
    75-8; puddling pits, 10; orientation of
    particles, 282; separation, 78
clay cutans, 68, 69

Clay-with-Flints, 71
climatic change; chronology, 3; influence on
    sediments, 138-9; *see also* pluvials
coding systems, 12
collophane, 71
colluvial deposits, 69, 77, 225, 265, 268, 319
colour; recording, 15; of soils, 69; of
    sediments, 205, 280, 297-8
computer recording systems, *see* information
    recovery
consistency of sediments, 15, 70
Crayford, 330
cryoturbation, 43

Danebury, 9
Dannewerke, 52
data recovery, *see* information recovery
dating methods; isotopic, 3; relative, 234-6;
    by pollen analysis, 347; *see also* chronology,
    radiocarbon dating
density separation, 79-80, 116
diagenesis, of clays, 81; of sands, 83-4; in cave
    sediments, 137
diatoms, 162, 220-4
Dibsi Faraj, 276-87
Die Kelders, 289-314
disaggregation, 73
dispersion methods, 72
Dragonby, 62
Drama, plain of, 256-65

Ealing, 330
ecosystems, models for, 24-5
electron microscopy, 83-4, 91, 289, 307-10
environment, coastal, 175, 204-10;
    reconstruction, 5, 66, 144, 289-93, 204-10,
    213-14, 386-8; cultural-environmental
    information systems, 23; data recovery
    from, 28; evidence from buried soils, 68; of
    deposition of sediments, 34, 65-6, 83-4,
    204-10, 213-14, 324; contexts for cultural
    entities, 24; significance of minerals, 80-1
epoxy resins, 67
eustatic changes, *see* sea level change
Ewe Crag Slack, 359
excavation, 9-10, 23, 37-9; *see also* under site
    names and areas